Springer Series in Statistics

Advisors:
P. Bickel, P. Diggle, S. Fienberg, K. Krickeberg,
I. Olkin, N. Wermuth, S. Zeger

T0184063

Springer
New York
Berlin
Heidelberg
Hong Kong
London
Milan
Paris
Tokyo

Springer Series in Statistics

Andersen/Borgan/Gill/Keiding: Statistical Models Based on Counting Processes.
Atkinson/Riani: Robust Diagnostic Regression Analysis.
Berger: Statistical Decision Theory and Bayesian Analysis, 2nd edition.
Borg/Groenen: Modern Multidimensional Scaling: Theory and Applications.
Brockwell/Davis: Time Series: Theory and Methods, 2nd edition.
Chan/Tong: Chaos: A Statistical Perspective.
Chen/Shao/Ibrahim: Monte Carlo Methods in Bayesian Computation.
David/Edwards: Annotated Readings in the History of Statistics.
Devroye/Lugosi: Combinatorial Methods in Density Estimation.
Efromovich: Nonparametric Curve Estimation: Methods, Theory, and Applications.
Eggermont/LaRiccia: Maximum Penalized Likelihood Estimation, Volume I: Density Estimation.
Fahrmeir/Tutz: Multivariate Statistical Modelling Based on Generalized Linear Models, 2nd edition.
Fan/Yao: Nonlinear Time Series: Nonparametric and Parametric Methods.
Farebrother: Fitting Linear Relationships: A History of the Calculus of Observations 1750-1900.
Federer: Statistical Design and Analysis for Intercropping Experiments, Volume I: Two Crops.
Federer: Statistical Design and Analysis for Intercropping Experiments, Volume II: Three or More Crops.
Ghosh/Ramamoorthi: Bayesian Nonparametrics.
Glaz/Naus/Wallenstein: Scan Statistics.
Good: Permutation Tests: A Practical Guide to Resampling Methods for Testing Hypotheses, 2nd edition.
Gouriéroux: ARCH Models and Financial Applications.
Gu: Smoothing Spline ANOVA Models.
Györfi/Kohler/Krzyżak/ Walk: A Distribution-Free Theory of Nonparametric Regression.
Haberman: Advanced Statistics, Volume I: Description of Populations.
Hall: The Bootstrap and Edgeworth Expansion.
Härdle: Smoothing Techniques: With Implementation in S.
Harrell: Regression Modeling Strategies: With Applications to Linear Models, Logistic Regression, and Survival Analysis.
Hart: Nonparametric Smoothing and Lack-of-Fit Tests.
Hastie/Tibshirani/Friedman: The Elements of Statistical Learning: Data Mining, Inference, and Prediction.
Hedayat/Sloane/Stufken: Orthogonal Arrays: Theory and Applications.
Heyde: Quasi-Likelihood and its Application: A General Approach to Optimal Parameter Estimation.
Huet/Bouvier/Poursat/Jolivet: Statistical Tools for Nonlinear Regression: A Practical Guide with S-PLUS and R Examples, 2nd edition.
Ibrahim/Chen/Sinha: Bayesian Survival Analysis.
Jolliffe: Principal Component Analysis, 2nd edition.

(continued after index)

Samaradasa Weerahandi

Exact Statistical Methods for Data Analysis

With 10 figures

Springer

Samaradasa Weerahandi
P.O. Box 58
Milburn, NJ 07041
USA

Library of Congress Cataloging-in-Publication Data
Weerahandi, Sam.
Exact statistical methods for data analysis / Sam Weerahandi.
p. cm. -- (Springer series in statistics)
Includes bibliographical references and index.
1. Mathematical statistics. I. Title. II. Series
QA276.W435 1994 9428642
519.5--dc20

ISBN 0-387-40621-2 Printed on acid-free paper.

First softcover printing, 2003.

© 1995 Springer-Verlag New York, Inc.
All rights reserved. This work may not be translated or copied in whole or in part without the written permission of the publisher (Springer-Verlag New York, Inc., 175 Fifth Avenue, New York, NY 10010, USA), except for brief excerpts in connection with reviews or scholarly analysis. Use in connection with any form of information storage and retrieval, electronic adaptation, computer software, or by similar or dissimilar methodology now known or hereafter developed is forbidden.
The use in this publication of trade names, trademarks, service marks, and similar terms, even if they are not identified as such, is not to be taken as an expression of opinion as to whether or not they are subject to proprietary rights.

Printed in the United States of America.

9 8 7 6 5 4 3 2 1 SPIN 10947616

www.springer-ny.com

Springer-Verlag New York Berlin Heidelberg
A member of BertelsmannSpringer Science+Business Media GmbH

To Dilani, Himali, Nuwan, and my parents

Preface

This book covers some recent developments as well as conventional methods in statistical inference. In particular, some rather important theory and methods in generalized p-values and generalized confidence intervals are included. Exact statistical methods based on these developments can be found in Chapters 5-10; they are presented as extensions (not alternatives) to conventional methods. They can be utilized with or without deviating from the conventional philosophy of statistical inference. Conventional methods alone do not always provide exact solutions to even simple problems involving nuisance parameters such as that of comparing the means of two exponential distributions. As a result, practitioners often resort to asymptotic results in search of approximate solutions even when such approximations are known to perform rather poorly with typical sample sizes.

The new theory and methods presented in this book play a vital role in linear models, in particular. For instance, with this approach, a practitioner can perform ANOVA with or without the assumption of equal error variances, an assumption usually made for simplicity and mathematical tractability rather than anything else. In typical applications, violation of this assumption is now known to be much more serious than the assumption of normality. When the assumption of homoscedasticity is not reasonable, the conventional F-test frequently leads to incorrect conclusions and often fails to detect significant experimental results. Such a lack of power of testing procedures can result in serious repercussions in applications, especially in biomedical research where each sample point is so vital and expensive.

The book is suitable as a professional reference book on statistical methods or as a textbook for an intermediate course in statistics. The book will be useful to industrial statisticians and to researchers and experimenters, including those in the agricultural and biomedical fields, who utilize statistical methods in their work. It will also be useful to students and teachers of statistics. I have tried to make this book easy to read, concise, and yet self-contained as much as possible. The mathematical level of the book is kept rather low and very little prior knowledge of statistical notions and concepts is assumed.

As a special feature of this book, exact statistical methods are provided for each application considered including inferences about parameters of the binomial, exponential, and normal distributions. The methods are exact in the sense that the tests and the confidence intervals developed in the book are based on exact probability statements rather than on asymptotic approximations; inferences based on them can be made with any desired accuracy, provided that assumed parametric model and/or other assumptions are correct. To make this possible in the case of discrete distributions, and to draw conclusions which are more general and informative than fixed level tests, hypothesis testing in this book is often carried out on the basis of observed levels of significance (p-values). Bayesian counterparts are developed for some important problems. In most applications, under noninformative priors, the Bayesian results in interval estimation and testing one-sided hypotheses are numerically equivalent to the generalized confidence intervals and generalized p-values presented in this book. There is readily available computer software to implement these exact statistical methods, and yet to my knowledge this is the first book which covers recent developments in statistical inferences based on generalized confidence intervals and generalized p-values.

Practitioners who prefer to carry out tests in the setting of the Neyman-Pearson theory can also find approximate fixed level tests based on the p-values. Exact p-values given for such problems as the ANOVA and regression under unequal error variances provide excellent approximations for fixed level tests. According to simulation studies reported in the literature, Type I error and power performance of these approximations are usually better than the performance of more complicated approximate tests. Therefore, exact significance tests and generalized confidence intervals reported in this book are useful even for those practitioners who insist on fixed level tests and conventional confidence intervals.

The book is written with both the student and the practitioner in mind. For the benefit of practitioners, a fairly comprehensive set of formulae is given, with or without proofs, for all important results under the topics covered in this book. A large number of illustrative examples and exercises is provided. They range from simple numerical examples to extended versions of theoretical results presented in the text. The problems given at the end of each chapter are intended to further illustrate the concepts and methods covered in the chapter and to stimulate further thoughts about results covered in the book. Many problems are given with outlines of the steps involved in deriving solutions. More emphasis is placed on concepts and methods than on mathematical rigor. Whenever possible, the methods and concepts are developed without decision theoretic notions and definitions.

I have attempted as far as possible to make the chapters of the book dealing with methods and applications easy to read for anyone who has taken an introductory course in mathematical statistics. A previous course in advanced statistical methods is helpful but not essential in order to study this book. Some knowledge of probability and distribution theory with an introduction to statistical inference is required, however. In particular, familiarity with special statistical distributions and the underlying problems of statistical inference concerning parameters of special distributions is required. I have taken extra effort to keep the mathematical level of the text as low as possible, often giving intuitive arguments without regard to their mathematical rigor. For the benefit of those readers who might be interested in formal proofs of theorems and further details of results, a faily comprehensive set of references is provided.

The first two chapters of the book are devoted to providing a fairly comprehensive, and yet a concise and easy-to-read introduction to the theory of statistical inference, both classical and Bayesian. Most textbooks in statistics take the Neyman-Pearson approach to hypothesis testing and so Chapter 2 is devoted to presenting concepts in hypothesis testing the way Fisher treated the problem. In Chapter 3, with a view towards demonstrating the simplicity and generality of Fisher's method of significance testing based on p-values, the theory is applied to simple problems in statistical inference involving the normal, binomial and Poisson distributions. Since most textbooks do not provide formulae for computing exact p-values for testing parameters of even widely used distributions, some are presented in this chapter in a concise manner. Chapter 4 presents nonparametric methods based on the classical notions presented in the first two chapters. Chapters 5 and 6 present some new concepts, notions, and theories in significance testing and interval estimation. The usefulness of these developments is evident from a variety of applications undertaken in the last six chapters. Practitioners who are interested in exact statistical methods in linear models rather than their derivations can find them in chapters 7-10. For the benefit of such readers, in Chapters 8-10 I have given the results before their derivations. A large number of numerical examples are provided to illustrate applications of methods based on the normal theory. Inferences concerning parameters of exponential distributions with some numerical illustrations can be found in Chapters 1, 2, 5, and 6.

The computations of the major exact parametric methods presented in this book can be performed with the XPro statistical software package. The exact nonparametric methods presented in Chapter 4 and the first half of Chapter 3 can be carried out using software packages such as StatXact and TESTIMATE. Major statistical software packages such as BMDP, SAS, SPSS, and SPlus all provide tools to carry out some of the computations of parametric and nonparametric methods. At back of this book there is a page describing how to get more information about XPro, StatXact, and TESTIMATE, software packages which specialize in exact methods.

I am very grateful to a number of people who read the manuscript and made useful comments and suggestions. Especially, I express my thanks and appreciation to D. Amaratunga, M. Ananda, J. D. Healy, M. Koschat, A. Mcintosh, and K. Samaranayake for reading the manuscript. Thanks are also due to S. R. Dalal, R. E. Hausman, A. Sadrian of Bell Communications

Research, and M. Koschat of Yale University for encouraging me to write this book. I extend my gratitude also to P. W. Epasinghe, R. A. Dayananda, R. A. Johnson, K. Perera, X. Meng, V. A. Samaranayake, T. Spacek, K. W. Tsui, and especially to J. Kettenring of Bell Communications Research and J. V. Zidek of the University of British Columbia, for the help, guidance, and support they provided at various occasions. I also thank Martin Gilchrist, my editor at Springer-Verlag, for his cooperation, patience, and support during the course of this project. Finally, I thank my family for their endurance and encouragement during the preparation of the manuscript.

Millington, New Jersey
September, 1994

Sam Weerahandi

Contents

Chapter 1

Preliminary Notions

1.1. Introduction

In this book we are concerned with the basic theory and methods of *statistical inference* and major applications such as Regression and Analysis of Variance. We will be dealing with statistical problems where it is of interest to make statistical statements about certain parameters of a population from which a set of data have been generated. It is assumed that the characteristics of interest in the population can be represented by a set of random variables and the parameters of interest are related to the distribution of the random variables. Except in Chapter 4, we shall assume that the form of the probability distribution of the random variables is known up to a set of unspecified parameters, including the parameters of interest. The statistical methods based on this assumption are called *parametric* methods. The methods that do not require the specification of the underlying distribution are known as *nonparametric*, or *distribution-free* methods; this is the subject of Chapter 4.

The principal focus of this book is on testing of statistical hypotheses and interval estimation concerning the parameters of statistical distributions. In important applications, however, we will deal with the problem of point estimation as well. In this chapter we present, in a concise manner, preliminary notions common to all types of statistical inference and the notions and methods in interval estimation and in point estimation.

Let $\mathbf{X}$ be an observable random vector representing some population characteristics and let Ξ denote the range of values that $\mathbf{X}$ can assume. Sometimes Ξ will be referred to as the *sample space*. In many applications $\mathbf{X} = (X_1, \ldots, X_n)$ is a random sample taken from the population, where n is the sample size and $X_1, \ldots, X_n$ are independently and identically distributed random variables. Let $F_{\mathbf{X}}(\mathbf{x}; \boldsymbol{\theta})$ be the cumulative probability distribution function of $\mathbf{X}$ which is sometimes abbreviated as *cdf* of $\mathbf{X}$, where $\boldsymbol{\theta}$ is a vector of unknown parameters. The set Θ of all possible values of the parameter vector $\boldsymbol{\theta}$ is called the *parameter space*. When $\mathbf{X}$ is a discrete random vector, its probability mass function will be denoted by $f_{\mathbf{X}}(\mathbf{x})$. Sometimes, the probability mass function is abbreviated as *pmf*. We will use the same notation to denote the probability density function (sometimes abbreviated as *pdf*) of a continuous random vector.

1.2. Sufficiency

The concept of sufficiency introduced by Fisher (1922) and further developed by Neyman (1935) is of interest in itself and is practically useful in a wide variety of problems of statistical inference. Although it deserves a thorough investigation, this is beyond the scope of this book. For the sake of self-containment, however, we will give a brief definition and outline a method to find sufficient statistics. The reader is referred to Huzurbazar (1976) for a comprehensive exposition of theory and methods dealing with sufficiency.

In many inference problems that we will encounter, we will be able to summarize the information in the data by a few summary statistics. In other words we will be able to find some functions of $\mathbf{X}$, say a vector $\mathbf{T}(\mathbf{X})$, that provide us with as much information about $\boldsymbol{\theta}$ as the original data itself. Such a set of real-valued functions is called a vector of sufficient statistics for $\boldsymbol{\theta}$. For a number of formal statements concerning the notion that one can simplify a statistical decision-making problem by means of sufficiency, the reader is referred to Huzurbazar (1976). A formal definition of the concept is as follows.

Definition 1.1. Let $\mathbf{X}$ be a random vector whose distribution depends on the unknown parameter $\boldsymbol{\theta}$. A set of statistics $\mathbf{T}(\mathbf{X})$ is said to be *sufficient* for $\boldsymbol{\theta}$ if the conditional distribution of $\mathbf{X}$, given $\mathbf{T}(\mathbf{X})=\mathbf{t}$, does not depend on $\boldsymbol{\theta}$.

A set of sufficient statistics is said to be *minimal sufficient* if and only if it is a function of every other set of sufficient statistics.

Of course, if $X_1, \ldots, X_n$ is a simple random sample from $f_X(x;\theta)$, the sample itself is sufficient for θ. Further, the set of order statistics $(X_{(1)}, \ldots, X_{(n)})$ is also sufficient for θ. In many situations, however, these are not the minimal sufficient statistics and we can actually reduce the dimension of an inference problem by searching for solutions based on minimal sufficient statistics.

This definition of sufficient statistics is not at all convenient in most applications. Not only does one have to guess what nontrivial functions of $\mathbf{X}$ are sufficient, but one also has to go through what is typically a tedious derivation of the required conditional distribution. Fortunately, the following criterion, a theorem due to Neyman (1935), makes the derivation of a sufficient statistic a trivial task.

The factorization criterion.

Let **X** be a random vector from an absolutely continuous distribution or a discrete distribution with the probability density function or the probability mass function $f(1 boldx| \ \theta)$. The vector of statistics $\mathbf{T}=\mathbf{t}(\mathbf{X})$ is sufficient for θ if and only if $f(\mathbf{x}| \ \theta)$ can be factored as

$$f(\mathbf{x}| \ \boldsymbol{\theta}) = g(\mathbf{t}(\mathbf{x}), \boldsymbol{\theta}) \ h(\mathbf{x}) \ , \tag{1.1}$$

where the function g depends on the observed random vector x only through the value of **t**, and the function h may depend on **x** but not on $\boldsymbol{\theta}$.

Example 1.1. Exponential Distribution

Let $X_1, \ldots, X_n$ be a random sample of size n from an exponential distribution with parameter θ. The probability density function of the underlying random variable X is

$$f_x(x) = \frac{1}{\theta} e^{-\frac{x}{\theta}} \quad \text{for } x > 0,$$

where θ is the mean of the distribution. We shall show that $T = \sum X_i$ is a sufficient statistic for θ. Since $X_1, \ldots, X_n$ are independent random variables , the joint probability density function of $\mathbf{X} = (X_1, \ldots, X_n)$ can be written as

$$f_X(\mathbf{x}) = \prod_{i=1}^{n} \frac{1}{\theta} e^{-\frac{x_i}{\theta}} = \frac{1}{\theta^n} e^{-\frac{\sum x_i}{\theta}} \quad \text{for } \mathbf{x} > \mathbf{0}$$

Hence, according to the factorization criterion (in this example $h(\mathbf{x}) \equiv 1$), $T = \sum X_i$ is a sufficient statistic for θ.

Example 1.2. Uniform distribution

Let $X_1, \ldots, X_n$ be a random sample from a uniform distribution having the density function

$$f_X(x) = \frac{1}{\theta} \quad \text{for } 0 \le x \le \theta \, .$$

That X has a uniform distribution over the interval $[0,\theta]$ will be denoted as $X \sim U(0,\theta)$. To enable the application of the factorization criterion in this problem, in which the sample space depends on θ, the density function needs to be expressed in the form

$$f_{\mathbf{X}}(\mathbf{x}) = \prod_{i=1}^{n} \frac{1}{\theta} I_{[0,\theta]}(x_i) = \frac{1}{\theta^n} I_{[0,\theta]}(max(\mathbf{x})) ,$$

where

$$I_{[0,\theta]}(y) = \begin{cases} 1 & \text{if } 0 \le y \le \theta \\ 0 & \text{otherwise} \end{cases}$$

and $max(\mathbf{x})$ is the largest value of x_i, $i=1,\dots,n$. Since $f_{\mathbf{X}}(\mathbf{x})$ is a function of $max(\mathbf{x})$ and θ only, it is now clear that $T(\mathbf{X}) = max(\mathbf{X})$ is a sufficient statistic for θ.

1.3. Complete Sufficient Statistics

A concept which plays a vital role particularly in the theory of point estimation and in hypothesis testing is the *completeness* of a sufficient statistic. This property is stronger than the minimality of a sufficient statistic. As we will clarify in Example 1.5, in many applications involving two dimensional minimal sufficient statistics for a single parameter, a set of minimal sufficient statistics need not be complete.

Definition 1.2. Suppose $\mathbf{T}(\mathbf{X})$ is a sufficient statistic for a parameter θ. The statistic $\mathbf{T}$ and the family of distributions of $\mathbf{T}$ parameterized by θ are said to be *complete* if for any given statistic $h(T)$,

$$E(h(T)) \equiv 0 \text{ for all } \theta \in \Theta$$

implies that

$$Pr[h(T) = 0] \equiv 1 \quad ,\text{for all } \theta \in \Theta , \tag{1.2}$$

where Θ is the parameter space.

A complete sufficient statistic is necessarily minimal. The converse is not necessarily true, however. For a proof of the former and for in-depth results and details the reader is referred to Lehmann (1983).

Example 1.3. Geometric distribution

Let X be an observation from the geometric distribution with probability mass function

$$f_X(x;\theta) = (1-\theta)\theta^x \quad x = 0, 1, 2, \dots ,$$

where $0 < \theta < 1$. Consider any real-valued function $h(X)$ of X such that

$$E(h(X)) = \sum_{x=0}^{\infty} h(x)(1-\theta)\theta^x \equiv 0 \text{ for all } \theta \in (0,1) .$$

In order for this convergent power series to be identically 0, each coefficient of θ^x, $x = 0, 1, 2, \ldots$ must also be 0. Since $0 < \theta < 1$, this implies that $h(x) \equiv 0$. Therefore, $Pr[h(X) = 0] \equiv 1$ for all $0 < \theta < 1$, implying that X is a complete statistic. Also, $f_X(x;\theta)$ form a complete family of distributions.

Example 1.4. Uniform distribution over a unit interval

Let $X_1, \ldots, X_n$ be a random sample from the uniform distribution with density function

$$f_X(x) = 1 \qquad \text{for } \theta \le x \le \theta + 1 ,$$

where θ is an unknown parameter that can assume any real number. The joint density function of $\mathbf{X} = (X_1, \ldots, X_n)$ can be expressed as

$$f_{\mathbf{X}}(\mathbf{x}) = I_{(-\infty,\theta+1]}(max\ (\mathbf{x})\ )\ I_{[\theta,\infty)}(min\ (\mathbf{x})\ )$$

By applying the factorization criterion, it is now easily seen that $\mathbf{T} = (min(\mathbf{X}), max(\mathbf{X}))$ is sufficient for θ. Although it can be shown that this statistic is minimal sufficient as well, it is not complete. To see this, consider the random variable $Y = X - \theta \sim U(0,1)$ (the uniform distribution over $[0,1]$) and let e be the expected value of $max(\mathbf{Y})$. (In fact $e = n/(n+1)$; see Exercise 1.1). Since $X = Y + \theta$, $E(max(\mathbf{X}))$ can be expressed in terms of e as $E(max(\mathbf{X})) = e + \theta$. Moreover, it follows from the fact that $1 - Y \sim U(0,1)$, $E(min(\mathbf{X})) = 1 - e + \theta$. Hence, for instance, if we define $h(\mathbf{T}) = max(\mathbf{X}) - min(\mathbf{X}) - 2e + 1$, we get $E(h(\mathbf{T})) \equiv 0$ for all θ. Therefore, $\mathbf{T}$ is not complete.

1.4. Exponential Families of Distributions

Most of the parametric families of distributions that we will encounter in this book are members of what is known as the exponential family of distributions. A number of important theorems related to the notions discussed in the preceding section can be stated in general without any reference to particular members of this class. We consider here only the distributions which are either discrete or continuous. A reader interested in definitions and results pertaining to more general distributions is referred to Lehmann (1983 & 1986).

Definition 1.3. A family of distributions with probability mass function or probability density function $f(x;\boldsymbol{\theta})$ is said to belong to the *exponential family of distributions* if $f(x;\boldsymbol{\theta})$ can be expressed in the form

$$f(x;\boldsymbol{\theta}) = a(\boldsymbol{\theta})\ b(x)\ e^{\sum_{j=1}^{k} c_j(\boldsymbol{\theta}) d_j(x)} \qquad (1.3)$$

for a suitable choice of functions $a()$, $b()$, $\mathbf{c}()$ and $\mathbf{d}()$, where $\boldsymbol{\theta}$ is a vector of parameters.

Let $\mathbf{X} = (X_1, \ldots, X_n)$ be a random sample of size n from an exponential family of distributions. By applying the factorization criterion it is seen that the vector of statistics

$$\mathbf{T}(\mathbf{X}) = \left(\sum_{i=1}^{n} d_1(X_i), \ldots, \sum_{i=1}^{n} d_k(X_i) \right) \tag{1.4}$$

is sufficient for the parameter vector $\boldsymbol{\theta}$. If the parameters are redefined as $c_j(\boldsymbol{\theta}) = \theta_j$, $j=1, \ldots, k$, the exponential family is said to have its *natural parametrization.* In this case, it can be shown (see, for instance Lehmann (1983)) that $\mathbf{T}$ is a complete sufficient statistic for $(\theta_1, \ldots, \theta_k)$.

Example 1.5. Beta distribution

Suppose $X_1, \ldots, X_n$ form a random sample from the beta distribution with density function

$$f(x;\alpha,\beta) = \frac{\Gamma(\alpha+\beta)}{\Gamma(\alpha)\Gamma(\beta)} x^{\alpha-1}(1-x)^{\beta-1} \quad \text{for } 0 < x < 1 ,$$

where $\alpha > 0$ and $\beta > 0$ are parameters of the distribution; this distribution will be denoted as $B(\alpha,\beta)$. Notice that the density function can be rewritten as

$$f(x;\alpha,\beta) = \frac{\Gamma(\alpha+\beta)}{\Gamma(\alpha)\Gamma(\beta)} \frac{1}{x(1-x)} e^{\alpha log(x) + \beta log(1-x)} \quad \text{for } 0<x<1 .$$

With this representation α and β can be treated as the natural parameters of the distribution. Moreover, with the notation in (1.3) $d_1(x) = log(x)$ and $d_2(x) = log(1-x)$. Therefore, the statistic $\mathbf{T} = \left(\sum_{i=1}^{n} \log(X_i), \sum_{i=1}^{n} \log(1-X_i) \right)$, is complete sufficient for (α,β).

Consider an exponential family of distributions with its natural parametrization. As explicitly specified in Theorem 1.1, the distribution of the complete sufficient statistic $\mathbf{T}$ also belongs to an exponential family. A reader interested in the proof of this theorem is referred to Lehmann (1986).

Theorem 1.1. Let $\mathbf{T} = (\sum d_1(X_i), \cdots, \sum d_k(X_i))$ be the complete sufficient statistic based on a random sample from the exponential family (1.3) with $c_j(\boldsymbol{\theta}) = \theta_j$, $j = 1, \ldots, k$. The joint probability mass function or the density function of $\mathbf{T}$ is of the form

$$f_{\mathbf{T}}(\mathbf{t};\boldsymbol{\theta}) = A(\boldsymbol{\theta}) \, B(\mathbf{t}) \, e^{\sum_{j=1}^{k} \theta_j t_j} , \tag{1.5}$$

where $t_j = \sum_{i=1}^{n} d_j(X_i)$.

Let $\mathbf{T} = (\mathbf{U},\mathbf{V})$ be a partition of the set of complete sufficient statistics. Let $\boldsymbol{\theta} = (\boldsymbol{\theta}_u,\boldsymbol{\theta}_v)$ be the corresponding partition of the parameter vector. The following useful corollary concerning the conditional distributions of one given the other (and the marginal distribution of each component) can be deduced from Theorem 1.1. (See Appendix B)

Corollary 1.1. The conditional distribution of $\mathbf{U}$ given $\mathbf{V} = \mathbf{v}$ forms an exponential family. Moreover, this distribution is independent of parameters $\boldsymbol{\theta}_v$.

1.5. Invariance

The concept of sufficiency is just one method by which an investigator can reduce the dimension of a statistical problem. The principle of invariance is another method which is often useful in further simplifying statistical inference problems. Many statistical problems that we will encounter in applications possess various natural symmetries. Then, it may be reasonable to search for solutions which are in some sense consistent with the symmetries. This in turn will impose restrictions on statistical procedures (thus reducing eligible candidates) to help us reduce the complexity of a problem. The notion of invariance is a mathematical way of giving a meaning to the symmetry of a statistical problem.

To motivate the idea, consider the problem of estimating the mean life time of a certain brand of light bulbs. Suppose the life time, X, is exponentially distributed with mean θ. Note that in this example θ is a scale parameter. Suppose that based on an observation, x, an investigator estimates θ by $\hat{\theta}(x)$, when x is measured in years. If the units are to be specified in months instead, the investigator should be willing to estimate the parameter by $12\hat{\theta}(x)$; for any other choice of change of scale, say k, the estimate should be revised in a similar manner. This imposes a natural restriction on the form of the estimator; namely, $\hat{\theta}()$ should satisfy the equation

$$\hat{\theta}(kX) = k\,\hat{\theta}(X) \tag{1.6}$$

where, for instance, $k=12$. An estimator having this property is called a scale invariant estimator. The underlying statistical problem as well as the family of distributions are also called scale invariant.

In order to define the invariance of a family of distributions, consider a random variable X (or a random vector $\mathbf{X}$) with its distribution function $F(X;\,\theta)$ parameterized by a parameter (or a vector of parameters) θ. Let Θ be the parameter space of possible values of θ. Let G be a group of transformations of the sample space Ξ. A member of G is denoted by g. For example, in the case of scale transformations, the group can be defined as $g(X) = kX$, where k is a positive constant. Assume that the transformations are one-to-one and onto Ξ itself. Further assume that, given that X is a random variable, $g(X)$ is also a random variable.

Definition 1.4. If for any given $g \in G$ and $\theta \in \Theta$, there exists a unique $\theta' \in \Theta$ such that the distribution function of $g(X)$ is $F(g(X); \theta')$, then the family of distributions $F(X; \theta)$, $\theta \in \Theta$ is said to be *invariant* under the group G.

The family of distributions is said to be *location invariant* if the property holds for the group $g(X+c) = g(X) + c$, where c is any given real number. It is called *scale invariant* if the property holds for the group defined by $g(kX) = k\, g(X)$. Finally, if $g(a + bX) = a + b\, g(X)$, a family invariant under g will be called an *affine invariant* family. The group of transformations $\overline{G}$ defined by $\overline{g}(\theta) = \theta'$ for $\theta \in \Theta$ is called the induced group of transformations of the parameter space, where $\theta' \in \Theta$ is as defined by Definition 1.4 for a given $g \in G$. It can be shown that each $\overline{g} \in \overline{G}$ is a one-to-one transformation of Θ onto itself; that is, Θ is invariant under g.

Example 1.6. Uniform distribution of unit length and mean θ

Suppose that X is a random variable uniformly distributed over the interval $[\theta-\frac{1}{2},\theta+\frac{1}{2}]$; that is $X \sim U(\theta-\frac{1}{2},\theta+\frac{1}{2})$, where θ is any real number. For any given constant c, the distribution of $X + c$ is $X + c \sim U(\theta + c - \frac{1}{2}, \theta + c + \frac{1}{2})$. This means that the distribution of $X + c$ belongs to the same family of uniform distributions as that of X. Therefore, the family of uniform distributions is location invariant. For a given constant c, the induced transformation of the parameter space is given by $\overline{g}(\theta) = \theta + c$.

The invariance of statistical problems will be defined later depending on the type of inference problem. Until then, the invariance of a statistic can be defined without any reference to the particular statistical problem in which the statistic is employed. Let $\mathbf{X}$ be a random vector with the distribution $F(\mathbf{X}: \boldsymbol{\theta})$. Assume that the distribution is invariant under a group G of transformations; that is, for a given $g \in G$ the distribution of $g(\mathbf{X})$ is $F(g(\mathbf{X}): \overline{g}(\boldsymbol{\theta}))$, where $\overline{g}$ is the induced transformation corresponding to g. Let $\mathbf{T}(\mathbf{X})$ be a statistic (possibly vector valued) based on $\mathbf{X}$.

Definition 1.5. A statistic $\mathbf{T}(\mathbf{X})$ is said to be *invariant under a group G* if

$$\mathbf{T}(g(\mathbf{X})) = \mathbf{T}(\mathbf{X}) \quad \text{for } all \ \mathbf{X} \in \Xi \text{ and } g \in G. \tag{1.7}$$

A statistic $\overline{\mathbf{T}}(\mathbf{X})$ is said to be *maximal invariant* if it is invariant and if

$$\overline{\mathbf{T}}(\mathbf{X}_1) = \overline{\mathbf{T}}(\mathbf{X}_2) \text{ implies } \mathbf{X}_2 = \mathbf{g}(\mathbf{X}_1) \text{ for some } \mathbf{g} \in G. \tag{1.8}$$

Example 1.7. Location invariance

Let $\mathbf{X} = (X_1, \ldots, X_n)$ be a vector of real numbers, where $n > 2$. Consider the group G of location transformations defined by

$$g(\mathbf{X}) = \mathbf{X} + c,$$

where c is any real number. Obviously, any difference of two components of $\mathbf{X}$ such as $X_2 - X_1$ is invariant. But, it is not maximal invariant. However, the statistic $\mathbf{T}(\mathbf{X}) = (X_2 - X_1, X_3 - X_1, ..., X_n - X_1)$ is maximal invariant. To see this suppose that $\mathbf{T}(\mathbf{X}) = \mathbf{T}(\mathbf{X}')$. This implies that $X_i - X_1 = X'_i - X'_1$ for $i = 1, ..., n$. Letting $c = X_1 - X'_1$ we get $X'_i = X_i + c$ thus proving that $\mathbf{T}$ is maximal invariant. Obviously, the maximal invariant is not unique; there are other maximal invariant $(n-1)$-tuples. The statistic $(X_1 - \bar{X}, \ldots, X_n - \bar{X})$ is also a maximal invariant, where $\bar{X} = \sum X_i / n$.

Example 1.8. Scale and affine invariance

Let $\mathbf{X} = (X_1, \ldots, X_n)$ be a vector of real numbers, where $n > 2$. Consider the group G of scale transformations defined by

$$g(\mathbf{X}) = k\,\mathbf{X},$$

where $k > 0$. Any ratio such as X_1 / X_2 which is not affected by the change of scale is scale invariant. Maximal invariants can be defined as before by forming independent $(n-1)$-tuples. To define a useful maximal invariant, let $\bar{X}$ and $S^2 = \sum (X_i - \bar{X})^2 / n$ be, respectively, the mean and the variance of the data $X_1, \ldots, X_n$. Then, it can be shown that, in particular, the statistic $T_1(\mathbf{X}) = (X_1/S, \ldots, X_n/S)$ is maximal invariant under the scale transformation. It is also evident from foregoing results that the statistic

$$T_1(\mathbf{X}) = \left(\frac{X_1 - \bar{X}}{S}, \ldots, \frac{X_n - \bar{X}}{S}\right)$$

is maximal invariant under the affine transformation defined by $\mathbf{X} \to k\,\mathbf{X} + c$, where $k > 0$ and $-\infty < c < \infty$.

The following theorem will be useful in testing of hypotheses and in interval estimation; a proof of the theorem can be found in Appendix B.

Theorem 1.2. Let $\bar{\mathbf{T}}(\mathbf{X})$ be a maximal invariant under a group of transformations, say G. Then, a function $\mathbf{T}(\mathbf{X})$ is invariant under G if and only if $\mathbf{T}$ depends on $\mathbf{X}$ only through $\bar{\mathbf{T}}$.

1.6. Maximum Likelihood Estimation

The problem of *point estimation* of parameters of a distribution is the subject of the next 3 sections; the Bayesian counterpart of the problem is presented in Appendix A. In this book we will confine our attention to widely used estimation methods such as the methods of maximum likelihood estimation and mimimum variance unbiased estimation. We will not undertake an exposition of optimum properties of point estimators. Nor will will attempt to characterize or try to find what can be considered as the 'best' estimator for a given situation unless it is evident from the criteria on which the definition of the estimation method is based. Rather, we will search for reasonable and natural estimators which will play a role in other types of inference problems as well. A reader interested in other types of estimation methods, their optimum properties, decision theoretic elements such as estimation loss functions, etc, is referred to Lehmann (1983), and Mood, Graybill, and Boes (1974).

Let $X_1, \ldots, X_n$ be a random sample from a distribution $F_X(x; \theta)$, where the form of the distribution is known but θ is an unknown parameter. In general the distribution may consist of a vector $\boldsymbol{\theta}$ of unknown parameters which need to be estimated. In some applications we might be interested in only a subset of components of $\boldsymbol{\theta}$; the other parameters are treated as nuisance parameters. In dealing with a function of parameters of a certain distribution, we will assume that the distribution $F_X()$ has been reparameterized so that the function of interest appears as a component of $\boldsymbol{\theta}$. The problem in point estimation is to find a statistic $\hat{\boldsymbol{\theta}}(\mathbf{X})$, a function of $X_1, \ldots, X_n$, to be used as an estimator of $\boldsymbol{\theta}$. Although we will not follow any convention strictly, often we will call the random variable representing this statistic an *estimator* and a value, say $\hat{\boldsymbol{\theta}}(\mathbf{x})$, that it takes on an *estimate*. Let $f_{\mathbf{X}}(\mathbf{x};\boldsymbol{\theta})$ be the probability density function or the probability mass function of $\mathbf{X} = (X_1, \ldots, X_n)$. The components of $\mathbf{X}$ are not necessarily independent although this may be the case in many applications.

Now we are in a position to define a simple and convenient method of finding point estimators, namely the method of maximum likelihood. This method typically yields good estimates of intuitive appeal and has the advantage of not requiring the specification of an optimization objective function (e.g. an estimation loss function). We shall first define what is meant by a likelihood function on which the method is based.

Definition 1.6. When treated as a function of $\boldsymbol{\theta}$, the joint pdf or pmf, say $f_{\mathbf{X}}(\mathbf{x}; \boldsymbol{\theta})$ is said to be the *likelihood function* based on $\mathbf{x} = (x_1, \ldots, x_n)$. It is denoted by $L(\boldsymbol{\theta}; \mathbf{x})$.

If $X_1, \ldots, X_n$ is a set of independent random observations from distributions with densities $f_i(x_i; \boldsymbol{\theta})$, $i = 1, \ldots, n$, the likelihood function can be obtained as

$$L(\boldsymbol{\theta}; x_1, \ldots, x_n) = f_1(x_1; \boldsymbol{\theta}) \ldots f_n(x_n; \boldsymbol{\theta}) \ .$$

In particular, if $X_1, \ldots, X_n$ is a random sample from a distribution with pdf or pmf $f(x; \boldsymbol{\theta})$, the likelihood function can be computed as

$$L(\boldsymbol{\theta}; x_1, \ldots, x_n) = \prod_{i=1}^{n} f(x_i; \boldsymbol{\theta}) \ . \tag{1.9}$$

Definition 1.7. Let $L(\boldsymbol{\theta}; \mathbf{X})$ be a likelihood function based on a sample $\mathbf{X} = (X_1, \ldots, X_n)$ from a distribution $F(\mathbf{X}; \boldsymbol{\theta})$, where $\boldsymbol{\theta}$ is a vector of unknown parameters taking values in a parameter space Θ. If $\hat{\boldsymbol{\theta}} = \hat{\boldsymbol{\theta}}(\mathbf{X})$ is a vector of statistics that maximizes $L(\boldsymbol{\theta}; \mathbf{X})$ subject to $\hat{\boldsymbol{\theta}} \in \Theta$, then $\hat{\boldsymbol{\theta}}(\mathbf{X})$ is said to be a *maximum likelihood estimator* of $\boldsymbol{\theta}$.

Given a particular sample $\mathbf{x} = (x_1, \cdots, x_n)$, $\hat{\boldsymbol{\theta}}(\mathbf{x})$ is called the maximum likelihood estimate of $\boldsymbol{\theta}$. This expression is often abbreviated as MLE. The reasoning that motivated MLE is that the set of observations $(x_1, \cdots, x_n)$ is most likely to have come from the distribution with pdf or pmf $f_{\mathbf{X}}(\mathbf{x}; \hat{\boldsymbol{\theta}})$, where $\hat{\boldsymbol{\theta}} = \hat{\boldsymbol{\theta}}(\mathbf{x})$ is the MLE. In other words, the observed sample is most likely to occur when $\boldsymbol{\theta} = \hat{\boldsymbol{\theta}}(\mathbf{x})$. The Bayesian counterpart of this notion can be found in Appendix A.

Example 1.9. MLE of the parameter of exponential distribution

Let $\mathbf{X} = (X_1, \ldots, X_n)$ be a random sample from the exponential distribution with pdf

$$f_X(x) = \alpha e^{-\alpha x} \quad \text{for } x > 0 \,, \tag{1.10}$$

where $\alpha > 0$ is an unknown parameter. In order to find the MLE of α, the likelihood function can be obtained as

$$L(\alpha; \mathbf{x}) = \prod_{i=1}^{n} f_X(x_i) = \alpha^n e^{-\alpha \sum_{i=1}^{n} x_i} \,. \tag{1.11}$$

It is convenient in this case to maximize the log likelihood function

$$L^*(\alpha; \mathbf{x}) = n \log \alpha - \alpha \sum_{i=1}^{n} x_i \,. \tag{1.12}$$

It is now easily seen by taking the first two derivatives that $\hat{\alpha} = n / \sum x_i$, maximizes L^* and the likelihood function. Therefore, $1/\bar{x}$ is the MLE of α, where $\bar{x}$ is the mean of the sample taken from the exponential distribution.

The MLE of the mean of the exponential distribution, namely $\mu = 1/\alpha$ can be derived in a similar manner or deduced from the last result with the aid of what is commonly known as the *invariance property of maximum likelihood estimators*. This property states that if $\hat{\theta}$ is the MLE of θ, then $g(\hat{\theta})$ is the MLE of $g(\theta)$ provided that $g(.)$ has a single-valued inverse. Now it is clear that the MLE of the mean of the exponential distribution is $\hat{\mu} = \bar{x}$.

Example 1.10. Monotone likelihood functions

In some applications the likelihood function can be increasing or decreasing over the parameter space. This example illustrates the method of finding the MLE in such a situation.

Let $X_1, \ldots, X_n$ be a random sample from the density function

$$f_X(x) = \frac{2x}{\alpha^2} \quad \text{for } 0 \le x \le \alpha . \tag{1.13}$$

The likelihood function can be expressed as

$$L(\alpha) = \frac{2^n \prod_{i=1}^{n} x_i}{\alpha^{2n}} \quad \text{for } 0 \le max(\mathbf{x}) \le \alpha . \tag{1.14}$$

Since this is a decreasing function α, in order to maximize L, α needs to be chosen as small as possible subject to the constraint $0 \le max(\mathbf{x}) \le \alpha$. Now it is evident that the MLE of α is $\hat{\alpha} = max(\mathbf{x})$.

1.7. Unbiased Estimation

A class of estimators which plays an important role in statistical methods involving normal distributions, in particular, is that of unbiased estimators. For instance, in analysis of variance and in regression the unbiased estimators of the underlying parameters naturally enter into other types of inferences as well.

Consider again a random sample $X_1, \ldots, X_n$ from a distribution $F(X; \theta)$ and an estimator $\hat{\theta}(\mathbf{X})$ of θ, a parameter of interest. Before the sample is observed $\hat{\theta}$ is a random variable. If the mean of the distribution of this random variable is θ, then we say that $\hat{\theta}$ is an *unbiased estimator* of θ. For many estimation problems usually there is a class of unbiased estimators. For instance, estimators such as X_1, $(X_1 + X_2)/2$, $(2X_1 + X_2)/3$, and $\bar{X}$, the sample mean, are all unbiased estimators of the mean of the distribution. In such situations one would prefer an unbiased estimator having the smallest possible sampling variation defined in an appropriate manner. These considerations provide the motivation for the following definition.

Definition 1.8. An estimator $\hat{\theta}(\mathbf{X})$ of θ is said to be a ***minimum variance unbiased estimator*** (sometimes abbreviated MVUE) of θ if it has the following two properties:

(i) $E(\hat{\theta}(\mathbf{X})) = \theta$; that is $\hat{\theta}$ is an unbiased estimator

(ii) $Var(\hat{\theta}(\mathbf{X})) \le Var(\theta^*(\mathbf{X}))$ for any other estimator $\theta^*(\mathbf{X})$, which is also unbiased for θ; that is $E(\theta^*(\mathbf{X})) = \theta$.

In some applications it is easy to guess the minimum variance unbiased estimator although it may not be easy to show that it satisfies the second property of Definition 1.8. This is particularly true when the MVUE can be obtained from a scale or a location transformation of the MLE of the parameter. In other situations it is not even easy to guess what the MVUE is. Of the following two theorems, the first is useful in the latter case and the second is useful in the former case.

Theorem 1.3 (Rao-Blackwell). Let $\mathbf{T}(\mathbf{X})$ be a sufficient statistic based on $X_1, \ldots, X_n$ and let $\theta^*(\mathbf{X})$ be an unbiased estimator of θ. Consider the statistic defined by the following conditional expectation

$$\hat{\theta}(\mathbf{t}) = E(\,\theta^*(\mathbf{X}) \mid \mathbf{T}=\mathbf{t}\,), \tag{1.15}$$

provided it exists. Then,

(i) $E(\,\hat{\theta}(\mathbf{T})\,) = \theta$; that is, $\hat{\theta}$ is also an unbiased estimator of θ,

(ii) $Var(\,\hat{\theta}(\mathbf{X})\,) \leq Var(\,\theta^*(\mathbf{X})\,)$; that is $\hat{\theta}$ has equal or smaller variance than θ^*.

Theorem 1.4 (Lehmann-Scheffe). Let $\mathbf{T}(\mathbf{X})$ be a complete sufficient statistic. If $\hat{\theta}(\mathbf{T})$ is an unbiased estimator of θ based on $\mathbf{T}$, then $\hat{\theta}$ is a minimum variance unbiased estimator.

A simple proof of Theorem 1.3 can be found in Mood, Graybill, and Boes (1974); Theorem 1.4 is an immediate consequence of Theorem 1.3 and Definition 1.2 of a complete statistic. The following examples illustrate the usefulness of these theorems.

Example 1.11. MVUE of the parameter of a Uniform distribution

Let $X_1, \ldots, X_n$ be a random sample from the uniform distribution $X \sim U(0,\theta)$. Recall from Example 1.2 that $T = \max(\mathbf{X})$ is a sufficient statistic for θ. This statistic is in fact complete sufficient for θ. In view of this result we can expect to find the MVUE as a function of T. It can be easily shown that (see Exercise 1.1) the mean of the distribution of θ is given by

$$E(T) = \frac{n}{n+1}\theta\,.$$

This means that $T = max(\mathbf{X})$ is not quite unbiased for small samples, and yet $T(n+1)/n$ is unbiased. Since this is a complete statistic, it follows from the Lehmann-Scheffe theorem that $\hat{\theta} = \dfrac{n+1}{n}\,max(X_1, \ldots, X_n)$ is the MVUE of θ

Example 1.12. Estimating a survivor probability when the lifetime is exponentially distributed

Let $X_1, \ldots, X_n$ be a random sample of exponentially distributed lifetimes. Let μ be the mean lifetime so that the distribution of the underlying random variable is $X \sim G(1,\mu)$, the gamma distribution with the shape parameter 1 and the scale parameter μ. It was shown in Example 1.9 that the sample mean $\bar{X}$ is the MLE of μ. Since the expected value of the sample mean is the population mean, this estimator is unbiased as well. Moreover, it follows from results concerning exponential families of distributions that this statistic as well as $S = \sum X_i$ are complete. Consequently, $\bar{X}$ is the MVUE of μ.

Let us now turn to the more interesting problem of constructing the MVUE of a survivor probability of the form $\rho = Pr(X \geq x_0) = e^{-x_0/\mu}$, the probability that the unit having the exponential life time distribution will survive at least x_0 units of time. In order to utilize the Rao-Blackwell theorem, consider first the simple estimator defined by

$$\rho^* = \begin{cases} 1 & \text{if } X_1 \geq x_0 \\ 0 & \text{if } X_1 < x_0 \end{cases}.$$

It immediately follows from this definition that ρ^* is an unbiased estimator of ρ, because

$$E(\rho^*) = 1\, Pr(X_1 \geq x_0) + 0\, Pr(X_1 < x_0) = e^{-x_0/\mu}. \quad (1.16)$$

Therefore, we can find the MVUE as $\hat{\rho} = E(\rho^*(X_1) | S = s))$. To evaluate this conditional expectation notice that X_1 and $S - X_1$ are independently distributed as

$$X_1 \sim G(1,\mu) \quad \text{and} \quad S - X_1 \sim G(n-1,\mu)$$

respectively. Consequently, their sum and the ratio are also independently distributed. In fact

$$R = \frac{X_1}{S} \sim B(1,n-1) \quad \text{and} \quad S \sim G(n,\mu) \quad (1.17)$$

are independently distributed. Hence, the required conditional expectation can be found as

$$\hat{\rho} = Pr(X_1 \geq x_0 \mid S = s) = Pr(R \geq \frac{x_0}{s})$$
$$= \int_{x_0/s}^{1} (n-1)(1-r)^{n-2}\, dr = (1 - \frac{x_0}{s})^{n-1} \quad (1.18)$$

for $s > x_0, n > 1$; if these conditions are not satisfied the estimate is zero. Since, $S = \sum X_i$ is a complete sufficient statistic it is now evident that $\hat{\rho} = (1 - x_0/\sum x_i)^{n-1}$ is the MVUE of $\rho = exp(-x_0/\mu)$, thus completing the derivation.

1.8. Least Squares Estimation

A method of estimation particularly useful in linear models that we undertake in Chapters 8 through 10 is the method of least squares. To describe this method with the aid of a particular application, consider the problem of estimating a linear relationship between two variables; this is a particular case of the regression models that we will study in Chapter 10.

In investigating an association between two variables, two concepts that are extensively entertained in scientific research are *correlation* and *regression*. The role of the former is to find out if there is a linear relation between the two variables which the latter is concerned with the task of establishing that relationship; more precise definitions of the concepts will be given below. Consider the association between two variables, Y and X. Let (y_i, x_i) $i=1,\ldots, n$ be a set of observations from the distribution of (Y,X). The pair of sample means and the sample variances of these data are denoted by $(\bar{x},\bar{y})$ and (s_x^2, s_y^2), respectively. A plot of these paired data is called a *scatter diagram*. When X and Y tend to be linearly related, the points on a scatter diagram will tend to cluster around a straight line. A measure of how well the relationship is characterized by a straight line is provided by the *coefficient of correlation* defined as

$$r = \frac{s_{xy}}{s_x s_y}, \text{ where } s_{xy} = \frac{\sum_{i=1}^{n}(x_i - \bar{x})(y_i - \bar{y})}{n} \tag{1.19}$$

is the *sample covariance* between X and Y. It is easily seen that this measure tends to be 0 when there is no linear association between the two variable and tends to be 1 or -1 depending on whether there is a positive or a negative linear relationship between them.

After examining the scatter diagram and the correlation coefficient, an investigator may wish to fit a straight line to the data so that Y can be predicted from X. Then, Y is called the *dependent variable* or the *response variable* of the regression and X is called the *independent variable* or the *explanatory variable*. To obtain the *linear regression* of Y on X, consider the model

$$y_i = \alpha + \beta x_i + e_i, \tag{1.20}$$

where α and β are parameters of the model and e_i is the *residual* or the *error* representing the discrepancy between y_i and the fit $\hat{y}_i = \alpha + \beta x_i$ given by the model. Without any distributional assumptions the parameters of the model can be estimated by minimizing the *error sum of squares* $q = \sum_{i=1}^{n} e_i^2$ with respect to the parameters. This method of parameter estimation is called the *method of least squares* . In estimating the parameters of the simple linear regression of Y on X, the intercept α and the slope β of the straight line are chosen so as to minimize

$$q = \sum_{i=1}^{n} [y_i - (\alpha + \beta x_i)]^2.$$

By differentiating the above equation with respect to α and β and equating the derivatives to zero we get

$$\frac{dq}{d\alpha} = -2\sum_{i=1}^{n}[y_i - \alpha - \beta x_i] = 0$$

and

$$\frac{dq}{d\beta} = -2\sum_{i=1}^{n}[y_i - \alpha - \beta x_i]x_i = 0.$$

By evaluating the second derivatives it is easily seen that the solution of these two equations do indeed minimize the error sum of squares q. Therefore, solving the two equations for α and β we get the *least squares estimates*

$$\hat{\beta} = \frac{\sum_{i=1}^{n} x_i y_i - n\bar{x}\bar{y}}{\sum_{i=1}^{n} x_i^2 - n\bar{x}^2} \quad \text{and} \quad \hat{\alpha} = \bar{y} - \hat{\beta}\bar{x}. \tag{1.21}$$

The straight line $\hat{\alpha} + \hat{\beta}x$ obtained in this manner is called the *least squares regression line* of Y on X. The regression of X on Y can be estimated from (1.21) by interchanging the roles of X and Y.

To illustrate the application of (1.21) consider the data set in Table 1.1. The table gives the scores on the final tests in mathematics and statistics obtained by fifteen students who took both courses in a certain semester. Let us study the association between the two sets of scores. For this purpose let x and y stand for scores in mathematics and statistics, respectively.

Table 1.1. Test scores in mathematics and in statistics

Mathematics	48 52 55 61 66 69 70 70 74 75 78 82 85 90 95
Statistics	54 55 50 67 65 70 74 68 65 73 85 80 85 89 90

A scatter plot of these data are given in Figure 1.1.

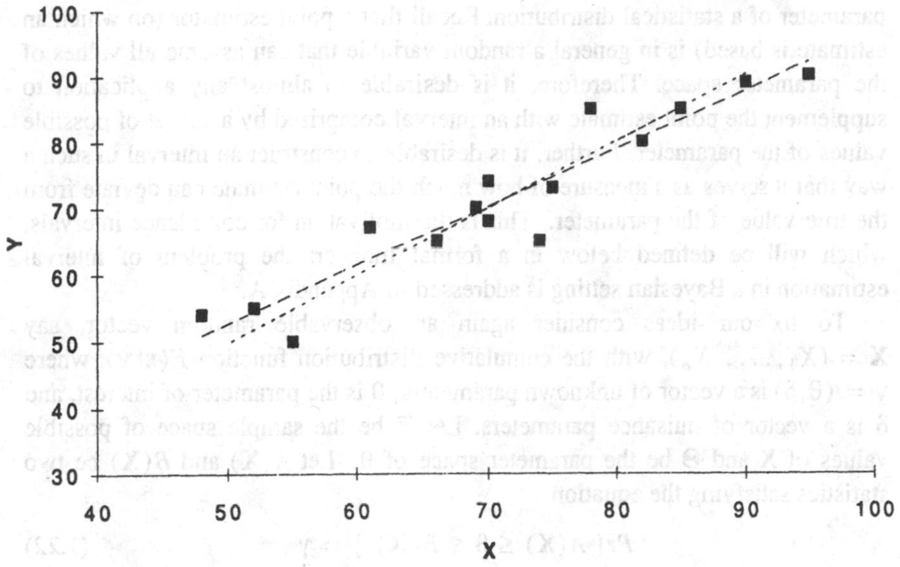

Figure 1.1. Regression of Y on X and regression of X on Y.

The sample means and sample variances of the two sets of data are

$$\bar{x} = 71.33,\ \bar{y} = 71.33,\ s_x^2 = 173.56, \text{ and } s_y^2 = 150.22.$$

The sample covariance of the data is $s_{xy} = 152.16$. Hence, the correlation coefficient is $r = 152.16/\sqrt{(173.56\times150.22)} = .9423$. This suggests a strong positive linear relation between scores in mathematics and statistics. In this application it makes sense to talk about the regression of X on Y as well as the regression of Y on X, which can be employed to predict one score from the other. The two regressions computed using (1.21) are

$$x = -.918 + 1.013y \quad \text{and} \quad y = 8.80 + .8767x,$$

respectively. For example, for a student who scored 50 in mathematics, the predicted score in Statistics is

$$\hat{y} = 8.8 + .8767\times50 = 52.6.$$

The two regression lines are also shown in Figure 1.1.

1.9. Interval Estimation

In the previous two sections we dealt with the problem of point estimation of a parameter of a statistical distribution. Recall that a point estimator (on which an estimate is based) is in general a random variable that can assume all values of the parameter space. Therefore, it is desirable in almost any application to supplement the point estimate with an interval comprised by a subset of possible values of the parameter. Further, it is desirable to construct an interval in such a way that it serves as a measure of how much the point estimate can deviate from the true value of the parameter. This is the motivation for confidence intervals, which will be defined below in a formal manner; the problem of interval estimation in a Bayesian setting is addressed in Appendix A.

To fix our ideas consider again an observable random vector, say $\mathbf{X} = (X_1, \ldots, X_n)$, with the cumulative distribution function $F(\mathbf{x}|\, \nu)$, where $\nu = (\theta, \delta)$ is a vector of unknown parameters, θ is the parameter of interest, and δ is a vector of nuisance parameters. Let Ξ be the sample space of possible values of $\mathbf{X}$ and Θ be the parameter space of θ. Let $A(\mathbf{X})$ and $B(\mathbf{X})$ be two statistics satisfying the equation

$$Pr[\, A(\mathbf{X}) \le \theta \le B(\mathbf{X})\,] = \gamma\,, \tag{1.22}$$

where γ is a prespecified constant between 0 and 1. If the observed values of the two statistics are $a = A(\mathbf{x})$ and $b = B(\mathbf{x})$, then it is said that $[a,b]$ is a *confidence interval* for θ with the *confidence coefficient* γ. If $A(\mathbf{X})$ is found such that

$$Pr[\, A(\mathbf{X}) \le \theta\,] = \gamma\,,$$

then $[a,\infty) \cap \Theta$ is said to be a right-sided confidence interval for θ, where Θ is the parameter space. Similarly, left-sided confidence intervals can be constructed using a probability statement of the form $Pr[\, B(\mathbf{X}) \ge \theta\,] = \gamma$. Depending on the application, a left-sided confidence interval, a right-sided confidence interval, a shortest possible interval, or some other interval might be desirable.

In general we may be interested in a vector of parameters, say $\boldsymbol{\theta}$. In such situations we will be interested in confidence regions for $\boldsymbol{\theta}$ and the probability statements leading to such regions will be of the form

$$Pr[\, \mathbf{X} \in C_\gamma(\boldsymbol{\theta})\,] = \gamma\,,$$

where $C_\gamma(\boldsymbol{\theta})$ is a subset of the sample space Ξ.

Typically γ is a quantity that must be set to be equal to a large number on the scale $[0,1]$ so that one can be highly confident that the interval $[a,b]$ would contain the true value of the parameter θ. The nominal values of γ typically used in many applications are .9, .95 and .99. For instance if γ=.95, the interval $[a,b]$ obtained in this manner is called a 95% confidence interval.

Once the interval $[a,b]$ is computed using a particular sample, the interval may or may not cover the true value of the parameter. Nevertheless, the probability statement on which the confidence interval is based has some implications in repeated sampling. To describe this, consider an investigator who constructs, for instance, 95% confidence intervals in all situations of interval estimation. Then, it follows from the law of large numbers that approximately 95% of intervals would cover the true values of underlying parameters. Of course this is also the case if the same interval estimation problem is repeated with samples of fixed size taken from the same population. These properties are known as the repeated sampling properties of confidence intervals.

When there exists a confidence interval with a certain confidence coefficient, usually there is a family of confidence intervals. This is the case even when one is interested in a one-sided confidence interval. For instance, if $X_1, \ldots, X_n$ is a random sample from the uniform distribution over $[0,\theta]$, it is possible to construct left-sided confidence intervals for θ on the basis of X_1, $\bar{X}$, $max(\mathbf{X})$, etc. In such situations we prefer a confidence interval with the shortest possible expected length. When the underlying distribution is free of nuisance parameters one can usually find a confidence interval with desirable properties by means of a sufficient statistic. In this case as well as when nuisance parameters are present the process of constructing confidence intervals can be facilitated by searching for a function of sufficient statistics with the following property.

Definition 1.8. Let $R = r(\mathbf{X};\boldsymbol{\theta})$ be a real-valued function of $\mathbf{X}$ and $\boldsymbol{\theta}$, the parameters of interest. If the distribution of the random variable R does not depend on θ, then R is called a *pivotal quantity*.

If a pivotal quantity is available, a confidence interval with confidence coefficient γ can be be found by solving the inequalities

$$r_1 \leq r(\mathbf{x};\, \boldsymbol{\theta}) \leq r_2 \tag{1.23}$$

for θ, where the real numbers r_1 and r_2 are chosen such that

$$Pr(r_1 \leq r(\mathbf{X};\, \boldsymbol{\theta}) \leq r_2) = \gamma\,.$$

When there is no simple sufficient statistic, it may be still possible to facilitate the construction of a pivotal quantity with the aid of the *probability integral transform*, which states that the cdf of a continuous random variable has a uniform distribution over $[0,1]$. As a result, if $X_1, \ldots, X_n$ is a random sample from a certain population with cdf $F_X(x;\, \theta)$, then, $U = F_X(X)$ has a uniform distribution and $-log(U)$ has an exponential distribution with mean 1; that is $-log(F_X(X)) \sim G(1,1)$. Consequently,

$$R(\mathbf{X};\, \theta) = -\sum_{i=1}^{n} \log F_X(X_i;\, \theta) \sim G(n,1)\,. \tag{1.24}$$

Hence, R is a pivotal quantity for θ. These procedures are illustrated by the following examples.

Example 1.13. Confidence intervals for the mean of an exponential distribution

Let $X_1, \ldots, X_n$ be a random sample from an exponential population with mean μ; that is $X \sim G(1,\mu)$. Let us construct a right-sided confidence interval and an equal-tail confidence interval for μ. It was shown in Example 1.1 that $T = \sum X_i$ is a sufficient statistic. In order to find a pivotal quantity based on this sufficient statistic, notice that $T \sim G(n,\mu)$, the gamma distribution with shape parameter n and the scale parameter μ. Consequently the random variable $2T/\mu$ has a chi-square distribution with $2n$ degrees of freedom, a distribution free of unknown parameters. Therefore, $2T/\mu$ is a pivotal quantity for μ.

A right-sided 95% confidence interval for μ can be constructed by means of the probability statement

$$Pr\left(\frac{2\sum X_i}{\mu} \le \chi^2_{2n,.95}\right) = .95 ,$$

where $\chi^2_{2n,.95}$ is the .95 quantile of the chi-square distribution with 2n degrees of freedom. Now it is clear that

$$\frac{2\sum X_i}{\chi^2_{2n,.95}} \le \mu \tag{1.25}$$

is the right-sided 95% confidence interval based on the sufficient statistic. Similarly, the equal-tail confidence intervals based on T can be found using probability intervals of the form

$$Pr\left(\chi^2_{2n,.025} \le \frac{2\sum X_i}{\mu} \le \chi^2_{2n,.975}\right) = .95 ,$$

which in turn yield confidence intervals of the form

$$\frac{2\sum x_i}{\chi^2_{2n,.975}} \le \mu \le \frac{2\sum x_i}{\chi^2_{2n,.025}} . \tag{1.26}$$

Since the gamma distribution is not symmetric about its mean, an explicit formula for the shortest 95% confidence interval based on T is not available; this can be accomplished by numerical methods, however.

Example 1.14. Confidence intervals for the median of Cauchy distribution

Let $X_1, \ldots, X_n$ be a random sample from the Cauchy distribution with probability density function

$$f_X(x) = \frac{1}{\pi(1 + (x-\theta)^2)} ,$$

where θ is the median of the distribution. Obviously, for θ, there is no nontrivial sufficient statistic that can reduce the complexity of the problem. The cumulative distribution function of X is

$$F_X(x) = \frac{1}{2} + \frac{1}{\pi} \tan^{-1}(x - \theta) .$$

It follows from equation (1.24) that

$$R(\mathbf{X}; \theta) = -\sum_{i=1}^{n} \log\left[\frac{1}{2} + \frac{1}{\pi} \tan^{-1}(X_i - \theta)\right] \sim G(n,1) ,$$

which can be employed as a pivotal quantity. Since $R(\mathbf{X}; \theta)$ is an increasing function of θ, confidence intervals of θ can be constructed using (1.23). The procedure needs to be implemented by numerical methods as (1.23) does not yield explicit solutions in this case.

Exercises

1.1. Let $X_1, \ldots, X_n$ be a random sample from the uniform distribution with density function

$$f_X(x) = \frac{1}{\theta} \qquad \text{for } 0 \le x \le \theta .$$

(a) Show that the cdf of $T = max(\mathbf{X})$ is

$$F_T(t) = \left(\frac{t}{\theta}\right)^n \qquad \text{for } 0 \le t \le \theta .$$

(b) Hence deduce the pdf of T. (c) Show that $E(T) = \frac{n}{n+1}\theta$. (d) Hence or otherwise find the expected value of the random variable $Y = min(\mathbf{X})$, the minimum of all the observations. (e) Find the cdf and the pdf of Y. (f) Examine whether the uniform family of distributions parameterized by θ is an exponential family.

1.2. Let $X_1, \ldots, X_n$ be a random sample from the uniform distribution with density function

$$f_X(x) = \frac{1}{\theta} \qquad \text{for } 0 \le x \le \theta .$$

Perform the following inference problems on the basis of the minimal sufficient statistic for θ:

(a) find the MLE of θ;

(b) find the left-sided 99% confidence interval for θ;

(c) find the equal-tail 99% confidence interval for θ;

(d) find the shortest 99% confidence interval for θ.

1.3. Let $X_1, \ldots, X_n$ be a random sample from the uniform distribution with density function

$$f_X(x) = \frac{1}{\theta} \qquad \text{for } -\theta \le x \le 0 .$$

Examine whether the family of distributions parameterized by θ is affine invariant. Find a complete sufficient statistic for θ. Show that $T = -min(\mathbf{X})$, the negative of the minimum of all the observations, is the maximum likelihood estimator of θ. Find the minimum variance unbiased estimator of θ.

1.4. Let $X_1, \ldots, X_n$ be a random sample from the density function

$$f_X(x) = \frac{2x}{\alpha^2} \quad \text{for } 0 \le x \le \alpha .$$

(a) Show that $3X_1/2$ is an unbiased estimator of α. (b) Does the MLE exist? If so find it. (c) Does the MVUE exist? If so find it. (d) Find a $100\gamma\%$ right-sided confidence interval for α based on the statistic $T = max(\mathbf{X})$.

1.5. Let $X_1, \ldots, X_n$ be a random sample from the two-parameter exponential distribution

$$f_X(x) = \frac{1}{\beta} e^{-(x-\alpha)/\beta} \quad \text{for} \quad x > \alpha , \quad \beta > 0$$

Is this an exponential family of distributions? Show that the distribution is affine invariant. Find the the induced group of transformation on the parameter space.

1.6. Let $X_1, \ldots, X_n$ be a random sample from the exponential distributions with parameter α. The mean of the distribution is α^{-1}. (a) Find the MVUE of α. (b) Find the MLE of $\sigma^2 = Var(X)$. (c) Find the MVUE of σ^2

1.7. Let $X_1, \ldots, X_n$ be a random sample from the exponential distribution with mean θ

(a) Is there a natural conjugate family of distributions (see Appendix A) for θ? If so find the form of the posterior distribution of θ. (b) Find the posterior mean estimator and the posterior mode estimator under the noninformative prior of θ. (c) Find a $100\gamma\%$ Bayesian interval estimator of θ. (d) Find the posterior distribution of θ under the improper prior $\pi(\theta) = 1$. (e) Find the posterior mean estimator of $Var(X)$ given by that prior.

1.8. Let $X_1, \ldots, X_m$ and $Y_1, \ldots, Y_n$ be two sets of random samples from exponential distributions with means θ and β respectively. (a) Show that that the two sample means are independently distributed according to gamma distributions. (b) In terms of the sample means define a pivotal quantity which is distributed as an F-distribution. (c) Hence construct a left-sided 95% confidence interval for the ratio of the two means θ and β, namely $\rho = \theta/\beta$. (d) Find the MLE and the MVUE of ρ.

1.9. Let $X_1, \ldots, X_n$ be a random sample from the logistic distribution

$$F_X(x) = \frac{1}{1 + e^{-\theta x}} .$$

Assuming the improper prior $\pi(\theta) = 1$, establish procedures for constructing classical confidence intervals and Bayesian credible intervals for θ.

1.10. Suppose that X and Y are two random variables having geometric distributions with parameters α and β respectively. Assume that the two random variables are independently distributed. Given the data $x = 10$ and $y = 8$ from the two distributions, construct a 95% credible interval for the parameter $\theta = \alpha - \beta$ under noninformative priors.

1.11. Let $X_1, \ldots, X_n$ be a set of independent observations from the negative binomial distribution with probability mass function

$$f(x) = \binom{r+x-1}{x}(1-p)^r p^x \quad \text{for } x = 0, 1, 2, \cdots,$$

where r is a known positive integer and p is an unknown parameter. (a) Examine whether this an exponential family of distributions. (b) Show that $\sum X_i$ is a complete sufficient statistic for p. (c) Find the noninformative prior of p. (d) If the prior of p is a beta distribution with parameters α and β, find the posterior distribution of p. (e) Find the posterior mean estimator of p. (e) Find the generalized maximum likelihood estimator of p. (f) Establish the form of 95% left-sided credible intervals for p.

1.12. Let $X_1, \ldots, X_n$ be a random sample from the uniform distribution over the interval $[0, \theta]$. Suppose that θ has an exponential prior distribution with hyper parameter α. Show that the posterior distribution is a truncated gamma distribution. Find the generalized maximum likelihood estimate of θ.

1.13. Let $X_1, \ldots, X_n$ be a random sample from the normal distribution with pdf

$$f_X(x) = \frac{1}{\sqrt{2\pi}} e^{-\frac{1}{2}(x-\mu)^2},$$

where μ is the mean of the distribution. (a) Show that the noninformative prior of μ is $\pi(\mu) = 1$. (b) Show that the posterior distribution of μ is a normal distribution with mean $\bar{x}$ and variance $1/n$. (c) Find the 95% left-sided Bayesian confidence interval for μ. (d) Find the 95% left-sided classical confidence interval for μ. (e) Find the 95% HPD interval for μ.

Chapter 2

Notions in Significance Testing of Hypotheses

2.1. Introduction

In this book often we perform hypothesis testing on the basis of what is defined later as *p-value*, as opposed to testing at a fixed nominal level. The p-value is sometimes called the 'observed level of significance' or the 'significance value'. As a data dependent quantity, it serves as a measure of evidence in favor of or against a certain null hypothesis. The practice of making a judgement on a certain hypothesis with a p-value is often called significance testing. It is good statistical practice to report p-values in any study involving tests of hypotheses, especially in biomedical research. This way, the judgment of the adequacy of evidence and significance of study findings can be left to the experts and decision makers. Fixed level testing is useful, however in a few applications such as statistical quality control, and therefore we will provide fixed-level tests as well when they are available. With the present state of modern computing and statistical software, which allows a practitioner to compute the probability coverage under standard distribution in a matter of seconds, one can hardly find an excuse to do otherwise.

From a historical perspective, ironically, it was the idea of significance testing, and not the decision-theoretic fixed-level testing, that was practiced by such pioneers as Pearson(1900), Gossett (1908) and Fisher (1956). Unfortunately, the notion was not well defined and as a result fixed level testing in the form of making an 'accept/reject' decision became popular after the celebrated Neyman-Pearson fundamental lemma which was followed by a series of elegant and important theories. Moreover, before modern computing facilities became available, practitioners had to rely on tabulated quantiles of statistical distributions to perform tests of hypotheses.

Unlike conventional fixed-level testing, not only does significance testing provide more information to a practitioner, but also it requires no specification of a subjective nominal level such as .01 and .05. Testing of hypotheses with p-values as well as interval estimation are more general and informative than fixed-level testing. In particular, when the underlying family of distributions is discrete, the tests with size at nominal levels are not usually attainable and so one has to resort to fixed-level tests based on asymptotic approximations; interval estimation is also not always possible with discrete distributions. On the other hand, with standard discrete distributions, there is no difficulty computing exact p-values. In applications such as those in biomedical experiments fixed-level tests are not appealing even when they do exist, because the sufficiency of evidence in favor of or against a hypothesis would depend upon the prevailing circumstances and what is being tested, and should be left for the other experts and decision makers to judge.

The limited applicability of fixed-level testing is much more serious than would at first appear. Except for very special cases, the nonexistence of fixed-level tests is a typical problem in multi-parameter testing situations. For instance, as illustrated in Chapter 5, this is the case even in a simple testing situation of comparing the means of two exponential distributions. In statistical design of experiments, practitioners often make the undesirable assumption of equal variances just for the sake of mathematical tractability of the underlying fixed-level tests. When the variances are quite different, the classical F-test may lead to serious repercussions. In Chapter 8 we will demonstrate how to perform exact significance testing in this situation without the assumption of equal variances.

2.2. Test Statistics and Test Variables

Suppose $\mathbf{X}$ is an observable random vector from a certain family of distributions parameterized by a vector of unknown parameters $\zeta = (\theta, \delta)$, where θ is the parameter of interest and δ is a vector of nuisance parameters. Let $F_X(\mathbf{x}; \zeta)$ be the distribution function of $\mathbf{X}$. Let Θ be the parameter space of possible values of θ. The underlying hypothesis specifies subsets of Θ as regions to which θ belong. Let Ξ be the sample space of possible values of $\mathbf{X}$ and let $\mathbf{x}$ be the observed value of $\mathbf{X}$. Statistical inferences on θ are to be based on $\mathbf{x}$.

The observed significance levels are defined on the basis of what is known as data-based extreme regions defined on Ξ. Typically an extreme region corresponds to the tails of a distribution with boundaries determined by the observed value of the random variable. A proper definition of an extreme region requires a stochastic ordering of the sample space according to the magnitude of the parameter of interest. In many applications this can be accomplished by the use of test statistics defined below. Other more general methods of ordering the sample space will be discussed in Chapter 5.

Definition 2.1. A statistic (a real-valued function of **x**), $T(\mathbf{x})$ is said to be a *test statistic* for θ if it has the following two properties:

Property 1. The distribution of $T = T(\mathbf{X})$ is free of the nuisance parameter δ.

Property 2. The cumulative probability distribution function of T, namely $F_T(t) = Pr(T \le t)$, is a monotonic function of θ for any given t.

If $Pr(T > t)$ is a nondecreasing function of θ, we say that T is *stochastically increasing* in θ. We call an observable random quantity a test statistic only if it has each of these two properties. Property 1 is imposed so that probabilities of T are computable when the parameter of interest is specified. This property is the counterpart to the notion of *similarity* which in turn is related to the notion of *unbiasedness* in fixed-level testing. Property 2 requires that T is stochastically increasing or decreasing in θ to enable an ordering of the sample space.

A real-valued function of **X** and θ of the form $T = T(\mathbf{X};\ \theta)$ is called a *test variable* if $T(\mathbf{X};\ \theta_0)$ is a test statistic when the value of the parameter is known to be θ_0. The quantity $t_{obs} = T(\mathbf{x};\ \theta)$ will be referred to as the *observed value* of the test variable, where **x** is the observed value of **X**. It will become clear that the existence of a test variable is sufficient to define measures of how well the data support a certain hypothesis.

In many statistical problems of practical importance, the search for a test statistic can be confined to a set of *sufficient statistics*. With no loss of information, usually we can base statistical inferences including hypotheses testing on a set of minimal sufficient statistics. In problems involving no nuisance parameters, it may be possible to find a single statistic (a real-valued function of data), say $T(\mathbf{x})$, which is sufficient for θ, the parameter of interest. In this case we can merely check whether T has Property 2 of a test statistic. If the underlying distribution of **X** has a number of unknown parameters the problem of finding a test statistic or a test variable is somewhat complicated. Some methods that address this problem will be discussed in the following sections and in Chapter 5, in particular.

Example 2.1. Exponential distribution

Let $X_1, \ldots, X_n$ be a random sample of size n from an exponential distribution with parameter θ. The probability density function of the underlying random variable X is

$$f_x(x) = \frac{1}{\theta} e^{-\frac{x}{\theta}} \quad \text{for } x > 0,$$

where θ is the mean of the distribution.

It was shown in Example 1.1 that $T = \sum X_i$ is a sufficient statistic for θ.

To show that T is indeed a test statistic, notice that, being a sum of exponential random variables, T has a gamma distribution with parameters n and θ. This we write as $T \sim G(n,\theta)$. The probability density function of T is

$$f_T(t) = \frac{1}{\Gamma(n)\,\theta^n} t^{n-1} e^{-\frac{t}{\theta}} \quad \text{for } t > 0 . \tag{2.1}$$

It is clear from this density function that $W = T/\theta \sim G(n, 1)$. Therefore, the cumulative distribution function of T can be expressed in the form

$$\begin{aligned} F_T(t) &= Pr(W \le t/\theta) \\ &= F_W(t/\theta) \quad , \end{aligned}$$

where F_W is the cumulative distribution function of W. It is evident now that $F_T(t)$ is a decreasing function of θ. This means that T is stochastically increasing in θ. Hence, T is a test statistic for testing θ.

Example 2.2. Beta distribution

Suppose that $X_1, \ldots, X_n$ are independently and identically distributed according to a particular beta distribution with density function

$$f_X(x) = \theta\, x^{\theta-1} \quad , \quad \text{for} \quad 0 < x \le 1,$$

where θ is a positive parameter. We saw in Example 1.5 that $T(\mathbf{X}) = \prod X_i$ is a sufficient statistic for θ. To establish that T is also a test statistic for θ, note first of all that the random variables $Y_i = X_i^\theta$,$i = 1, \ldots, n$ have independent uniform distributions on the $(0 , 1]$ interval. In terms of these uniform random variables the distribution function of T can be expressed as

$$\begin{aligned} F_T(t) &= Pr(\textstyle\prod X_i \le t) \quad \text{for } 0 < t \le 1 \\ &= Pr(\textstyle\prod Y_i \le t^\theta) \\ &= F_W(t^\theta), \end{aligned}$$

where F_W is the cumulative distribution function of the random variable $W = \prod Y_i$. Since the distribution of W is free of θ, it is now clear that the distribution of T is a decreasing function of $\theta > 0$, thus indicating that T is a test statistic.

We have assumed so far that when θ is the only unknown parameter of the underlying family of distributions, the information in a random sample taken from the distribution can be summarized by a single sufficient statistic. In some applications this may not be the case, and yet it may be still possible to find an appropriate test statistic with the aid of the probability integral transform.

Let $X_1, \ldots, X_n$ be a random sample from a continuous distribution $F(x; \theta)$ with no nuisance parameters. Then, $U = F(X)$ has a uniform distribution and $-log(U)$ has an exponential distribution with parameter 1. Consequently,

$$T(\mathbf{X};\, \theta) = -\sum_{i=1}^{n} \log F(X_i;\, \theta) \sim G(n,1)\ . \qquad (2.2)$$

The random quantity T defined in (2.2) forms a test variable with a known distribution if the cdf $F(x)$ of X is a monotonic function of θ. The procedure of finding a test variable by this approach is illustrated by Example 2.3.

Example 2.3. Cauchy distribution

Let $X_1, \ldots, X_n$ be a random sample from the Cauchy distribution discussed in Example 1.14. The cumulative distribution function of X is

$$F_X(x) = \frac{1}{2} + \frac{1}{\pi} \tan^{-1}(x - \theta)\ ,$$

which is a decreasing function of θ. Therefore,

$$T(\mathbf{X}) = -\sum_{i=1}^{n} \log \left(\frac{1}{2} + \frac{1}{\pi} \tan^{-1}(X_i - \theta) \right)$$

is a test variable with a known distribution.

2.3. Definition of p-Value

In this book we consider one-sided null hypotheses and point null hypotheses only. These are the most important types of hypotheses that one would usually encounter in practical applications. Let us first consider one-sided null hypotheses for which the definition of the p-value is simple; the definition of the p-value for point null hypothesis will be given later in this section. To be specific, consider the problem of testing left-sided null hypotheses of the form

$$H_0\!: \theta \le \theta_0 \quad \text{against} \quad H_1\!: \theta > \theta_0\ , \qquad (2.3)$$

where θ is the parameter of interest, θ_0 is a specified value of the parameter, and H_1 is the alternative hypothesis.

In the literature on significance testing, the p-value is often defined as the probability of observing a point in the sample space which can be considered as extreme as, or more extreme than, the observed sample. A precise statement of these ideas requires a well defined stochastic ordering of the sample space. In this chapter we confine our exposition to those situations where such an ordering is provided by a test statistic or a test variable of the form $T(\mathbf{X};\, \theta)$. It should be emphasized that there are more general ways of ordering the sample space and, therefore, the problem will be revisited in Chapter 5. The Bayesian treatment of the hypothesis testing problem is presented in Appendix A.

Assume without loss of generality that T is stochastically increasing in θ; that is $Pr(T > t)$ is a nondecreasing function of θ. Notice in this case that the larger values of T can be considered as extreme values of the distribution of T

under H_0. Now we are in a position to define what is meant by an extreme region of the sample space, and in turn the p-value.

Definition 2.2. If the test variable $T = T(\mathbf{X}, \theta)$ is stochastically increasing, then a *data-based extreme region* for H_0: $\theta \le \theta_0$ is defined as the subset $C_{\mathbf{x}}$ of the sample space given by

$$C_{\mathbf{x}} = \{\mathbf{X} \in \Xi : T(\mathbf{X},\theta) \ge T(\mathbf{x},\theta)\} .$$

Definition 2.3. If $C_{\mathbf{x}}$ is an extreme region, its p-value, (the observed level of significance) is defined as

$$p = \sup_{\theta \le \theta_0} Pr(\mathbf{X} \in C_{\mathbf{x}} | \theta) .$$

If $T = T(\mathbf{X},\theta)$ is the test variable on which an extreme region is based, the p-value can be conveniently computed as

$$p = Pr(T \ge t_{obs} \mid \theta = \theta_0) , \tag{2.4}$$

provided that T is stochastically increasing in θ, where $t_{obs} = T(\mathbf{x},\theta_0)$. The functions

$$\pi(\theta) = Pr(T(\mathbf{X},\theta) \ge t_{obs} \mid \theta) \tag{2.5}$$

and

$$\pi_0(\theta) = Pr(T(\mathbf{X},\theta_0) \ge t_{obs} \mid \theta)$$

are called the *data-based power functions* of the test variable $T(\mathbf{X},\theta)$ and the test statistic $T(\mathbf{X},\theta_0)$, respectively. Of these two definitions of power functions, while the latter is the one that can be used in comparisons of alternative extreme regions, the former is useful in stating the implications of p-values in fixed-level testing and in interval estimation. If the distribution of $T(\mathbf{X};\theta_0)$ does not depend on nuisance parameters, usually $\pi(\theta) = \pi_0(\theta)$ when θ is a location parameter or a scale parameter.

Since the optimality of most powerful tests in fixed-level testing is readily incorporated in the definition of test variables based on sufficient statistics and that of p-values, power functions do not play a very important role in significance testing. However, they are particularly useful in presenting propositions without specifying and restricting to specific hypotheses.

If T is stochastically decreasing in θ, the observed level of significance is given by $p = Pr(T \le t_{obs} \mid \theta = \theta_0)$. Moreover, it is easily seen that the p-value for testing

$$H_0\colon \theta \ge \theta_0 \quad \text{against} \quad H_1\colon \theta < \theta_0$$

can be computed using the formula $p = Pr(T \le t_{obs} | \theta = \theta_0)$, or the formula $p = Pr(T \ge t_{obs} | \theta = \theta_0)$ according as T is stochastically increasing or decreasing in θ.

In each of these cases the p-value serves as a measure (on the scale [0,1]) of how well the data support or discredit the null hypothesis; the smaller the p-value, the greater the evidence against the null hypothesis. In other words, smaller values of the observed level of significance favor the alternative hypothesis and its larger values favor the null hypothesis. As the null hypothesis is treated as the hypothesis being tested, given a choice of p-values based on a number of test variables, the test variable with the smallest p-value is preferred as the measure providing most evidence against the null hypothesis (see Thompson (1985) for a discussion of these ideas and for some related results); if there are two test statistics with the same p-value, the one with the larger power function for all $\theta > \theta_0$ (if one such exists) is preferred. In the case of test variables, the comparisons can be made only in terms of the power function $\pi_0(\theta)$. In many important problems involving no nuisance parameters an optimal test statistic in this sense is provided by a minimal sufficient statistic. In such situations the optimality of the conventional likelihood ratio procedure in fixed-level testing is readily imbedded in our definition of a p-value based on the minimal sufficient statistic. The methods of dealing with significance testing situations with nuisance parameters will be addressed in the following three sections and in Chapter 5.

Example 2.4. Exponential distribution (continued)

Suppose that $X_1, \ldots, X_n$ form a random sample from an exponential population with mean θ. It is seen in Example 2.1 that $T = \sum X_i$ is a test statistic and that T is stochastically increasing in θ. Let $t_{obs} = \sum x_i$ be the observed value of the test statistic. Consider the following hypotheses:

$$H_0: \theta \le \theta_0,\ H_1: \theta > \theta_0 .$$

Since $T \sim G(n,\theta)$, the random variable $2T/\theta$ has a chi-square distribution with $2n$ degrees of freedom; that is the gamma distribution $G(n,2)$. The appropriate p-value in this application can be computed as

$$\begin{aligned} p(t_{obs}) &= Pr(T \ge t_{obs} \mid \theta = \theta_0) \\ &= 1 - C_{2n}(2t_{obs} / \theta_0) \quad , \end{aligned} \tag{2.6}$$

where the function C_{2n} is the cdf of χ^2 distribution with 2n degrees of freedom. Hence, significance tests on the above hypotheses can be based on the p-value given by (2.6). The corresponding data-based power function is, of course, $\pi_0(t_{obs};\theta) = 1 - C_{2n}(2t_{obs} / \theta)$. A Bayesian treatment of this problem can be found in Appendix A.

Example 2.5. Testing the parameter of a uniform distribution

Suppose n independent observations are available from the uniform distribution with density function

$$f_X(x) = \frac{1}{\theta} \quad \text{for } 0 \le x \le \theta .$$

The hypotheses

$$H_0: \theta \ge \theta_0, \quad H_1: \theta < \theta_0 \tag{2.7}$$

are to be tested using a random sample, say $\mathbf{X} = (X_1, \ldots, X_n)$. We saw in Example 1.2 that $T(\mathbf{X}) = max\ (\mathbf{X})$ is a sufficient statistic for θ.

The cumulative distribution function of T is

$$\begin{aligned} F_T(t) &= Pr\left[max\ (\mathbf{X}) \le t \right], \quad 0 \le t \le \theta \\ &= Pr(X_1 \le t, \ldots, X_n \le t) \\ &= \prod_{i=1}^{n} Pr\ (X_i \le t) \\ &= \left[\frac{t}{\theta} \right]^n , \end{aligned}$$

which is a decreasing function of θ. In other words, T is stochastically increasing in θ. Since the null hypothesis in (2.7) is right-tailed, the appropriate extreme region corresponds to the left tail of the distribution of T. Therefore, (2.7) can be tested on the basis of the observed level of significance

$$p(t_{obs}) = \left[\frac{t_{obs}}{\theta_0} \right]^n ,$$

where t_{obs} is the observed value of $max\ (\mathbf{x})$.

Example 2.6. Testing the parameter of a geometric distribution

Bernoulli trials with probability of a success p are repeated until a success is obtained. If 14 failures are obtained before the first success, test the null hypothesis,

$$H_0: p \le .02 \quad \text{against} \quad H_1: p > .02 \tag{2.8}$$

Let X denote the random variable representing the number of failures which occur before the first success is obtained. Clearly X has a geometric distribution with the probability mass function

$$f_X(x) = p\,(1-p)^x \quad \text{for } x = 0,1,2,\ldots.$$

Tests concerning p can be based on X itself. The cumulative distribution function of X is

$$\begin{aligned} F_X(x) &= Pr\,(X \le x) \\ &= \sum_{i=0}^{x} pq^i \\ &= p\,\frac{(1-q^{x+1})}{(1-q)} \\ &= 1 - q^{x+1} \;, \end{aligned}$$

where $q = 1-p$. Obviously, $F_X(x)$ is an increasing function of p, and therefore, X is a test statistic for p. Since X is stochastically decreasing in p, the left tail of the distribution corresponds to the extreme region on which significance tests of (2.8) can be based. Therefore, the observed level of significance corresponding to the observed value $x = 14$ is

$$\begin{aligned} p &= F_x(14 \mid p = .02) \\ &= 1 - .98^{15} \\ &= .26 \end{aligned}$$

According to this p-value, $x = 14$ can not be considered as an extreme value under H_0. There is almost no evidence against the null hypothesis. A Bayesian treatment of the testing problem pertaining to the geometric distribution can be found in Appendix A.

The case of point null hypotheses

Now consider hypotheses of the form

$$H_0 \colon \theta = \theta_0 \quad H_1 \colon \theta \neq \theta_0 \,, \tag{2.9}$$

where θ_0 is a specified constant. Suppose $T = T(\mathbf{X})$ is a test statistic satisfying properties 1 and 2. When the distribution of T is not symmetric about θ there is no common agreement about the definition of the observed level of significance. Gibbons and Pratt (1975) discussed (also see Cox and Hinkley (1974)) alternative definitions of p-values on which significance tests on H_0 can be based. In this book we use a natural extension of Definition 2.3 when there exists a function of T which tends to take larger values for greater discrepancies between θ and θ_0.

Definition 2.4. Let T be a test statistic of θ and let C_t be a subset of τ, the sample space of T. Suppose C_t has the following property:

Property 3. Given any fixed t and δ, the probability, $Pr(T \in C_t)$ is a nondecreasing function of (i) $\theta - \theta_0$ when $\theta \ge \theta_0$, and (ii) $\theta_0 - \theta$ when $\theta < \theta_0$; more or less, this means that there exists a function of T which is stochastically increasing in $|\theta - \theta_0|$.

Sometimes, the following milder property is considered adequate for defining an extreme region; it should be noted, however, that a test having Property 3 is equally or more powerful than a test with the following property.

Property 3'. Given any fixed t and δ, $Pr(T \in C_t | \theta) \geq Pr(T \in C_t | \theta_0)$ for all $\theta \in \Theta$.

A subset of the sample space of T having at least Property 3' is considered as an extreme region for testing (2.9). The p-value for testing H_0: $\theta = \theta_0$ against H_1: $\theta \neq \theta_0$ based on an extreme region $C_{t_{obs}}$ is defined as

$$p = Pr(T \in C_{t_{obs}} | \theta = \theta_0), \tag{2.10}$$

where t_{obs} is the observed value of the T statistic. The corresponding data-based power function is

$$\pi_0(\theta; t_{obs}) = Pr(T \in C_{t_{obs}} | \theta).$$

When θ is a location parameter of the underlying family of distribution one can easily find, with the aid of Theorem 2.1, an extreme region based on T satisfying Property 3, provided that T has a unimodal distribution. This procedure will be illustrated in the next example. When θ is a scale parameter, the method can be applied perhaps upon logarithmic transformations of appropriate quantities. In other situations one may be able to define an appropriate p-value for a point null hypothesis by exploiting special properties of the underlying distribution such as the monotone likelihood ratio property. The proofs of the following results are given in Appendix B.

Theorem 2.1. Let T be a continuous random variable with the unimodal probability density function $f(t)$ with the mode at μ. Let M be the parameter space and let τ be the support of $f(t) = f(t;\mu)$. Assume that, given any $w > 0$ such that $t = \mu + w \in \tau$, there exists $k(w) > 0$ such that $\mu - k(w) \in \tau$ and that $f(\mu - k(w)) = f(\mu + w)$. Then, given $\mu_0 \in M$, the probability $Pr(-k(w) \leq T - \mu_0 \leq w)$ is a nonincreasing function of $|\mu - \mu_0|$.

Corollary 2.1. Suppose X is an observable random variable and $Y = X - \theta$ is a continuous random variable with a unimodal probability density function $f_Y(y)$, which does not depend on the parameter θ. Given an observed value x of X, if $\tilde{x}$ is chosen such that $f_Y(\tilde{x} - \theta) = f_Y(x - \theta)$, then $Pr(\tilde{x} - \theta \leq Y \leq x - \theta \mid \theta = \theta_0)$ is a nonincreasing function of $|\theta - \theta_0|$.

Example 2.7. One parameter normal distribution

Suppose that $X_1, \ldots, X_n$ form a random sample from a normal distribution with probability density function

$$f_X(x) = \frac{1}{\sqrt{2\pi}} e^{-\frac{1}{2}(x-\mu)^2},$$

where μ is the mean of the distribution. It is desired to test the following hypotheses:

$$H_0: \mu = \mu_0 \quad , \quad H_1: \mu \neq \mu_0 \quad ,$$

where μ_0 is a specified value of μ.

The joint probability density function of $X_1, \ldots, X_n$ can be written in the form

$$f_{\mathbf{X}}(\mathbf{x}) = \frac{1}{(2\pi)^{n/2}} e^{-\sum x_i^2/2} e^{-n\mu^2/2 + \mu \sum x_i}.$$

It is evident from this factorization that $T = \sum X_i$ is a sufficient statistic for μ. It is well known that the distribution of T is normal with mean $n\mu$ and variance n. Consequently, T is stochastically increasing in μ thus implying that T is indeed a test statistic. Under H_0, the distribution of T/n is a bell-shaped curve symmetric about μ_0. Therefore, according to Theorem 2.1, significance tests of H_0 can be based on the random variable $Y = |\bar{X} - \mu_0|$, because Y tends to take larger values for any departure of μ from μ_0, where $\bar{X} = \sum X_i / n$ is the sample mean. Hence, H_0 can be tested on the basis of the p-value

$$\begin{aligned} p &= Pr\left[\, |\bar{X} - \mu_0| \geq |\bar{x} - \mu_0| \,\right] \\ &= 1 - Pr\left(-\sqrt{n}(\bar{x} - \mu_0) < Z < \sqrt{n}(\bar{x} - \mu_0)\right) \\ &= 2\,\Phi\left[-\sqrt{n}\,(\bar{x} - \mu_0)\right], \end{aligned} \tag{2.11}$$

where Z is a standard normal random variable and Φ is the cumulative distribution function.

2.4. Generalized Likelihood Ratio Method

So far we have considered hypothesis testing situations where the parameter of interest is the only unknown parameter of the underlying family of distributions and there is a single statistic which is sufficient for the parameter. The purpose of this section is to provide a simple criterion which will enable us to find test statistics for many situations where such conditions are not met.

Let $X_1, \ldots, X_n$ be a random sample from a family of distributions $F_X(x;\zeta)$, where $\zeta = (\theta, \boldsymbol{\delta})$, θ is the parameter of interest, and δ is a vector of nuisance parameters. Let $f_X(x;\zeta)$ be the density function (mass function in the discrete case) of X. Then the likelihood function of ζ is

$$L(\theta,\delta;\ \mathbf{x}) = \prod_{i=1}^{n} f_X(x_i:\zeta)\ .$$

Then the generalized likelihood ratio (sometimes abbreviated as GLR) is defined as the function of **x** and θ given by

$$G(\mathbf{X};\ \theta) = \frac{\underset{\delta}{Sup}\ L(\theta,\delta;\ \mathbf{x})}{\underset{\zeta}{Sup}\ L(\theta,\delta;\ \mathbf{x})}\ . \tag{2.12}$$

Suppose the function G depends on the data and θ only through a quantity $T(\mathbf{X};\ \theta)$. Then we can merely test whether T has the properties of a test variable. In many situations of practical importance this is indeed the case and the resulting test variable is the same as the one with optimum properties described in the following two sections.

As in conventional fixed-level testing, the generalized likelihood procedure is not quite based on any clearly defined optimum criteria or desirable properties. Nevertheless, this method often leads to the same test variables that one can derive by more complicated procedures based on optimum considerations and desirable properties. Furthermore, the GLR procedure has various asymptotic optimum properties as discussed, for instance, in Cox and Hinkley (1974).

Example 2.8. Two-parameter exponential distribution

Let $X_1, \ldots, X_n$ be a random sample from the two-parameter exponential distribution; that is

$$f_X(x) = \frac{1}{\beta}\, e^{-(x-\alpha)/\beta} \qquad \text{for} \quad x > \alpha\,, \quad \beta > 0$$

where α and β are unknown parameters. Suppose the following hypotheses are to be tested:

$$H_0\!: \beta \le \beta_0 \qquad H_1\!: \beta > \beta_0\ , \tag{2.13}$$

where β_0 is a prespecified constant. In order to find a sufficient statistics for (α,β), the joint probability density function of $X_1, \ldots, X_n$ can be written in the form

$$\begin{aligned} f_{\mathbf{X}}(\mathbf{x}) &= \prod_{i=1}^{n} \frac{1}{\beta}\ e^{-(x_i-\alpha)/\beta}\ I_{(\alpha,\infty)}(x_i) \\ &= \frac{1}{\beta^n}\ e^{-n(\bar{x}-\alpha)/\beta}\ I_{(\alpha,\infty)}(min(\mathbf{x}))\ , \end{aligned}$$

where $\bar{x}$ is the sample mean and *min* (**x**) is the smallest value of x_i ; $i = 1, \ldots, n$.

It is now evident that $\bar{X}$ and *min* (**X**) are minimal sufficient statistics for

(α, β). Since there is no single sufficient statistic for β, let us find a test variable for β by applying the generalized likelihood ratio method. It can be easily shown (Exercise 2.5) that the maximum likelihood estimators of α and β are $X_{(1)} = min\,(\mathbf{X})$ (regardless of whether β is known or unknown) and $\bar{X} - X_{(1)}$ respectively. Hence, the generalized likelihood ratio can be expressed as

$$GLR = \frac{\max_{\alpha} L(\alpha,\beta;\mathbf{x})}{\max_{\alpha,\beta} L(\alpha,\beta;\mathbf{x})} = \frac{\beta^{-n}\, e^{-n(\bar{x} - x_{(1)})/\beta}}{(\bar{x} - x_{(1)})^{-n}\, e^{-n}}$$

Clearly, GLR depends on the data and β only through the random variable

$$T(\mathbf{X};\beta) = \frac{\bar{X} - X_{(1)}}{\beta} . \tag{2.14}$$

It can be shown (Exercise 2.5) that the random variable T defined by (2.14) has the gamma distribution

$$T \sim G\,(\,n-1\,,n^{-1})$$

The cumulative distribution function of $T_0 = (\bar{X} - X_{(1)})/\beta_0$ is

$$\begin{aligned} F_{T_0}\,(t_0) &= Pr\,(T\beta/\beta_0 \le t_0) \\ &= Pr(\,2nT \le 2n\,t_0\,\beta_0/\beta\,) \\ &= C_{2(n-1)}(2n\,t_0\,\beta_0/\beta)\,, \end{aligned} \tag{2.15}$$

where $2nT \sim \chi^2_{2(n-1)}$ and $C_{2(n-1)}$ is its cumulative distribution function. It follows immediately from (2.15) that T_0 is a test statistic and that T is a test variable. Therefore, significance tests of (2.13) can be carried out using the observed level of significance

$$p = 1 - C_{2(n-1)}(2n\,(\bar{x} - x_{(1)})/\beta_0)\,. \tag{2.16}$$

It is easily seen in this example that the power functions of T and T_0 are $\pi(\beta) = \pi_0(\beta) = 1 - C_{2(n-1)}(2n\,(\bar{x} - x_{(1)})/\beta)$.

2.5. Invariance in Significance Testing

The purpose of this section and the next section is to study some desirable properties in significance testing. Once a problem of finding an appropriate test statistic has been reduced by the principle of sufficiency, these properties will enable us to reduce further the number of statistics on which a test statistic can be based.

Consider a significance testing situation where the underlying family of distributions is invariant under a group G of transformations on the sample space Ξ. Let $\overline{G}$ be the induced group of transformations on the parameter space Θ. Consider the problem of testing the null hypothesis

$$H_0\colon \theta \in \Theta_0 \quad \text{against} \quad H_1\colon \theta \in \Theta_1 ,$$

where Θ_0 and Θ_1 are subsets of Θ. The testing problem is said to be invariant under group G if the hypotheses are not affected by the transformation of the distribution; that is, more precisely, if

$$\overline{g}(\Theta_0) = \Theta_0 \quad \text{and} \quad \overline{g}(\Theta_1) = \Theta_1 \tag{2.17}$$

for all $\overline{g} \in \overline{G}$. This commonly used definition is actually rather too restrictive and has very limited applicability. In Chapter 5 we will study a simple way of relaxing the restrictive nature of the definition.

In significance testing of an invariant hypothesis, it is desirable that the observed level of significance be not affected by the group of transformations. For instance, in a scalė invariant problem, regardless of whether the variable x is measured in feet or in meters we like to get the same p-value so that it serves as a true measure of how well the data support the null hypothesis. Moreover, this property is desired regardless of how the two hypotheses specify the parameter θ. According to (2.17) the p-value can possibly be affected only by G and not by $\overline{G}$. The invariance of the data-based power function, as specified in the following definition, satisfies these requirements.

Definition 2.5. A data-based power function $\pi(\mathbf{x};\theta)$ is said to be invariant under G if $\pi(g(\mathbf{x});\theta) = \pi(\mathbf{x};\theta)$ for all $\mathbf{x} \in \Xi$, $\theta \in \Theta$, and $g \in G$.

When the testing problem is invariant under a certain group of transformations, the following theorem is useful in reducing the number of statistics on which a test variable can be based. It assures that our search for invariant tests can be confined to the class of test variables which depend on the data only through maximal invariants. We defer the proof of this theorem until Chapter 5 where a more general version of the theorem is given.

Theorem 2.2. Suppose that the testing problem is invariant under a group, G, of transformations on the sample space Ξ. Let $\phi(\mathbf{X})$ be a set of maximal invariants with respect to that group. If $T = T(\phi)$ is a continuous random variable, then any invariant data-based power function $\pi(\mathbf{x};\theta)$ can be obtained using T.

Example 2.9. Comparing two exponential distributions

Suppose that $X_1, \ldots, X_m$ and $Y_1, \ldots, Y_n$ form two sets of random samples from exponential distributions with means μ_x and μ_y respectively. Suppose that it is desired to test hypotheses of the form

$$H_0\colon \mu_x \le \mu_y \quad \text{against} \quad H_1\colon \mu_x > \mu_y . \tag{2.18}$$

It follows from Example 1.1 that $U = \sum X_i$ and $V = \sum Y_i$ are sufficient statistics for μ_x and μ_y respectively. Furthermore, these random variables are independently distributed as

$$U \sim G(m,\mu_x),\ V \sim G(n,\mu_y) \tag{2.19}$$

In order to find a single statistic on which tests can be based, note that the family of joint distributions of these random variables is invariant under the group of common scale transformations

$$(U,V) \to (k\,U, k\,V) \quad ,(\mu_x, \mu_y) \to (k\,\mu_x, k\,\mu_y)$$

where k is a positive constant. Further, this transformation leaves the hypotheses in (2.18) invariant. Therefore, any scale invariant test of (2.18) can be constructed using the maximal invariant

$$T(U,V) = \frac{U}{V}$$

Let $\theta = \mu_x/\mu_y$ be the parameter of interest. Since

$$\frac{2U}{\mu_x} \sim \chi^2_{2m} \quad \text{and} \quad \frac{2V}{\mu_y} \sim \chi^2_{2n}$$

are independent random variables, the random variable $nT/m\theta$ has an F distribution with $2m$ and $2n$ degrees of freedom. In terms of θ, (2.18) can be rewritten as

$$H_0: \theta \le 1 \text{ against } H_1: \theta > 1 .$$

It is now clear that T is indeed a test statistic for θ, because $Pr\ (T > t)$ is an increasing function of θ. Therefore, the hypothesis in (2.18) can be tested using the p-value

$$p = Pr\left(T > \frac{\sum x_i}{\sum y_i} \right)$$

$$= 1 - F_{2m,2n}\left(\bar{x}/\bar{y} \right) , \tag{2.20}$$

where $F_{2m,2n}$ is the cdf of the F distribution with 2m and 2n degrees of freedom, and $\bar{x}$ and $\bar{y}$ are the means of the two samples from respective exponential populations.

2.6. Unbiasedness and Similarity

In this section we briefly describe the notions of unbiasedness and similarity in the context of significance testing. For the definitions and more details of these notions in fixed-level testing the reader is referred to Lehmann (1986). To describe the nature of the related concepts, first consider one-sided hypotheses of the form

$$H_0\colon \theta \le \theta_0 \ , \quad H_1\colon \theta > \theta_0 \ .$$

Let $T = T(\mathbf{X};\ \theta)$ be a test variable which is stochastically increasing in θ. Then, for $\theta_1 \le \theta_2$, we have

$$Pr(T \ge t \mid \theta_1) \le Pr(T \ge t \mid \theta_2) \ ,$$

and consequently, the p-value, $p(t)$, satisfies the inequality $p(t) \le \pi(t \mid \theta)$ for all $\theta > \theta_0$, where $\pi(t \mid \theta) = Pr(T \ge t \mid \theta)$ is the power function and $p(t) = \pi(t \mid \theta_0)$.

There is a similar inequality for other types of hypotheses as well and this is a consequence of Property 2, Property 3, or Property 3' that we have required from a test statistic. We call this property the p-unbiasedness, or merely unbiasedness, of the resulting test. Also notice that Property 1 of a test variable implies that the p-value is a constant independent of the nuisance parameter δ. This property will be referred to as p-similarity, or just similarity as this will cause no confusion in this book. This means that we have readily incorporated the notions of unbiasedness and similarity in our definitions of test variables and p-values. So far we have focused on a single parameter of interest and so the properties that we imposed on test variables have not been too restrictive (see Exercise 2.4 for a variable which does not satisfy Property 2 and yet provides an unbiased testing procedure) or difficult to verify. This may not be the case when we need to deal with a hypothesis concerning a vector of parameters of interest, say $\boldsymbol{\theta}$. In such situations we can merely impose the unbiasedness and similarity in place of Properties 1 to 3.

In order to give a more general definition of the two notions, consider the problem of testing the following two hypotheses:

$$H_0\colon \theta \in \Theta_0 \ , \qquad H_1\colon \theta \in \Theta_1 \ . \tag{2.21}$$

Let $\Theta_B = \overline{\Theta}_0 \cap \overline{\Theta}_1$ be the boundary of the hypotheses. In this book we consider only hypotheses with a single (and possibly vector valued) boundary, say $\Theta_B = \boldsymbol{\theta}_0$. Let $C(\mathbf{x})$ be an extreme region designed for testing (2.21) and let

$$\pi(\mathbf{x};\ \boldsymbol{\theta}) = Pr(\ \mathbf{X} \in C(\mathbf{x} \mid \boldsymbol{\theta})\ ) \tag{2.22}$$

be its power function. The definitions of the notions can now be given with these notations.

Definition 2.6. A test based on an extreme region $C(\mathbf{x})$ is said to be *p-unbiased* if

$$\pi(\mathbf{x};\ \boldsymbol{\theta}) \ge p(\mathbf{x}) \quad \text{for all } \ \boldsymbol{\theta} \in \Theta_1 \ , \tag{2.23}$$

where

$$p(\mathbf{x}) = \pi(\mathbf{x};\ \boldsymbol{\theta}_0) \tag{2.24}$$

is defined to be the p-value.

Definition 2.7. A test based on an extreme region $C(\mathbf{x})$ is said to be *p-similar* (on the boundary) if, given any $\mathbf{x} \in \Xi$,

$$\pi(\mathbf{x};\, \boldsymbol{\theta}_0) = p \tag{2.25}$$

is a constant independent of $\boldsymbol{\delta}$, where $p = p(\mathbf{x})$ is the p-value.

In Chapters 7 through 10 we will undertake hypothesis testing in situations involving a number of parameters of interest. In this section we utilize these notions to derive tests in multi-parameter problems with a single parameter of interest. Notice that a p-unbiased test based on a data-based power function continuous in $\boldsymbol{\theta}$ is p-similar. Once a p-similar test is derived one can simply check whether the test is unbiased as well. Therefore, in the rest of this section we focus our attention on tests which are similar on the boundary.

For given $\theta = \theta_0$, let $\mathbf{S}$ be a sufficient statistic for $\boldsymbol{\delta}$, the vector of nuisance parameters. Suppose, conditional on $\mathbf{S}$, one can find a p-similar extreme region; that is

$$Pr(\mathbf{X} \in C(\mathbf{x}) \mid S; \boldsymbol{\theta}_0) = p\,. \tag{2.26}$$

Define

$$I_c = \begin{cases} 1 & \text{if } \mathbf{X} \in C(\mathbf{x}) \\ 0 & \text{if otherwise} \end{cases}.$$

With this notation, note that

$$\begin{aligned} Pr(\mathbf{X} \in C(\mathbf{x}) \mid \boldsymbol{\theta}_0) &= E(\, I_c\,) \\ &= E(\, E(\, I_c \mid \mathbf{S}\,)\,) \\ &= p\,. \end{aligned} \tag{2.27}$$

This means that the unconditional test is also p-similar. The following theorem (see the Appendix B for a proof) deals with the converse of this result.

Theorem 2.3. If $\mathbf{S}$ is a complete sufficient statistic for $\boldsymbol{\delta}$ when $\theta = \theta_0$, then every similar test has the structure (sometimes called *Neyman structure*) in (2.26).

The foregoing results are useful in deriving similar tests in testing situations where there are nuisance parameters. In view of Theorem 2.3, we can base our search for similar tests on sufficient statistics for θ and its conditional distribution given $\mathbf{S}$. This approach is especially useful in dealing with exponential families of distributions, for which Corollary 1.1 can be employed to find the required conditional distribution. The procedure is illustrated by the following example.

Example 2.10. Comparing two exponential distributions (continued)

Let $X_1, \ldots, X_m$ and $Y_1, \ldots, Y_n$ be two sets of independent random samples from exponential distributions with the joint distribution function

$$f(\mathbf{x},\mathbf{y}) = \alpha^m \beta^n e^{-\alpha \sum x_i - \beta \sum y_j},$$

where α and β are unknown parameters. It is desired to test the null hypothesis

$$H_0: \theta \geq 0 \quad \text{against} \quad H_1: \theta < 0 ,$$

where $\theta = \alpha - \beta$.

Notice that this is equivalent to the hypothesis testing problem considered in Example 2.9. Recall that that $U = \sum X_i$ and $V = \sum Y_i$ are sufficient statistics for the unknown parameters. In terms of these variables and θ, the above distribution can be expressed as

$$f = (\theta+\beta)^m \beta^n e^{-\theta u - \beta(u+v)} .$$

It is evident from this equation that, for any given θ, $S = U+V$ is a sufficient statistic for β, which can be treated as the nuisance parameter of the problem. Therefore, according to Theorem 2.3, the p-value for testing the above hypotheses can be found using the conditional distribution of U given S. Recall that U and V are independently distributed as $U \sim G(m, 1/\alpha)$ and $V \sim G(n, 1/\beta)$, and therefore, the random variables U/S and S are independent. Hence, the p-value can be computed as

$$\begin{aligned} p &= Pr(U \geq u \mid S=s) \\ &= Pr(\frac{U}{S} \geq \frac{u}{s} \mid S=s) \\ &= Pr(\frac{U}{S} \geq \frac{u}{s}) \\ &= Pr(T \geq \frac{u}{v}) , \end{aligned}$$

where $T = U/V$. This is the same test statistic derived in Example 2.9 and therefore the p-value is given by (2.20).

In this example the principle of invariance and the notion of similarity give rise to the same test statistic. In Chapter 4 we will study situations where similarity can provide solutions when the problem can not be reduced by invariance. In Chapter 6 we will study situations where the two principles complement each other.

2.7. Interval Estimation and Fixed-Level Testing

When the underlying family of distributions is continuous, the notions in significance testing have implications both for interval estimation and for fixed-level testing. If T is a continuous test variable for a parameter θ, then one can construct confidence intervals of θ on the basis of T. The confidence intervals based on T can be deduced also from data-based power functions (based on T) when they are readily available.

To illustrate these, assume without loss of generality that the test variable

$T = T(\mathbf{X};\theta)$ is stochastically increasing in θ. Let $F_T(t)$ be the cumulative distribution function of T. Assume that $F_T(t)$ is free of nuisance parameters. It follows from the probability integral transform that the random variable $U = F_T(T)$ has a uniform distribution over the interval $[0,1]$. Therefore,

$$Pr(\gamma_1 \le F_T(T) \le \gamma_2) = \gamma_2 - \gamma_1 ,$$

and consequently, $[\gamma_1 \le F_T(t_{obs}) \le \gamma_2]$ will provide a $100(\gamma_2-\gamma_1)\%$ confidence region for θ, where t_{obs} is the observed value of the test variable T. Let $\pi(t;\theta)$ be a readily available data-based power function corresponding to a certain one-sided null hypothesis. Recall that

$$F_T(t) = \pi(t;\theta) \quad \text{or} \quad 1 - \pi(t;\theta) \tag{2.28}$$

according as the null hypothesis is right-sided or left-sided. In either case, $\pi(T;\theta)$is a uniform random variable and as a result,

$$[\gamma_1 \le \pi(t_{obs};\theta) \le \gamma_2] \tag{2.29}$$

leads to an exact $100(\gamma_2-\gamma_1)\%$ confidence region for θ.

Next consider the problem of testing hypotheses at a fixed desired nominal level, say α. To be specific, consider hypotheses of the form

$$H_0\!: \theta \le \theta_0 \ , \quad H_1\!: \theta > \theta_0 \ .$$

Let $p(t) = \pi(t;\ \theta=\theta_0)$ be an observed level of significance based on a continuous test variable. Then, it follows from (2.28) that the random variable $V = p(T)$ has a uniform distribution over the unit interval $[0,1]$. An exact fixed-level test of size α follows immediately from this result:

$$\text{reject } H_0 \text{ if } \ p(t_{obs}) < \alpha \ . \tag{2.30}$$

In general, if $p(T)$ has a uniform distribution, then the decision rule (2.30) provides an exact fixed-level test for any null hypothesis for which $p(t)$ serves as an observed level of significance. Fixed level tests of hypotheses including one-sided and point null hypotheses can also be deduced from $100(1-\alpha)\%$ confidence intervals constructed using (2.29). In this approach, the null hypothesis is rejected at α level if the $100(1-\alpha)\%$ confidence interval does not contain θ_0, the boundary point of the hypotheses.

Example 2.11. Two-parameter exponential distribution (continued)

Let $X_1, \ldots, X_n$ be a random sample from the two-parameter exponential distribution

$$f_X(x) = \frac{1}{\beta}\, e^{-(x-\alpha)/\beta} \qquad \text{for} \quad x > \alpha\ , \quad \beta > 0$$

where α and β are unknown parameters. Consider the problem of constructing an equal-tail 95% confidence interval for β and the problem of testing the hypotheses

$$H_0: \beta \le \beta_0 \qquad H_1: \beta > \beta_0 , \tag{2.31}$$

at .05 level, where β_0 is a specified constant.

It was shown in Example 2.8 that

$$p(\bar{x}, x_{(1)}) = 1 - C_{2(n-1)} \ (2n\,(\bar{x} - x_{(1)})/\beta_0) \tag{2.32}$$

is the observed level of significance for testing (2.31). Since the test variable defined in (2.14) is a continuous random variable, H_0 can be rejected at .05 level if $p(\bar{x}, x_{(1)}) < .05$; that is if

$$\frac{2n(\bar{x} - x_{(1)})}{\beta_0} > q_{.95} ,$$

where q_γ is the γth percentile of the chi-square distribution with $2(n-1)$ degrees of freedom.

To construct confidence intervals of β, we can deduce from (2.32) or derive directly from the test variable T that

$$\pi(t;\, \beta) = 1 - C_{2(n-1)} \ (2n\,(\bar{x} - x_{(1)})/\beta), \tag{2.33}$$

where $t = (\bar{x} - x_{(1)})/\beta$. By (2.33), an equal tail 95% confidence interval of β can be found by solving the equation

$$0.025 \le \pi(t;\, \beta) \le 0.975 . \tag{2.34}$$

It is now evident from (2.33) and (2.34) that

$$\frac{2n(\bar{x} - x_{(1)})}{q_{.975}} \le \beta \le \frac{2n(\bar{x} - x_{(1)})}{q_{.025}}$$

is a 95% confidence interval of β.

Exercises

2.1. Let $X_1, \ldots, X_n$ be a random sample from the uniform distribution with density function

$$f_X(x) = \frac{1}{\theta} \qquad \text{for } -\theta \le x \le 0 .$$

Show that $T = min(\mathbf{X})$, the minimum of all the observations, is a test statistic for θ. Find the observed level of significance based on the observed value $\min(\mathbf{x})$ for testing the hypotheses

$$H_0: \theta \le \theta_0, \quad H_1: \theta > \theta_0 .$$

Deduce from the power function of the test, or find otherwise,

(a) a left-sided 95% confidence interval for θ,

(b) an equal tail 95% confidence interval for θ

(c) a fixed-level test of size .05 for testing the above hypotheses.

2.2. Let $X_1, \ldots, X_n$ be a random sample from the beta distribution with density function

$$f_X(x) = \theta\, x^{\theta - 1} \qquad \text{for } 0 < x \le 1 .$$

Find a test statistic for θ using the formula (2.2). In terms of the cdf of the gamma distribution find its p-value for testing each of the following pairs of hypotheses:

$$(a) \; H_0: \theta \le \theta_0, \; H_1: \theta > \theta_0 ,$$

$$(b) \; H_0: \theta \ge \theta_0, \quad H_1: \theta < \theta_0 .$$

2.3. Let $X_1, \ldots, X_n$ be a random sample from the exponential distribution with density function

$$f_X(x) = \alpha\, e^{-\alpha x} \qquad \text{for } x > 0 .$$

Establish a procedure for significance testing of the hypotheses

$$H_0: \alpha = \alpha_0, \quad H_1: \alpha \ne \alpha_0 .$$

2.4. Let $X_1, \ldots, X_n$ be a random sample from the density function

$$f_X(x) = \frac{2x}{\alpha^2} \qquad \text{for } 0 \le x \le \alpha .$$

Show that $T = \max(\mathbf{X})$ is a sufficient statistic and yet it is not a test variable for α. Is T a complete statistic? Show that the extreme region given by the right tail of T is unbiased and similar for testing the hypotheses

$$H_0: \alpha \le \alpha_0, \quad H_1: \alpha > \alpha_0 ,$$

where $\alpha_0 > \max(\mathbf{x})$, the observed value of T.

2.5. Let $X_1, \ldots, X_n$ be a random sample from the two-parameter exponential distribution

$$f_X(x) = \frac{1}{\beta} e^{-(x-\alpha)/\beta} \quad \text{for} \quad x > \alpha, \ \beta > 0$$

where α and β are both unknown parameters.

(a) Show that $X_{(1)}$ and $\bar{X} - X_{(1)}$ are the maximum likelihood estimators of α and β respectively, where $\bar{X}$ is the sample mean and $X_{(1)}$ is the smallest of all n observations.

(b) Show that $\bar{X} - X_{(1)}$ has a gamma distribution. Find the distribution of $X_{(1)}$.

(c) By the method of generalized likelihood ratio, find a test variable for α.

(d) Find the form of the p-value based on the test variable for testing one-sided hypotheses on α.

(e) Assuming the improper prior $\pi(\alpha,\beta) = 1$, find the posterior probability of the null hypothesis.

2.6. In Exercise 5, show that the statistics $\bar{X} - X_{(1)}$ and $X_{(1)}$ are independently distributed. Hence, derive procedures based on these sufficient statistics for testing each of the following parameters by exploiting an appropriate invariance property of the family of distributions:

(a) for β when α is the nuisance parameter;

(b) for α when β is the nuisance parameter;

(c) for the mean of the distribution when β is the nuisance parameter;

(d) Find a left-sided confidence interval for α.

2.7. Let $X_1, \ldots, X_m$ and $Y_1, \ldots, Y_n$ be two sets of random samples from exponential distributions with parameters α and β respectively. The means of the two distributions are α^{-1} and β^{-1} respectively. Suppose that it is desired to test hypotheses of the form

$$H_0: \theta \le \theta_0 , \quad H_1: \theta > \theta_0 ,$$

where $\theta = \alpha - \beta$. Derive procedures for testing these hypotheses by each of the following methods:

(a) by the method of generalized likelihood ratio;

(b) by the method of invariance;

(c) by Bayesian method with the improper prior $\pi(\alpha,\beta) = 1/(\alpha\beta)$.

2.8. Deduce each of the following from the p-value obtained in Exercise 2.7:

(a) a 95% left-sided confidence interval of θ;
(b) a 95% equal tail confidence interval of θ;
(c) a fixed-level test of size .05 for testing the hypotheses in Exercise 2.7.

2.9. Let $X_1, \ldots, X_n$ be a random sample from the logistic distribution

$$F_X(x) = \frac{1}{1 + e^{-\theta x}} .$$

Establish procedures for testing one-sided hypotheses on θ.

2.10. Let $X_1, \ldots, X_n$ be a random sample from the uniform distribution on the interval $[\alpha, \beta]$. Show that $\min(\mathbf{X})$ and $\max(\mathbf{X})$ are jointly sufficient for the parameters of the distribution. By using the notion of similarity, find p-values for testing the following hypotheses:
(a) left-sided null hypotheses on α;
(b) right-sided null hypotheses on β.

2.11. Suppose that X and Y are two random variables having geometric distributions with parameters α and β respectively. Assume that the two random variables are independently distributed. By applying Theorem 2.3, find a p-similar test for testing the null hypothesis

$$H_0 \colon \alpha \leq \beta \quad \text{against} \quad H_1 \colon \alpha > \beta .$$

Show that the test is p-unbiased as well.

2.12. Let X be an observation from a geometric distribution with parameter α. Show that the noninformative prior of p given by Jeffreys' rule (see Appendix A) is

$$\pi(p) = p^{-1}(1-p)^{-1/2} .$$

Find the posterior distribution of p given $X = x$. Hence find the form of the posterior odds ratio for testing left-sided null hypotheses against right-sided alternative hypotheses.

Chapter 3

Review of Special Distributions

3.1. Poisson and Binomial Distributions

Most textbooks in statistics take the Neyman-Pearson decision theoretic approach to tackle problems of hypothesis testing. Therefore, they do not provide formulae for computing exact p-values even for testing parameters of special distributions. The purpose of this chapter is to develop procedures of significance testing for the parameters of widely used distributions, such as the binomial distribution. The problems of point estimation and interval estimation will be briefly discussed.

The binomial and Poisson distributions are the most widely used discrete distributions in statistics. Practical applications involving these distributions arise in a variety of areas including industrial experiments, statistical quality control, and biomedical research. The binomial distribution also plays an important role in nonparametric statistical methods. The purpose of this chapter is to develop methods for making inferences about parameters of these two distributions followed by a similar development for the normal distribution.

The starting point in defining a binomial distribution based on the outcomes of an experiment is to classify the outcomes into one of two categories, either a *success* or a *failure*; in using the word 'success' in this context one does not necessarily mean that the outcome is desirable. For example, an item inspected in a production line are either defective or nondefective, an experimental vaccine is either effective or not effective against a certain disease, the eldest child in a randomly selected family is either a female or a male, and so on. Suppose the experiment is repeated a number of times, say n trials, and the outcomes are recorded. Let X be the number of successes observed in repeating the experiment a fixed number of n trials. Obviously, the possible values of X

are $0,1,\ldots,n$. The random variable X is said to have a binomial distribution with parameters n and p if the following conditions hold:

1. The experiment is repeated a fixed number (n) of independent and identical trials.

2. The outcome of each trial can be classified either as a success or as a failure; that is, each trial is a *Bernoulli* trial.

3. The probability of a success, $p = Pr(success)$, remains constant from trial to trial. Moreover, $Pr(failure) = 1 - p$.

We shall denote a binomial random variable X with parameters n and p as

$$X \sim B(n, p).$$

The probability mass function of the random variable is given by the binomial formula

$$p_X(x) = Pr(X = x) = \binom{n}{x} p^x (1-p)^{n-x}, \quad x = 0,1,\ldots,n, \tag{3.1}$$

where $\binom{n}{x} = n!/(x!(n-x)!)$. The following alternative representations of the cumulative distribution function of X will prove to be useful in applications:

$$F_X(x) = Pr(X \le x) = \sum_{j=0}^{x} \binom{n}{j} p^j (1-p)^{n-j} \tag{3.2}$$

$$= 1 - n\binom{n-1}{x} \int_0^p y^x (1-y)^{n-x-1}\, dy$$

$$= 1 - F_B(p), \tag{3.3}$$

where F_B is the cdf of the beta random variable

$$B \sim Beta(x+1, n-x).$$

It can be shown (Exercise 3.1) that the mean and the variance of the binomial distribution are $E(X) = np$ and $Var(X) = np(1-p)$, respectively.

A probability distribution which is closely related to the binomial distribution is the Poisson distribution. When the probability of a rare event is small, the number of occurrences of the event in a large number of trials tends to be distributed as a Poisson random variable. It naturally arises more directly in the occurrences of a rare event (in a small time interval) over a given period of time. Typical examples of Poisson random variables are the number of telephone calls received at an exchange per hour, the number of customers handled by a bank teller per day, the number of vehicles arriving at a toll booth during a certain hour of a Sunday, the number of radioactive particles that decay per minute, the number of light bulbs that need replacing in an office building during a year, and so on.

To give the conditions under which the number of events that occur during a time interval will be distributed as a Poisson, consider a fixed period of time say $[0,T)$. Let X denote the number of events which occur over this period. If the following conditions hold, then X has a Poisson distribution with parameter $\lambda = \mu T$:

1. The probability that exactly one event occurs during a small time interval of length δt is equal to $\mu\delta t + o(\delta t)$, where $o(\delta t)/\delta t \to 0$ as $\delta t \to 0$ and $\mu > 0$.

2. The probability that two or more events occur during a small time interval of length δt is $o(\delta t)$ (as defined above).

3. The occurrences of events in one interval are independent of those in any other interval.

It is not difficult to derive (Exercise 3.2) from these conditions that the probability mass function of X is

$$p_X(x) = Pr(X = x) = \frac{e^{-\lambda}\lambda^x}{x!}, \quad x = 0, 1, 2, \ldots, \tag{3.4}$$

where $\lambda = \mu T$.

In general, any random variable having a probability mass function of the form (3.4) is said to be a Poisson random variable with parameter λ. It is denoted by

$$X \sim P(\lambda).$$

The cumulative probability distribution function of this random variable can be computed using any of the following alternative formulae:

$$F_X(x) = Pr(X \le x) = \sum_{j=0}^{x} \frac{e^{-\lambda}\lambda^j}{j!} \tag{3.5}$$

$$= 1 - \frac{1}{x!}\int_0^{\lambda} y^x e^{-y} dy$$

$$= 1 - F_G(\lambda), \tag{3.6}$$

where F_G is the cdf of the gamma random variable

$$G \sim Gamma(x+1,1).$$

It is easy to show (see Exercise 3.3) that the mean and the variance of a Poisson random variable, X, with parameter, λ, are both equal to λ; that is $E(X) = \lambda$ and $Var(X) = \lambda$.

3.2. Point Estimation and Interval Estimation

Before we address the problem of significance testing of hypotheses in this context let us briefly review the elements of point estimation, as they also play a role in other types of inferences including Bayesian inference. First consider the problem of estimating the parameter of a Poisson distribution. Let $X_1, \ldots, X_n$ be a random sample from $P(\lambda)$. The joint probability mass function is

$$f_{\mathbf{X}}(\mathbf{x}) = \prod_{i=1}^{n} \frac{e^{-\lambda}\lambda^{x_i}}{x_i!} = \frac{e^{-n\lambda}\lambda^{\sum x_i}}{\prod x_i!}. \tag{3.7}$$

Hence, it follows from the factorization criterion that $S = \sum x_i$ is a sufficient statistic for the parameter λ. The distribution of this random variable is (see Exercise 3.4)

$$S \sim P(n\lambda); \tag{3.8}$$

that is, S has a Poisson distribution with mean $n\lambda$.

The log likelihood of function given by (3.7) is

$$\log(L(\lambda;\, \mathbf{x})) = -n\lambda + \sum_{i=1}^{n} x_i \log(\lambda) - \sum_{i=1}^{n} \log(x_i!) \tag{3.9}$$

Differentiating the log likelihood function with respect to λ we obtain the first two derivatives of the log likelihood function, $-n + \sum x_i/\lambda$ and $-\sum x_i/\lambda^2$, respectively. Since the second derivative is negative, the maximum likelihood estimator of λ can be found by equating the first derivative to zero. Hence, the MLE of λ based on a random sample of size n is

$$\hat{\lambda} = \sum_{i=1}^{n} x_i/n = \bar{x}. \tag{3.10}$$

Since the sample mean is an unbiased estimator of the population mean, this estimator is unbiased. In fact, it follows from the Lehmann-Scheffe theorem (Theorem 1.4) and from (3.8) that this estimator is also the MVUE of λ.

Let us now turn to the binomial distribution. Since a binomial random variable denotes the number of successes in n trials, it is sufficient to consider a single binomial random variable, say $X \sim B(n,p)$, where p is an unknown parameter and n is assumed to be known, as usually is the case. In fact, the sample sum of a random sample of size k from this distribution is a sufficient statistic for p and denotes the number of successes obtained in nk independent trials of an experiment. Hence, $\sum x_i \sim B(nk,p)$, thus reducing to a problem involving a single binomial distribution. Also note that in view of (3.8) we could have obtained a similar reduction with Poisson distribution as well.

The likelihood function of p can be obtained directly from equation (3.1):

$$\log(L(p;x)) = \log\binom{n}{x} + x\log(p) + (n-x)\log(1-p).$$

It immediately follows from this log likelihood function that the MLE of the binomial parameter p is

$$\hat{p} = \frac{x}{n}. \tag{3.11}$$

Since $E(X) = np$, the estimator given by (3.11) is unbiased. Consequently, according to the Lehmann-Scheffe theorem, $\hat{p}$ is also the MVUE of p.

Exact interval estimators with conventional frequency interpretations are not available for parameters of Poisson and binomial distributions. Nevertheless, it is possible to find interval estimates to guarantee the desired confidence level. Moreover, there is no difficulty obtaining exact intervals in a Bayesian setting with noninformative priors or with natural conjugate priors. We shall undertake this task in Section 3.4.

For example, in the notation of (3.6), the $100\gamma\%$ upper confidence bound $\bar{\lambda}$ for the parameter λ of the Poisson distribution is computed as

$$Pr(X \le x \mid \bar{\lambda}) = 1 - F_G(\bar{\lambda}) = \gamma.$$

Other confidence intervals can be found using appropriate probability statements. The confidence bounds can be computed, for instance, using the S-Plus software package. The desired properties of the confidence interval $[0,\bar{\lambda}(x)]$ for λ are easily deduced from this representation. Since this interval is based on an exact probability statement, despite its unconventional interpretations, some statisticians consider this as an exact confidence interval. Confidence intervals for the binomial parameter can be found in a similar manner using (3.3). For example, if $x < n$, the $100\gamma\%$ upper confidence limit for p is the solution of the equation

$$Pr(X \le x \mid \bar{p}) = 1 - F_B(\bar{p}) = \gamma; \tag{3.12}$$

if $x = n$ is observed, the confidence bound is taken as 1. Other confidence intervals can be found in a similar manner with additional properties, if desired. Two-sided equal-tail confidence limits can be computed conveniently using the software packages such as StatXact and TESTIMATE. One-sided confidence intervals can also be computed using these software packages with the appropriate adjustment of the confidence coefficient.

3.3. Significance Testing of Parameters

Exact nonrandomized fixed level tests are also not available for binomial and Poisson parameters. On the other other hand, exact p-values and randomized fixed level tests can be easily obtained. Readers interested in randomized tests are referred to Ferguson (1967). Although randomized tests are of theoretical interest, they are not considered very appealing in practical applications. The derivation and the computation of exact p-values in this context are so simple that there is hardly any point to using the approximations one often finds in the literature. In this section we consider problems of testing hypotheses about parameters of binomial and Poisson distributions on the basis of p-values.

First consider the problem testing the parameter of a Poisson distribution. Consider left-sided null hypotheses and right-sided alternative hypotheses of the form

$$H_0: \lambda \le \lambda_0 \quad \text{versus} \quad H_1: \lambda > \lambda_0 . \tag{3.13}$$

We can confine our attention to tests based on the sufficient statistic $S = \sum X_i$, whose distribution is given by (3.8). From equation (3.6) we have

$$Pr(S \ge s) = F_{G_s}(n\lambda),$$

where F_{G_s} is the cdf of the gamma distribution with the shape parameter s and the scale parameter 1. Therefore, S is stochastically increasing in the parameter λ. So S serves as a test statistic and the right tail of its distribution defines the appropriate extreme region for testing hypotheses in (3.13). Hence, the observed level of significance for testing H_0 is

$$\begin{aligned} p &= Pr(S \ge s \mid \lambda = \lambda_0) \\ &= F_{G_s}(n\lambda_0), \end{aligned} \tag{3.14}$$

where s is the observed value of the sufficient statistic S. The p-value given by (3.14) in terms of the cdf of the gamma distribution can be easily computed, for instance, using the SPlus software package.

Next, turning to the binomial distribution consider the problem of testing hypotheses about the parameter p of $X \sim B(n,p)$ based on an observed value x of X. First consider the hypotheses

$$H_0: p \le p_0 \quad \text{versus} \quad H_1: p > p_0 . \tag{3.15}$$

It follows from equation (3.3) that

$$Pr(X \ge x) = F_{B_{x,n+1-x}}(p), \tag{3.16}$$

where $F_{B_{x,n+1-x}}$ is the cdf of the beta distribution

$$B_{x,n+1-x} \sim Beta(x, n+1-x).$$

It is evident from (3.16) that X is stochastically increasing in p. Consequently, X

is a test statistic suitable for constructing procedures for testing hypotheses about p. In particular, the right tail of its distribution defines unbiased extreme regions for testing left-sided null hypotheses. Therefore, the p-value for testing (3.15) is

$$\begin{aligned} p &= Pr(X \geq x \mid p = p_0) \\ &= F_{B_{x,n+1-x}}(p_0). \end{aligned} \tag{3.17}$$

Right-sided null hypotheses can be tested in a similar manner using the same test statistic. It is easily seen the that null hypothesis $H_0: p \geq p_0$ can be tested using the p-value

$$p = 1 - F_{B_{x+1,n-x}}(p_0). \tag{3.18}$$

Many statistical software packages (e.g. BMDP and TESTIMATE) provide procedures to perform this test.

There is no common agreement about the way to compute the p-value for point null hypotheses concerning the parameters of the binomial and Poisson distributions. Gibbons and Pratt (1975) discuss a number of alternative methods which may be suitable under various conditions. One desirable way of defining the p-value in these cases is to include in the extreme region each sample point which can be considered as extreme as or more extreme than the observed value of the test statistic. Often, the p-value is defined as the sum of probabilities of all X in either tail which do not exceed the probability $Pr(X = x)$.

Example 3.1. Testing the rate of defective soldering

An automated X-ray inspection system is used in a soldering process to detect defective solder connections in integrated circuits. Under controlled conditions the rate of defective connections is less than six per inspection period. Suppose in a certain inspection period, nine connections were found to be defective. Does this data suggest that the process is currently not in control?

To answer this question, let λ denote the current rate of defective connections and consider the hypotheses

$$H_0: \lambda \leq 6 \quad \text{versus} \quad H_1: \lambda > 6.$$

In this kind of application it is reasonable to assume that the underlying random variable is Poisson. Let X denote the number of defective connections detected during an inspection period. Then, we have $X \sim P(\lambda)$ and the observed value of X is $x = 8$. The p-value for testing the hypotheses can be computed as

$$\begin{aligned} p &= F_{G_8}(6), \\ &= .256. \end{aligned}$$

With this p-value there is no reason to doubt the controlled condition of the soldering process. The null hypothesis cannot be rejected with the available data.

Example 3.2. Testing the improved quality of a consumer durable

A statistician keeps track of the quality of electric appliances made by the company she works for. She knows that 2% of the microwave ovens sold in the past are reported to have some problems within the warranty period. The company introduces a new model with a claim that the quality of the product has improved. By the time 1000 sold units of the new model have passed their warranty period the statistician has received twelve problem reports. Has the quality of the product (measured in terms of the reported number of problems within the warranty period) really improved?

Consider the hypothesis H_0: $p \geq .02$ that the quality has not improved, where p is the probability that a sold microwave oven is reported to be defective within the warranty period. According to (3.18) this hypothesis can be tested based on the p-value

$$\begin{aligned} p &= 1 - F_{B_{13,988}}(.02) \\ &= .034. \end{aligned}$$

Therefore, the data provide fairly strong evidence in favor of the hypothesis that quality of the new model is better than the quality of the previous model. The hypothesis can be tested based on the appropriate confidence interval for p as well. The left-sided 95% confidence interval for p is $[0, .0194]$. In order to compute this upper confidence bound using StatXact, the confidence coefficient must be set at .9 as applied to its two-sided equal-tail interval. Since .02 does not fall in the confidence interval, it is evident from the interval also that the null hypothesis can be rejected at the .05 level.

3.4. Bayesian Inference

Let $X_1, \ldots, X_n$ be a random sample of size n from a Poisson distribution with parameter λ. The joint probability mass function of **X** is given by (3.7). It is evident from the form of (3.7) that the conjugate family of prior distributions for λ is the class of gamma distributions. So assume that

$$\lambda \sim Gamma(\alpha, \beta), \tag{3.19}$$

where α and β are the hyper parameters of the prior distribution. When a prior distribution is not available one can still carry out Bayesian inferences about λ using the noninformative prior

$$g(\lambda) = \lambda^{-1/2}. \tag{3.20}$$

The joint distribution of **X** and λ is

$$f(\mathbf{x},\lambda) = \frac{e^{-n\lambda}\lambda^{\sum x_i}}{\prod x_i!} \times \frac{\lambda^{\alpha-1}e^{-lambda/\beta}}{\Gamma(\alpha)\beta^{\alpha}}$$
$$= \frac{e^{-\lambda(n+1/\beta)}\lambda^{\sum x_i + \alpha - 1}}{\Gamma(\alpha)\beta^{\alpha}\prod x_i!}.$$

Since this expression as a function of λ is proportional to the density function of a gamma distribution with parameters

$$\tilde{\alpha} = \sum_{i=1}^{n} x_i + \alpha \quad \text{and} \quad \tilde{\beta} = \frac{1}{n + 1/\beta}$$

it is now clear that the posterior distribution of λ is

$$\lambda \sim Gamma(\tilde{\alpha},\tilde{\beta}). \tag{3.21}$$

It can be shown (Exercise 3.5) by taking a similar approach that, for the binomial problem with $X \sim B(n, p)$, the class of beta distributions provide a natural conjugate family and that if $p \sim Beta(\alpha,\beta)$ is the prior, then the posterior distribution of p given x is

$$p \sim Beta(\alpha + x, \beta + n - x). \tag{3.22}$$

It can be deduced from (3.21) or shown directly, that under the noninformative prior defined by (3.20) the posterior distribution becomes

$$\lambda \sim Gamma\left(\sum_{i=1}^{n} x_i + ½, \frac{1}{n}\right). \tag{3.23}$$

Inferences on the Poisson parameter can be based on the posterior distribution of λ. In particular, the Bayesian unbiased estimator of λ is

$$\hat{\lambda} = E(\lambda)$$
$$= \frac{\sum_{i=1}^{n} x_i + \alpha}{n + 1/\beta}.$$

One-sided hypotheses about λ can be tested simply, based on the posterior probability of the underlying null hypotheses. For example, the hypotheses in (3.13) can be tested based on the posterior probability $p = F_{\lambda}(\lambda_0)$, where F_{λ} is the cdf of the gamma posterior of λ. In particular, under the noninformative prior this probability reduces to

$$p = F_{G_{s+1/2}}(n\lambda_0), \tag{3.24}$$

where $s = \sum x_i$ and $F_{G_{s+1/2}}$ is the cdf of the gamma distribution with parameters $s + 1/2$ and 1. Notice that this posterior probability of the null hypothesis is closely related to (but, not quite equal to) the p-value given by (3.14). As illustrated by the following example, Bayesian treatment in this context is especially useful in obtaining interval estimates.

Example 3.3. Testing the rate of defective soldering (continued)

Consider again the problem in Example 3.1 concerning the X-ray inspection system to detect defective solder connections. Suppose it is desirable to obtain a 95% interval estimate of the current rate of defective connections. Recall that during the period in question, nine defects have been observed. Assuming the noninformative prior (3.20) given by Jeffrey's rule, we have the posterior distribution

$$\lambda \sim Gamma(9.5,1).$$

Since 2λ is distributed as chi-squared distribution with 19 degrees of freedom, Bayesian confidence intervals for λ can be constructed using the quantiles of this chi-squared distribution.

$$Pr(8.91 \leq 2\lambda \leq 32.9) = .95$$

and, therefore, $[4.455\ ,\ 16.45]$ is the 95% equal tail credible interval for the current rate of defective soldering.

3.5. The Normal Distribution

By far the most widely used probability distribution in a variety of statistical applications is the normal distribution. Many populations encountered in practical applications are found to be normally distributed to a sufficient degree of accuracy. In some situations this phenomenon can be justified by the central limit theorem. For example, the actual net weight of a canned product with a certain specified weight can be affected by many factors in the production process. The net deviation of the weight from the mean weight might be the net result of all the factors, thus helping the central limit theorem to work.

In this chapter we consider the problems of estimation and hypotheses testing concerning the mean and the variance of a normal distribution. The parameter of interest could well be a function of both parameters. Two such functions are also considered in this chapter. A number of later chapters will examine more complicated and practically important problems involving two or more normal distributions.

Suppose $X_1, X_2, ..., X_n$ is a random sample of size n taken from a normal distribution with the probability density function,

$$f_X(x) = \frac{1}{\sqrt{2\pi}\,\sigma} \exp\left(\frac{-(x-\mu)^2}{2\sigma^2} \right) \tag{3.25}$$

where μ and σ^2 are respectively the mean and the variance of the distribution. For notational simplicity we will write $X \sim N(\mu,\sigma^2)$ to indicate that X has a normal distribution with mean μ and variance σ^2; similar notations will be used for other commonly used distributions. The joint probability density function of

$X_1, X_2, \ldots, X_n$ and the likelihood function of (μ,σ^2) can be expressed as

$$f_{\mathbf{X}}(\mathbf{x};\mu,\sigma^2) = L(\mu,\sigma^2;\mathbf{x}) = \frac{\exp\left[-\dfrac{(n-1)s^2}{2\sigma^2} - \dfrac{n(\bar{x}-\mu)^2}{2\sigma^2}\right]}{(2\pi)^{\frac{1}{2}}\sigma^n}, \tag{3.26}$$

where $\bar{x}$ and $s^2 = \sum(x_i - \bar{x})^2 / (n-1)$ are, respectively, the sample mean and the sample variance. It is known that these random variables are independently distributed as

$$\bar{X} \sim N(\mu, \sigma^2/n), \text{ and } U = (n-1)S^2/\sigma^2 \sim \chi^2_{n-1}. \tag{3.27}$$

By $V \sim \chi^2_m$, we mean that the random variable V has a chi square distribution with m degrees of freedom and the probability density function,

$$f_V(v) = \frac{1}{2^{m/2}\Gamma(m/2)} v^{m/2-1} e^{-v/2}, \quad v > 0. \tag{3.28}$$

It follows from (3.26) that $\bar{x}$ and s^2 are sufficient statistics for the parameters of the distribution. Moreover, since the normal distribution is a member of the exponential family of distributions, these statistics as well as $(\sum x_i, \sum x_i^2)$ are jointly complete. Inferences on functions of μ and σ can be based on the random variables.

It can be conveniently deduced from the mean of the normal distribution and the mean of the chi square distribution appearing in (3.27) or shown directly that $\bar{X}$ and S^2 are the MVUEs of μ and σ^2, respectively. Moreover, it is clear from the likelihood function $L(\mu,\sigma^2)$ given by (3.26) that the MLEs of the two parameters are $\hat{\mu} = \bar{X}$ and $\hat{\sigma}^2 = (1 - 1/n)\, s^2$.

We shall show in the following sections that the usual one-sided and two-sided (depending on whether the underlying alternative hypothesis is one-sided or two-sided) critical regions given by the generalized likelihood ratio tests of μ and σ can be utilized to obtain appropriate extreme regions on which p-values can be based. Alternatively, one can derive the extreme region by invoking the principles of sufficiency, invariance and similarity as in the case of conventional theory of hypothesis testing based on the Neyman-Pearson Lemma. For the purpose of illustration we shall take different approaches in the following sections.

3.6. Inferences About the Mean

A case of known variance, namely $\sigma^2 = 1$, was considered in Example 2.7 and p-values for testing the mean were given in terms of the standard normal distribution. The corresponding result for a normal distribution with a known variance σ^2 can also be deduced from Example 2.7 by means of the transformations $X \to X/\sigma$ and $\mu \to \mu/\sigma$. In most practical applications both parameters of the normal distribution are unknown to the investigator and so here we drop the assumption of known variance.

Derivation of Test Statistic

Consider the problem of testing the mean when the variance is unknown. As in the construction of fixed level tests for this problem, test statistics leading to significance tests of μ can be found by invoking the principles of invariance, similarity, or unbiasedness. In the present situation, these methods yield equivalent test statistics and there is no need to impose all the requirements. We will confine our exposition in this section to the method of invariance and then show that the resulting tests do not depend on unknown parameters thus leading to similar (and unbiased) tests. The test statistic can be derived regardless of whether the alternative hypothesis is one-sided or two-sided. To fix ideas however first consider the particular one-sided null hypothesis, $H_0: \mu \le 0$. The null hypothesis is to be tested against the alternative hypothesis $H_1: \mu > 0$.

This testing problem remains invariant under the transformation of scale of the random variable X from which observations are made. The change of scale by a positive constant k results in similar scale transformations of sufficient statistics, namely, $\bar{X} \to k\bar{X}$, $S \to kS$. Clearly, a maximal invariant under these transformations is $\bar{X}/S$ or equivalently

$$T = \frac{\bar{X}}{S/\sqrt{n}}. \tag{3.29}$$

Consequently, according to Theorem 2.2, any invariant test based on the sufficient statistics $\bar{X}$ and S^2 can also be constructed using T alone.

To show that T is indeed a test statistic, rewrite (3.29) as

$$T = \frac{\sqrt{n}\,(\bar{X}-\mu)/\sigma + \delta}{S/\sigma} = \frac{Z+\delta}{\sqrt{U/(n-1)}}, \tag{3.30}$$

where $\delta = \sqrt{n}\,\mu/\sigma$, Z is a standard normal random variable, and U is the chi square random variable defined in (3.27). The following properties, including those which qualify T as a test statistic, follow immediately from (3.30).

1. T is stochastically increasing in μ; that is, more precisely, Pr $(T > t)$ is an increasing function of μ.

2. When $\mu = 0$, the distribution of T does not depend on any nuisance parameters; this means that tests based on T will be similar (and unbiased due to Property 1) indeed.

3. T is distributed according to the noncentral t distribution with $r = n-1$ degrees of freedom and noncentrality parameter δ. The probability density

function of T is

$$f_T(t) = \frac{\Gamma((r+1)/2)\,e^{-\delta^2/2}}{\Gamma(r/2)\sqrt{\pi r}}\left[1+\frac{t^2}{r}\right]^{-(r+1)/2} h(t)\,, \tag{3.31}$$

where

$$h(t) = \sum_{i=0}^{\infty}(\delta t\sqrt{2}/r)^i\,\frac{\Gamma((r+i+1)/2)}{\Gamma((r+1)/2)\,i!}\left[1+\frac{t^2}{r}\right]^{-i/2}.$$

Let $t_{obs}=\sqrt{n}\,\bar{x}/s$ denote the observed value of the test statistic T, where $\bar{x}$ and s are observed values of $\bar{X}$ and S respectively. The larger values of t_{obs} tends to discredit $H_0: \mu \le 0$. Since T is stochastically increasing in μ the extreme region corresponds to the right tail of the distribution of T. It is of interest to note that this is also the critical region suggested by the generalized likelihood ratio method. Hence, the p-value for testing H_0 can be computed as

$$p(t_{obs}) = Pr(T \ge \sqrt{n}\,\bar{x}/s \mid \mu=0) = 1-G_{n-1}(\sqrt{n}\,\bar{x}/s)\,, \tag{3.32}$$

where G_{n-1} is the cumulative distribution function of the Student's t distribution with $n-1$ degrees of freedom and the probability density function,

$$f_T(t) = \frac{\Gamma(n/2)}{\Gamma((n-1)/2)\sqrt{\pi(n-1)}}\left[1+\frac{t^2}{(n-1)}\right]^{-n/2}. \tag{3.33}$$

The power function corresponding to the extreme region used in (3.32) is $\pi_0(\delta) = 1-G_{n-1,\delta}(\sqrt{n}\,\bar{x}/s)$, where $G_{n-1,\delta}$ is the cumulative distribution function of the noncentral t distribution with $n-1$ degrees of freedom and the noncentrality parameter δ.

General one-sided and two-sided tests

Now consider one-sided hypotheses of the form

$$H_0: \mu \le \mu_0 \quad \text{versus} \quad H_1: \mu > \mu_0, \tag{3.34}$$

where μ_0 is a specified constant. The problem of testing (3.34) can be reduced to the preceding hypothesis testing problem by means of the location transformations $X_i \to X_i - \mu_0$, $i=1,\ldots, n$, and $\mu \to \mu - \mu_0$. Hence, it can deduced from (3.32) that the tests of significance of (3.34) can be based on the p-value

$$p(\bar{x},s) = 1-G_{n-1}(\sqrt{n}\,(\bar{x}-\mu_0)/s)\,. \tag{3.35}$$

The smaller the observed value of $p=p(\bar{x},s)$, the greater the evidence against the null hypothesis. The p-value can be easily computed using XPro, SPlus, or widely used statistical software packages such as BMDP, SAS, and SPSS.

It can be deduced from foregoing results or shown in a similar manner that tests of the right-sided null hypothesis H_0: $\mu \geq \mu_0$ can be performed on the basis of the p-value $p = G_{n-1}(\sqrt{n}(\bar{x} - \mu_0)/s)$.

In the case of point null hypotheses of the form H_0: $\mu = \mu_0$, the same test statistic, $T = \sqrt{n}(\bar{X} - \mu_0)/S = (Z + \delta)/\sqrt{U/(n-1)}$, can be utilized to define the appropriate extreme region. Since the distribution of T is $G_{n-1,\delta}(t)$ the random variable $|T|$ can be expected to be stochastically increasing in δ or in $|\mu - \mu_0|$. To see this from Theorem 2.1 applied to the standard normal distribution, the cumulative distribution function of $Y = |T|$ can be expressed as

$$F_Y(y) = Pr(|T| \leq y) = E(H_\delta(yV)),$$

where $V = \sqrt{U/(n-1)}$, $H_\delta(x) = Pr(-x \leq Z + \delta \leq x)$, and the expectation appearing in the last equation is taken with respect to the random variable V. From Theorem 2.1, $H_\delta(x)$ is a decreasing function of $|\delta|$. Hence, $F_Y(y)$ is also a decreasing function of $|\delta|$ thus implying the desired result. Therefore, under the null hypothesis, the appropriate extreme region in this case corresponds to the two tails of the underlying Student's t distribution. Hence the corresponding p-value is

$$p = Pr(|T| > |t_{obs}|) = 2G_{n-1}(-\sqrt{n}|\bar{x} - \mu_0|/s) \ ,$$

where $t_{obs} = \sqrt{n}(\bar{x} - \mu_0)/s$.

The paired t-test

The t tests developed above can also be employed to compare the means of two normal distributions when the available data consist of matched pairs. This situation arises when observations are taken from a bivariate normal distribution so that the distribution of the difference in the two random variables has a univariate normal distribution. Such paired data become available, for example, when a certain measurement is taken from a sample of subjects before and after a certain treatment is given.

Let $(u_1, v_1), \ldots, (u_n, v_n)$ be a random sample from a pair of random variables (U, V), which are not necessarily independent. Let μ_u and μ_v be the means of the two random variables. Let

$$X = U - V \quad \text{and} \quad \mu = \mu_u - \mu_v.$$

Assume that X is normally distributed. Since the mean of X is μ it is now clear that the two parameters μ_u and μ_v can be compared using p-values of the form (3.35) based on the sample mean and the sample variance of the data $(u_1 - v_1), \ldots, (u_n - v_n)$ from the distribution of X.

Confidence Intervals

Confidence intervals for μ can be deduced immediately from the p-value given by (3.35). Notice that when μ_0 is replaced by μ, equation (3.35) yields the power function of the test variable $(\bar{X}-\mu)/S$ whose distribution is independent of nuisance parameters. For instance, a left-sided 100γ% confidence interval for μ can be obtained using the formula $1-G_{n-1}(\sqrt{n}(\bar{x}-\mu)/s)\leq\gamma$ given by (2.29). Hence, the one-sided 100γ% confidence intervals for μ are

$$\mu \leq \bar{x} + \frac{s}{\sqrt{n}}t_\gamma \quad \text{and} \quad \mu \geq \bar{x} - \frac{s}{\sqrt{n}}t_\gamma \tag{3.36}$$

respectively, where $t_\gamma = t_{\gamma,n-1}$ is the γth quantile of the Student's t distribution with $n-1$ degrees of freedom. The symmetric (and shortest) 100γ% confidence interval for μ obtained using (2.29) is

$$\bar{x} - \frac{s}{\sqrt{n}}t_{(1+\gamma)/2} \leq \mu \leq \bar{x} + \frac{s}{\sqrt{n}}t_{(1+\gamma)/2} . \tag{3.37}$$

Example 3.4. An agricultural innovation

An agrarian research scientist claims that he has developed a hybrid seed of rice paddy which out performs a widely used seed by at least 30% in average yield under normal growing conditions. It is known that, on average a farmer can get 120 bushels of rice per plot with the currently popular seed. The researcher has obtained the following data from 14 experimental plots randomly selected from a farm with average soil conditions.

Table 3.1. Rice yield (in bushels) obtained from 14 experimental plots

148	160	176	178	159	153	163
164	156	148	167	163	157	159

Let us test whether this data supports the scientist's claim. Let μ be the true mean yield of rice that a farmer can obtain using the hybrid seed. The null hypothesis of less than 30% improvement in the yield is $H_0:\mu\leq 156$. This hypothesis is to be tested against the alternative hypothesis $\mu > 156$ as claimed. The sample mean and the sample variance of the data are $\bar{x}=160.8$ and $s^2=78.3$ respectively. The observed value of the T statistic is $t_{obs} = \sqrt{14}(\bar{x}-156)/s=2.04$. The observed significance level is $p = 1-G_{13}(2.04)=.031$. Therefore, the null hypothesis of false claim can be rejected as we have significant evidence in favor of the alternative hypothesis of more than 30% improvement. The 95% confidence interval given by (3.37) for the mean yield μ is

$$[\ \bar{x} - \frac{s}{\sqrt{14}} t_{.975}\ ,\ \bar{x} + \frac{s}{\sqrt{14}} t_{.975}\ ] = [\ 155.7\ ,\ 165.9\ ]\ ,$$

where the .975th quantile of t-distribution with 14 degrees of freedom is 2.16.

3.7. Inferences About the Variance

In this section , in the context of normal theory, we shall demonstrate the derivation of a test statistic using the generalized likelihood ratio method. As in the previous section, one can of course derive the same test statistic by exploiting the location invariance of the variance. We will first derive the test statistic by the likelihood ratio method and then show that it is location invariant.

The likelihood function of μ and σ^2 is given by (3.26). The maximum likelihood estimators of theses parameters are $\bar{X}$ and $(1-1/n)S^2$. It is evident that the generalized likelihood ratio depends on the data and σ^2 only through the potential test variable S^2/σ^2. The underlying statistic can be derived by invoking the notion of p-similarity as well. The procedure is indicated in Exercise 3.3.

It remains to be shown that S^2 is indeed a statistic appropriate for testing hypotheses concerning σ^2. Consider the null hypothesis

$$H_0\colon \sigma^2 \le \sigma_0^2 \ \text{ against } \ H_1\colon \sigma^2 > \sigma_0^2 \tag{3.38}$$

Notice that the function

$$Pr(S^2 > t) \ = \ = \ Pr(U > t(n-1)/\sigma^2) \ = \ 1 - C_{n-1}(t(n-1)/\sigma^2)\ , \tag{3.39}$$

is increasing in σ^2, where C_{n-1} is the cumulative distribution function of the χ^2 distribution with $n-1$ degrees of freedom. Therefore, S^2 is stochastically increasing in σ^2 and thus it is appropriate for testing (3.38). The hypotheses in (3.38) can be tested on the basis of the p-value

$$p(s^2) = 1 - C_{n-1}(s^2(n-1)/\sigma_0^2)\ . \tag{3.40}$$

Obviously, the hypothesis $H_0\colon \sigma^2 \ge \sigma_0^2$ versus $H_1\colon \sigma^2 < \sigma_0^2$ can also be tested using the same test statistic. The corresponding p-value $p = C_{n-1}(s^2(n-1)/\sigma_0^2)$. These p-values can be easily computed using XPro, SPlus, or widely used statistical software packages such as BMDP, SAS, and SPSS.

Confidence Intervals

Confidence intervals for the variance of a normal distribution can be deduced from (3.39) with t replaced by the observed sample variance s^2. For instance, the left-sided 100γ% confidence interval of σ^2 can be obtained using the formula $1 - C_{n-1}(s^2(n-1)/\sigma^2) \le \gamma$ given by (2.29). Clearly, the left-sided and right-sided $100\gamma\%$ confidence intervals of σ^2 are

$$\sigma^2 \le \frac{s^2(n-1)}{\chi^2_{1-\gamma}} \quad \text{and} \quad \sigma^2 \ge \frac{s^2(n-1)}{\chi^2_{\gamma}} \tag{3.41}$$

respectively, where $\chi^2_\gamma = \chi^2_\gamma(n-1)$ is the γth quantile of the chi-square distribution with $n-1$ degrees of freedom. The $100\gamma\%$ confidence interval for σ^2 given by equal tails of the chi-square distribution is

$$\frac{s^2(n-1)}{\chi^2_{(1+\gamma)/2}} \le \sigma^2 \le \frac{s^2(n-1)}{\chi^2_{(1-\gamma)/2}} . \tag{3.42}$$

Example 3.5. Variation of piston diameters

A factory manufactures pistons for a certain internal combustion engine. The diameters of pistons produced in the factory follow a normal distribution with mean 36 cm and variance .16 cm^2. An engineer make certain changes in the production process to reduce the variation in piston diameters. The engineer believes that the standard deviation of piston diameters is reduced at least by 25% due to the new design and that it has not affected the desired mean of 36 cm.

A sample of 20 pistons produced with the new design give a mean diameter of 36.4 cm and a standard deviation of .32 cm. Let us test whether this data is in agreement with the engineer's assertions. The underlying hypotheses concerning the new mean μ and the new standard deviation σ of piston diameters can be formulated as

$$H_0 : \sigma \le .3 , \quad H_1 : \sigma > .3 ,$$

$$H_{0'} : \mu = 36 , \quad H_{1'} : \mu \ne 36 .$$

The observed value of the test variables which define (3.40) and (3.35) are

$$t_1 = \frac{19s^2}{.09} = 21.62, \quad \text{and} \quad t_2 = \frac{\sqrt{20}\,|\bar{x}-36|}{s} = 1.25,$$

respectively. The corresponding significance values are $p(t_1) = 1 - C_{19}(21.62) = .30$ and $p(t_2) = 2G_{19}(-1.25) = .23$ respectively. These are quite typical values for the probabilities of the extreme regions under H_0. There is no significant evidence to doubt engineer's assertions. The new design seems to have reduced the variation of piston diameters with no adverse effect on the mean diameter. The 95% confidence interval of σ given by the formula (3.42) is

$$[\, 0.32(\frac{19}{30.14})^{1/2} ,\ 0.32(\frac{19}{10.12})^{1/2} \,] = [\, .25 , .44 \,] .$$

3.8. Quantiles of a Normal Distribution

Consider the problem of testing hypotheses about the parameter $\theta=\mu+k\sigma$, where μ and σ^2 are respectively the mean and the variance of the normal distribution under consideration and k is a specified constant. Note that θ is the kth quantile of $N(\mu,\sigma^2)$.

Statistical inferences concerning quantiles of a normal distribution arise in various applications including statistical quality control. To describe a particular application, suppose that a batch of products is considered satisfactory if the expected fraction of defectives in the batch does not exceed some standard tolerance level ρ. Whether or not an individual item in the batch is defective is determined on the basis of a quality characteristic X which is known to be normally distributed. The actual net weight of a canned product is an example of this kind of a random variable. Let μ be the target value of the characteristic (e.g. the specified weight of a jar of peanut butter) and let σ be the standard deviation of the actual value of the characteristic. An item is considered satisfactory if its quality characteristic x does not fall below some tolerance limit x_0 so that the probability of a defective item is $Pr(X\le x_0)=\Phi((x_0-\mu)/\sigma)$, where Φ is the cumulative distribution function of the standard normal distribution. Hence, the problem of accepting a batch of products can be formulated as that of testing the hypothesis

$$H_0:\Phi\left(\frac{x_0-\mu}{\sigma}\right)\le\rho.$$

Now it is evident that the underlying hypothesis testing problem is that of testing the quantile of a normal distribution. More precisely, the preceding hypothesis can be written as H_0: $x_0\le\mu+c\sigma$, where $c=\Phi^{-1}(\rho)$, the ρ quantile of the standard normal distribution. The hypothesis is to be tested on the basis of a random sample taken from a batch of products.

Consider the problem of testing the hypotheses

$$H_0:\ \theta\le\theta_0,\quad \text{versus}\quad H_1:\ \theta>\theta_0 \tag{3.43}$$

using a random sample of size n. Clearly this null hypothesis can also be transformed into a scale invariant hypothesis, namely to H: $\nu+k\sigma\le 0$, by means of the change of variables $Y_i=X_i-\theta_0$, where $\nu=\mu-\theta_0$. Then the problem reduces to that of testing H (against the natural alternative hypothesis) on the basis of the sample mean $\bar{Y}$ and the sample variance $S_y^2=S^2$ computed from the random sample available from $N(\nu,\sigma^2)$. It follows from Section 3.2 that scale invariant extreme regions for this problem can also be constructed from the statistic $T=\sqrt{n}\,\bar{Y}/S$. In terms of the original variables $T=\sqrt{n}\,(\bar{X}-\theta_0)/S$. We shall now show that T is an appropriate test statistic for testing (3.43).

Since $\sqrt{n}\,(\bar{X}-\theta_0)/\sigma$ = $\sqrt{n}\,(\bar{X}-\mu)/\sigma$ + κ, T has a noncentral t distribution with $n-1$ degrees of freedom and noncentrality parameter κ, where

$\kappa=\sqrt{n}\,[(\theta-\theta_0)/\sigma-k]$. Consequently, T is stochastically increasing in θ, the parameter of interest. Further, when $\theta=\theta_0$, this distribution is independent of unknown parameters. Hence T is indeed a test statistic leading to right-tailed extreme regions. Therefore, significance tests of (3.43) can be based on the p-value

$$p(\bar{x},s)=1-G_{n-1,-\sqrt{n}k}\left[\frac{\bar{x}-\theta_0}{s}\right]. \tag{3.44}$$

The power function of the test is $\pi_0(\theta)=1-G_{n-1,\kappa}((\bar{x}-\theta_0)/s)$.

As in the case of testing the mean of a normal distribution, right-sided null hypotheses of the form H_0: $\theta\geq\theta_0$ can be tested against H_1: $\theta<\theta_0$ using the p-value $p=G_{n-1,\kappa}((\bar{x}-\theta_0)/s)$. Finally, significance testing of a point null hypothesis, which specifies the desired quantile of the distribution to be θ_0, can employ the p-value $p=2G_{n-1,\kappa}(-|\bar{x}-\theta_0|/s)$.

The confidence intervals of θ can be deduced from (3.44) with the aid of equation (2.29).

Example 3.6. Statistical quality control

Under normal controlled operating conditions a factory turns out high-strength wire cables with less than 1% defectives. A cable is considered defective if its strength (in thousands of pounds per square inch) is less than 45. Table 3.2 gives the data on the strengths that were observed at a quality control experiment involving destructive tests on 16 cables.

Table 3.2. Strength of 16 wire cables: cable number i and its strength x_i.

i	1	2	3	4	5	6	7	8
x_i	50.9	56.3	63.8	55.8	49.7	55.0	48.5	58.2
i	9	10	11	12	13	14	15	16
x_i	46.3	57.6	50.5	58.8	43.9	51.4	45.8	55.3

Assume that the cable strength, X, is a normally distributed random variable. We are interested in testing the hypothesis that the production process is under control, that is H_0: $\theta \geq 45$, where θ is the .01 quantile of the distribution of X. More explicitly $\theta = \mu - 2.33\sigma$, where μ and σ^2 are the mean and the variance of X. Under our assumption that the distribution of X is normal, the testing can be based on the observed values of the sufficient statistics $\bar{x} = 52.99$ and $s^2=30.03$. The observed value of the test statistic is $t_{obs} = 4(\bar{x}-45)/s = 5.83$ and $\sqrt{n}k = -9.32$. Hence, the observed level of significance is $p = G_{15,9.32}(5.83) = .0079$. There is very strong evidence against the null hypothesis and so we can conclude without a reasonable doubt that the production process is not under control.

The Coefficient of Variation

The foregoing testing procedure can be readily applied to test hypotheses concerning the coefficient of variation $\rho = \sigma/\mu$ and the mean expressed in σ units, namely $\rho^{-1} = \mu/\sigma$. Consider, for instance, hypotheses of the form

$$H_0: \rho \geq \rho_0, \quad H_1: \rho < \rho_0,$$

where ρ_0 is a specified constant. Notice that, by defining $\theta = \mu - \rho_0^{-1}\sigma$ and $\theta_0 = 0$, these hypotheses can be transformed into hypotheses of the form in (3.43). It is now evident that the above hypotheses can be tested using the p-value

$$p = 1 - G_{n-1, \sqrt{n}\rho_0^{-1}}\left(\frac{\bar{x}}{s}\right).$$

Fixed-Level Tests

For each of the hypotheses considered above there exists a corresponding fixed level test. While these fixed level tests can be readily deduced from the p-values the converse is not possible without additional information. Therefore, it is always more informative to report p-values even in those hypothesis testing situations where 'accept/reject' type decisions are of primary interest.

If $p(\mathbf{x})$ is a p-value corresponding to a certain null hypothesis, say H_0, a corresponding test at a nominal level α (e.g. $\alpha = .05$) can be obtained either based on the extreme region on which the p-value is based or directly from the p-value. In each of the above cases the extreme regions on which the p-values are based give rise to critical regions on which UMP (uniformly most powerful) invariant (and unbiased) tests can be based. For instance, if the extreme region for testing H_0: $\theta \leq \theta_0$ is of the form $\{\mathbf{X} \mid T(\mathbf{X}) \geq t_{obs}\}$, the corresponding critical region is of the form $\{\mathbf{X} \mid T(\mathbf{X}) \geq t(\alpha)\}$, where $t(\alpha)$ is chosen such that $Pr(T \geq t(\alpha)) = \alpha$.

On the basis of the p-value the null hypothesis is rejected at level α if $p(\mathbf{x}) < \alpha$. Using the properties of the underlying test statistics the tests obtained in this manner can be easily expressed in terms of t_{obs}, that is, in terms of the sample mean $\bar{x}$ and sample variance s^2. For instance if the null hypothesis of interest is

$$H_{01}: \mu \leq \mu_0, \quad H_{02}: \mu \geq \mu_0, \quad or \quad H_{03}: \mu = \mu_0,$$

the corresponding UMP unbiased test of nominal level α is to reject H_0 if

$$t_\alpha < t_{obs}, \quad t_{obs} < -t_\alpha, \quad \text{and} \quad t_{\alpha/2} < |t_{obs}|$$

respectively, where t_α is the $1-\alpha$th quantile of the Student's t distribution with n-1 degrees of freedom and $t_{obs} = \sqrt{n}\,(\bar{x} - \mu_0)/s$.

3.9. Conjugate Prior and Posterior Distributions

Let $X_1, \ldots, X_n$ be a random sample from a normal distribution with mean μ and variance σ^2. Suppose, as usually the case, that both the mean and the variance are unknown. It is convenient, in this case, to specify the conjugate distributions in terms of the mean μ and $\tau = \sigma^{-2}$, where τ is the precision of the distribution. We shall show that if the conditional prior distribution of μ given σ^2 is a normal distribution, and if the marginal distribution of σ^{-2} is a gamma distribution, then the posterior distributions belong to the same family of distributions. In other words we shall show that the normal-gamma family of distributions is a conjugate family of distributions for (μ, σ^{-2}). Theorem 3.1 gives a precise statement of the result; a proof of this theorem is given in Appendix B.

Theorem 3.1. Suppose the joint prior distribution of (μ, σ^{-2}) is given by the following distributions:

$$\mu \mid \sigma^{-2} \sim N(\mu_0, \sigma^2/\eta_0), \quad \sigma^{-2} \sim G(\alpha_0, \lambda_0^{-1}). \tag{3.45}$$

Given a random sample $X_1, \ldots, X_n$ from $N(\mu, \sigma^2)$, the posterior distribution of (μ, σ^{-2}) is given by the following distributions:

$$\mu \mid \sigma^{-2}, \mathbf{x} \sim N(\mu_1, \sigma^2/\eta_1), \quad \sigma^{-2} \mid \mathbf{x} \sim G(\alpha_1, \lambda_1^{-1}), \tag{3.46}$$

where

$$\mu_1 = \frac{\eta_0 \mu_0 + n\bar{x}}{\eta_0 + n}, \quad \eta_1 = \eta_0 + n \tag{3.47}$$

and

$$\alpha_1 = \alpha_0 + n/2, \quad \lambda_1 = \lambda_0 + \frac{n-1}{2}s^2 + \frac{n}{2}\frac{\eta_0}{\eta_1}(\bar{x} - \mu_0)^2. \tag{3.48}$$

The marginal posterior distribution of μ can be obtained by integrating the joint posterior distribution with respect to σ^{-2}. As demonstrated by the proof of Theorem 3.2, the definition of the t-distribution in terms of normal and chi square random variables is a simpler way to establish the result.

Theorem 3.2. The marginal posterior distribution of μ is given by that of the standardized random variable $\nu = (\mu - \mu_1)(\eta_1 \alpha_1 / \lambda_1)^{1/2}$ having a t-distribution with $2\alpha_1$ degrees of freedom. That is, the probability density function of ν is

$$g_\nu(\nu) = \frac{\Gamma(\alpha_1 + 1)}{\Gamma(\alpha_1)\sqrt{2\pi\alpha_1}} \left[1 + \frac{\nu^2}{2\alpha_1} \right]^{-(\alpha_1 + 1/2)}. \tag{3.49}$$

Proof. From (3.46) we have $\sqrt{\tau\eta_1}(\mu - \mu_1) \mid \tau, \mathbf{x} \sim N(0,1)$. Since this distribution is free of τ, the unconditional distribution of $\sqrt{\tau\eta_1}(\mu - \mu_1)$ is also standard normal. It further follows from (3.46) that $2\tau\lambda_1 | \mathbf{x} \sim \chi^2_{2\alpha_1}$. Hence, by the definition of the t-distribution

$$\left(\frac{\eta_1 \alpha_1}{\lambda_1}\right)^{1/2} (\mu - \mu_1) | \mathbf{x} \sim t_{2\alpha_1}, \tag{3.50}$$

as claimed.

Of course, a similar result holds for the assumed prior distribution, namely,

$$\left(\frac{\eta_0 \alpha_0}{\lambda_0}\right)^{1/2} (\mu - \mu_0) \sim t_{2\alpha_0} .$$

It can be also seen from above results (by taking the limits $\lambda_0 \to 0$, $\eta_0 \to 0$, and setting the improper value $\alpha_0 = -\frac{1}{2}$ so that the improper prior $g(\mu,\tau) = \tau^{-1}$ is obtained) or shown directly that, if the improper prior distribution $g(\mu,\sigma) = \sigma^{-1}$ is assumed, then the posterior distribution becomes

$$\frac{\sqrt{n}(\mu - \bar{x})}{s} | \mathbf{x} \sim t_{n-1} . \tag{3.51}$$

It should be noted that this improper prior distribution is not quite the noninformative prior given by Jeffreys' method applied to Fisher information matrix. The noninformative matrix given by formula (A.1) is $g(\mu,\sigma) = \sigma^{-2}$.

3.10. Bayesian Inference About the Mean and the Variance

Assume again that $X_1, \ldots, X_n$ is a random sample from a normal distribution with mean μ and variance σ^2, where $n \geq 2$. Consider the posterior distribution of μ and σ^2 given by the results in Section 3.6. In this section we will suppress the implicit fact that the posterior distribution is a conditional distribution given the data **x**.

It is clear from (3.50) that the posterior distribution of μ is symmetric about μ_1, which is also the mean and the mode of the distribution. Therefore, the generalized maximum likelihood estimator of μ corresponding to the conjugate prior distribution (3.45) is

$$\hat{\mu} = \frac{\eta_0 \mu_0 + n\bar{x}}{\eta_0 + n} .$$

This is also the posterior mean, which is optimal under a variety of loss functions including the squared error loss.

In order to find the generalized maximum likelihood estimate of σ^2, let us express the posterior distribution of σ^{-2} given by (3.46) as

$$\log(g_2) = \log(c_2) + (\alpha_1 - 1)\log(\tau) - \lambda_1 \tau , \tag{3.52}$$

where $\tau = \sigma^{-2}$. Differentiating (3.52) with respect to τ we get

$$\frac{dg_2}{d\tau} = \frac{\alpha_1 - 1}{\tau} - \lambda_1 .$$

It is now clear that the generalized maximum likelihood estimate of σ^2 is

$$\hat{\sigma}^2 = \frac{\lambda_1}{\alpha_1 - 1} = \frac{\lambda_0 + \frac{n-1}{2}s^2 + \frac{n}{2}\frac{\eta_0}{\eta_0 + n}(\bar{x} - \mu_0)^2}{\alpha_0 + \frac{n}{2} - 1} .$$

It can be shown that the posterior mean of σ^2 is also given by this equation.

Credible intervals

Consider the random variable ν defined in Theorem 3.2. Since ν has a t-distribution, for instance, left-sided credible intervals of μ can be found as

$$Pr(\ \mu \le \mu_\gamma\ ) = Pr(\ \nu \le \left(\frac{\eta_1 \alpha_1}{\lambda_1}\right)^{1/2} (\mu_\gamma - \mu_1))$$

$$= G_{2\alpha_1}\left(\left(\frac{\eta_1 \alpha_1}{\lambda_1}\right)^{1/2} (\mu_\gamma - \mu_1)\right) ,$$

where $G_{2\alpha_1}$ is the cdf of the t-distribution with $2\alpha_1$ degrees of freedom and the parameters are as defined by (3.46). Let $t_\gamma(2\alpha_1)$ be the γth quantile of the t-distribution with $2\alpha_1$ degrees of freedom. Now it is evident that the left-sided $100\gamma\%$ Bayesian confidence interval (credible region) of μ is

$$\mu \le \mu_1 + \left(\frac{\lambda_1}{\eta_1 \alpha_1}\right) t_\gamma(2\alpha_1) .$$

Similarly the right-sided $100\gamma\%$ credible region of μ is

$$\mu \ge \mu_1 - \left(\frac{\lambda_1}{\eta_1 \alpha_1}\right) t_\gamma(2\alpha_1) .$$

Other types of credible regions can be found in a similar manner. In particular, the HPD credible region of μ is

$$\mu_1 - \left(\frac{\lambda_1}{\eta_1 \alpha_1}\right) t_{(1+\gamma)/2}(2\alpha_1) \le \mu \le \mu_1 + \left(\frac{\lambda_1}{\eta_1 \alpha_1}\right) t_{(1+\gamma)/2}(2\alpha_1) .$$

Notice that under the noninformative prior specified in Section 3.6 this interval is the same as the shortest $100\gamma\%$ confidence interval given by (3.37).

The credible regions of σ^2 can be derived using the result

$$\frac{2\lambda_1}{\sigma^2} \sim \chi^2_{2\alpha_1} . \tag{3.53}$$

Obviously, the one-sided 100γ% credible intervals of σ^2 by this posterior distribution are

$$\sigma^2 \leq \frac{2\lambda_1}{\chi^2_{1-\gamma}(2\alpha_1)} \quad \text{and} \quad \sigma^2 \geq \frac{2\lambda_1}{\chi^2_{\gamma}(2\alpha_1)} .$$

The equal-tail 100γ% credible interval of σ^2 is

$$\frac{2\lambda_1}{\chi^2_{(1+\gamma)/2}(2\alpha_1)} \leq \sigma^2 \leq \frac{2\lambda_1}{\chi^2_{(1-\gamma)/2}(2\alpha_1)} .$$

The HPD credible regions of this parameter can be found by numerical methods.

Posterior odds

Let us now turn to the problem of testing hypotheses about the parameters of a normal distribution. We shall consider only the problems of testing one-sided hypotheses under the natural conjugate priors and noninformative priors. In a Bayesian treatment, these priors are not quite appropriate for point null hypotheses. The readers interested in Bayesian testing of point null hypotheses are referred to Berger (1985).

Consider first the problem of testing

$$H_0 : \mu \leq k \quad \text{against} \quad H_1 : \mu > k ,$$

where k is a prespecified constant. These hypotheses are to be tested on the basis of the posterior distribution of μ given by (3.50). If the prior odds ratio is 1, then μ_0 is the same as k. The posterior probabilities of H_0 and H_1 are

$$\begin{aligned} p_0 &= Pr(\mu \leq k \mid \mathbf{x}) \\ &= Pr\left(\nu \leq \left[\frac{\eta_1 \alpha_1}{\lambda_1} \right]^{1/2} (k - \mu_1) \right) \\ &= G_{2\alpha_1}\left[\left(\frac{\eta_1 \alpha_1}{\lambda_1} \right)^{1/2} (k - \mu_1) \right] , \end{aligned} \tag{3.54}$$

and $p_1 = 1 - p_0$ respectively. Using (3.54) the posterior odds ratio can be computed as p_0 / p_1. The posterior odds ratio for testing right-sided null hypotheses is p_1 / p_0.

Consider now the situation of testing hypotheses about the variance, say $H_0 : \sigma^2 \leq \sigma_0^2$ versus $H_1 : \sigma^2 > \sigma_0^2$. We can use (3.53) in this case to compute the posterior probability of H_0 as

$$Pr(\sigma^2 \le \sigma_0^2) = Pr\left(\frac{2\lambda_1}{\sigma_0^2} \le \frac{2\lambda_1}{\sigma^2}\right)$$

$$= 1 - C_{2\alpha_1}(2\lambda_1/\sigma_0^2) ,$$

where $C_{2\alpha_1}$ is the cdf of the chi square distribution with $2\alpha_1$ degrees of freedom. The posterior probability of H_1 is of course $1-q_0$.

It is of interest to note that under the noninformative prior, p_0 and q_0 are the same as the p-values given by (3.35) and (3.40) respectively.

Example 3.7. Variation of piston diameters (continued)

Suppose the following prior distributions are available on the mean and the variance of the diameters of pistons produced according to the new design:

$$\mu | \tau \sim N(36,.08) , \quad \tau \sim G(1,.16^{-1}) .$$

Let us find the posterior probability of the hypothesis H_0: $\sigma \le .3$ that the engineer is interested in. The relevant posterior parameters are

$$\alpha_1 = 11 \quad \text{and} \quad \lambda_1 = \frac{.16+19}{2}(.32)^2 + 10(\frac{2}{22})(.4)^2 = 1.278 .$$

Therefore, the posterior probability of the null hypothesis is $q_0 = 1 - C_{22}(28.4) = .16$. Notice that mainly due to the prior mean .16 of σ^2 we have more evidence against the null hypothesis claimed by the engineer; the evidence is not very strong, however.

Exercises

3.1. Let X be a binomial random variable with parameters n and p. Show that the moment generating function of X is

$$\begin{aligned} M_X(t) &= E(e^{tX}) \\ &= (1 + p(e^t - 1))^n. \end{aligned}$$

Hence deduce that the mean and the variance of X are given by

$$E(X) = np \quad \text{and} \quad Var(X) = np(1-p),$$

respectively.

3.2. Consider the Poisson process presented in Section 3.1. Using the difference differential equation given by the conditions of the Poisson process $X(t)$, derive the formula (3.4) for its probability mass function.

3.3. Let X be a Poisson random variable with the parameter λ. Show that the moment generating function of X is

$$\begin{aligned} M_X(t) &= E(e^{tX}) \\ &= e^{\lambda(e^t - 1)}. \end{aligned}$$

Hence, deduce or show otherwise that the mean and the variance of X are given by

$$E(X) = \lambda \quad \text{and} \quad Var(X) = \lambda,$$

respectively.

3.4. Let $X_1, \ldots, X_n$ be a random sample from a Poisson distribution with parameter λ. Find the moment generating function of $S = \sum X_i$. Hence, deduce that S has a Poisson distribution with parameter $n\lambda$.

3.5. Let $X \sim \mathbf{B}(n, p)$ be a binomial random variable. Show that the class of beta distributions provide a natural conjugate family and that if $p \sim Beta(\alpha,\beta)$ is the prior, then the posterior distribution of p given x is

$$p \sim Beta(\alpha + x, \beta + n - x).$$

Show also that $g(p) = p^{-1/2}(1-p)^{-1/2}$ is the noninformative prior given by Jeffrey's rule.

3.6. Let $X_1, \ldots, X_n$ be a random sample from a normal population with mean μ and variance σ^2.

(a) Show that the maximum likelihood estimators of μ and σ^2 are $\bar{X}$ and $\sum (X_i - \bar{X})^2 / n$ respectively.

(b) By the method of GLR derive a statistic for testing hypotheses about μ.

(c) By the method of GLR derive a statistic for testing hypotheses about $\theta = \mu/\sigma$.

(d) Construct a left-sided 95% confidence interval for θ.

3.7. Let $X_1, \ldots, X_n$ be a random sample from a normal population with mean μ and variance σ^2. Consider the problem of testing the hypotheses

$$H_0: \mu=0, \quad H_1: \mu\neq 0.$$

(a) Show that p-similar tests of these hypotheses can be based on the conditional distribution of $\bar{X}$ given $W = \sum X_i^2$.

(b) Show that W and $\bar{X}/S$ are independent random variables.

(c) Hence, find a statistic for testing the above hypotheses and find its p-value.

3.8. In exercise 3.7, suppose it is desired to test the hypotheses

$$H_0: \sigma^2 \geq 1, \quad H_1: \sigma^2 < 1.$$

(a) Show that p-similar tests of these hypotheses can be based on the conditional distribution of $W = \sum X_i^2$ given $\bar{X}$.

(b) Show that S^2 and $\bar{X}$ are independent random variables.

(c) Hence or otherwise find a statistic for testing the above hypotheses and find its p-value.

(d) Establish a procedure for testing

$$H_0: \sigma^2 = 1, \quad H_1: \sigma^2 \neq 1.$$

3.9. Let $X_1, \ldots, X_n$ be a random sample from a normal population with mean μ_0 and variance σ^2, where μ_0 is a known constant. Show that the MLE of σ^2 is

$$\hat{\sigma}^2 = \frac{\sum_{i=1}^{n} (X_i - \mu_0)^2}{n}.$$

It is desired to test the following hypotheses:

$$H_0: \sigma^2 = 1, \quad H_1: \sigma^2 \neq 1.$$

Show that the testing problem is invariant under the change of location $X \to X + c$, where c is a constant. Find a single statistic which is maximal invariant under this transformation and show that the statistic has the properties of a test statistic. Hence derive the procedure for significance testing of the above hypotheses.

3.10. A sample of 100 families with children is selected at random. Out of the one hundred eldest children in these families fifty seven were found to be males. Test the hypothesis that first-born children are more likely to be males.

3.11. Consider the soldering inspection system considered in Example 3.1. In an inspection period eleven connections were found to be defective.

(a) Test the null hypothesis H_0: $\lambda \geq 6$ that the soldering process is out of control.

(b) Test the point null hypothesis H_0: $\lambda = 6$, that the soldering process is under control.

(c) Construct a 99% Bayesian confidence interval under the diffuse prior $g(\lambda) = \lambda$.

3.12. The diameters (in centimeters) of trunks of a sample of thirty-year-old trees selected at random from a certain forest are as follows:

64 33 39 26 48 21 56 74 34 81
24 45 94 31 44 51 33 51 25 42

Test the hypothesis that the mean diameter of trunks of thirty year old trees grown in this forest is more than 40 cm. Also, test the hypothesis that the standard deviation of the diameter is less than 20 cm. Construct a 95% confidence interval for the mean diameter of trunks.

3.13. A weight-loss diet manufacturer claims that (i) on average, regular users of this diet lose 20 pounds in weight after the first month on the diet, and (ii) more than 20% of regular diet users lose at least 30 pounds during the first month. The following data on weight loss are available from 16 users of this diet after a month on the diet:

14.0 3.0 24.9 15.8 3.1 30.6 22.8 19.1
20.7 17.7 18.9 18.4 43.2 32.8 1.6 2.0

Does this data support the manufacturer's claims?

3.14. One year ago, the mean age of people in a city watching a certain TV cartoon program was 15 years. A random sample of 100 people watching the cartoon show this year has a mean age of 16.2 years with a standard deviation of 3.6 years. Has the mean age of people watching the show increased significantly? Construct 99% confidence intervals for the mean age and the standard deviation of age of people in the city watching the show this year.

3.15. An experiment was conducted to compare a new brand of tires with a currently popular brand. A tire of the popular brand gives a mean mileage of

30,000 before it exceeds the recommended wear. A random sample of 30 new tires tested under normal conditions has a mean mileage of 36,000. The standard deviation of the sample is 3,000 miles. Test the hypothesis that the new brand yields at least 5,000 miles more than the popular brand. Construct a 95% right-sided confidence interval for the mean mileage of the new brand.

3.16. Let $X_1, \ldots, X_n$ be a random sample from a normal population with mean μ_0 and variance σ^2, where μ_0 is a known constant. Assume that the prior distribution of σ^{-2} is $G(\alpha,\beta)$.

(a) Find the posterior distribution of σ^{-2}.

(b) Deduce the posterior distribution of σ^2.

(c) Establish Bayesian procedures for testing one-sided hypotheses about σ^2.

(d) Establish Bayesian procedures for constructing credible intervals about σ^2.

3.17. In exercise 3.8 let μ and σ^2 be the mean and the variance of mileage that a new brand of tires yield. Construct the 95% HPD interval of μ under the priors, $\mu | \sigma^2 \sim N(30000, \sigma^2/2)$ and $\sigma^{-2} \sim G(1, 3000^{-2})$.

3.18. Let $X_1, \ldots, X_n$ be a random sample from the normal distribution with mean μ and variance σ^2. Find the noninformative prior distribution of (μ, σ^{-2}) and in turn find the posterior distribution of: (i)μ given the data and; (ii)σ^{-2} given the data.

Chapter 4
Exact Nonparametric Methods

4.1. Introduction

Most of this book deals with what are commonly referred as parametric methods, where we make assumptions about the underlying parametric family of distributions from which the observations are taken, and then make inferences about some of its unspecified parameters. In particular, the linear models in the last three chapters of the book are all based on the assumption that the underlying populations are normally distributed. While this assumptions may be justified and reasonable in some situations, this may not be the case in some other situations. In a loose sense, statistical procedures that are not based on the assumptions about the population distribution are referred to as *nonparametric*, or *distribution-free* methods. The terms 'parametric' versus 'nonparametric' basically arise from considerations in the context of hypothesis testing, as the latter allows comparisons and other tests of distributions regardless of whether they belong to a certain parametric family. It should be noted, however, that there is no common agreement among statisticians concerning whether certain statistical procedures can be classified clearly as parametric or nonparametric. For example, in many applications the binomial test that we studied in Section 3.3 which can be considered as both a parametric method and a nonparametric method.

The purpose of this chapter is to present widely used nonparametric tests in terms of exact p-values as opposed to approximate fixed-level tests as presented in many text books in statistics. In the previous chapters, except perhaps for some discrete distributions such as the binomial and Poisson, we basically dealt with parametric methods. Before we get into more complicated problems involving two or more parametric distributions, it is convenient to study problems involving one or more populations in the nonparametric setting. This is because exact as well as approximate methods in nonparametric methods are simpler, both analytically and conceptually. Moreover, the procedures in this context are intuitive and easy to understand. In any case, nonparametric methods are important in those situations where the underlying family of distributions is unknown, nonstandard, or when the available data consist of ranks rather than values of the underlying random variables.

Of course, nonparametric techniques have some disadvantages as well. The major disadvantage is that these methods tend to be less efficient compared to parametric methods when the underlying distributional assumptions are valid. For example, when a set of data actually provides sufficient evidence based on a parametric test to reject a null hypothesis at a given fixed-level, a nonparametric test applied to the same data set may fail to reject the hypothesis. Furthermore, nonparametric methods often do not take full advantage of the actual numerical value of an observation; they make only partial use of the information in the data in terms of relative ranks of observations, etc. Inefficiency of nonparametric methods is a consequence of this waste of information provided by the data.

Most of the exact nonparametric procedures are based on the idea of conditional inference introduced by Fisher (1925). In this approach, the nuisance parameters of the inference problem are eliminated by conditioning on certain functions of the observable random variables. For example, in contingency tables that we will study in Section 4.6, the conditioning is performed on one or both margins of the table, which contain the row and column sums of the data. Extreme regions obtained by this approach are also well-defined subsets of the sample space, and their conditional probabilities serve as a measure of how well the data supports or discredits the underlying hypotheses.

Until recently, most of the applications involving nonparametric tests were performed using asymptotic approximations. This is because, except for very small sample sizes, exact inferences in this context were computationally time-consuming or impractical even with modern computing facilities. This situation has now improved due to a variety of computationally efficient algorithms that were recently developed (see, for instance, Baker (1977), Mehta and Patel (1980, 1983, 1986), Pagano and Tritchler (1983), and Balmer (1988)). The computation of exact confidence intervals in this context is still a computationally prohibitive task and poses further challenges to methodology researchers to devise algorithms to make this possible with typical sample sizes which arise in applications. We shall not discuss available algorithms for computing exact p-values, as this is beyond the scope of this book. The readers interested in those algorithms are referred to the excellent surveys presented by Verbeek and Kroonenberg (1985), and Agresti (1992), and to the original articles.

In this book, we present some widely used nonparametric tests only. Those readers who are interested in other nonparametric techniques are referred to Agresti (1992), Gibbons (1985), Hollander and Wolfe (1973), Lehmann (1975), and Conover (1980).

4.2. The Sign Test

One of the simplest nonparametric tests is the two-sample sign test applied to data that come in matched pairs. The test applies equally well and was originally developed for testing a quantile (e.g. the median) of a population with an unknown continuous distribution. If the data are normally distributed, then we would compare the two distributions (or test the mean, in the case of a single distribution) using the t-test developed in Section 3.6. The sign test allows us to carry out a similar test when the normality assumption is not reasonable.

Let $(X_1,Y_1),\ldots,(X_n,Y_n)$ be a random sample of size n from a bivariate distribution. As in the case of the paired t-test, X could be thought of as the response of a subject (e.g. blood pressure of a patient) before a treatment (or treatment 1) and Y as the response after the treatment (or treatment 2). Let $p = Pr(X \le Y)$ and consider hypotheses of the form

$$H_0\colon p \ge p_0 \quad \text{versus} \quad H_1\colon p < p_0\,, \tag{4.1}$$

where p_0 is specified. On the other hand, in the case of testing the quantile of a distribution based on a single sample, say $X_1,\ldots,X_n$, the parameter of interest is a quantile q, or equivalently the probability $p = Pr(X \le q)$, where q is the unknown quantile of the distribution. In particular, if one wishes to test that treatment 1 is no more effective than treatment 2 or to test left-sided null hypotheses concerning the median of a distribution, then $p_0 = .5$. For simplicity, assume that there is zero probability that the pair of observations obtained from any subject is identical. In applications, one may omit any observed ties of X and Y from the available sample. When there are too many ties, however, this procedure may not be satisfactory; the reader is referred to Lehmann (1975), Conover (1980), and Gibbons (1985) for discussions about how to handle ties and for some variations of the sign test outlined here.

Let W denote the sign of $Y - X$ (or $q - X$), in the case of testing the quantile of a distribution. Obviously, the observations on W form a sequence of Bernoulli trials. The probability of a + sign is p. Consequently, the total number of + signs obtained in n independent trials, say U, follows a binomial distribution with parameters n and p. Therefore, the p-value of the sign test for testing hypotheses in (4.1) follows immediately from (3.3). With an observed number u of + signs, the null hypothesis is rejected if the observed level of significance

$$p = 1 - F_{B_u}(p_0) \tag{4.2}$$

is too small, where F_{B_u} is the cdf of the beta random variable

$$B \sim Beta(u+1, n-u).$$

In terms of the cdf of the F-random variable

$$W_u = \frac{n-u}{u+1}\frac{B}{1-B} \sim F_{2(u+1),2(n-u)}$$

the p-value can be computed conveniently as

$$p = 1 - H_{2(u+1),2(n-u)}\left\{\frac{n-u}{u+1}\frac{p_0}{1-p_0}\right\}$$

$$p = H_{2(n-u),2(u+1)}\left\{\frac{u+1}{n-u}\frac{1-p_0}{p_0}\right\}, \tag{4.3}$$

where $H_{r,s}$ is the cdf of the F-distribution with r and s degrees of freedom. Similarly, the p-value for testing left-sided null hypotheses is computed as

$$p = H_{2u,2(n-u+1)}\left\{\frac{n-u+1}{u}\frac{p_0}{1-p_0}\right\}. \tag{4.4}$$

Using the TESTIMATE software package p-values can be conveniently computed for both the one-sided and two-sided hypotheses.

Example 4.1. Testing the effectiveness of a course

In a study carried out to evaluate the effectiveness of a course designed to improve writing skills, essays written by fourteen students are scored before and after the course. Table 4.1 displays the results of the study; also given in the table are the sign of the difference between the after score and the before score.

Table 4.1. Scores before and after the course

Student	Before	After	Sign
A	78	84	+
B	65	61	-
C	67	63	-
D	58	62	+
E	52	59	+
F	85	78	-
G	60	72	+
H	48	55	+
I	55	63	+
J	72	76	+
K	80	75	-
L	63	70	+
M	57	65	+
N	49	45	-
O	62	75	+

Consider the hypothesis that the course is ineffective. In this application $n = 15$, $u = 10$, and $p_0 = .5$, and so the desired hypothesis can be tested using the p-value given by (4.4). The observed level of significance of the sign test is computed as

$$p = H_{20,12}(.6)$$
$$= .151.$$

Therefore, the evidence is not strong against the null hypothesis that the course is ineffective in improving writing skills.

4.3. The Signed Rank Test and the Permutation Test

Another test that can be used to test the median of a distribution is the signed rank test due to Wilcoxon (1945); the test is also known as the *Wilcoxon signed-rank test*. Just like the sign test, given a set of paired data it can be used for comparing the distributions of two correlated random variables as well. Unlike the sign test, not only does it depend on the signs of the differences of matched pairs, but it also uses the ranks of their absolute values. Hence, the Wilcoxon test tends to be more efficient in detecting differences in distributions.

Consider again a random sample $(X_1, Y_1), \ldots, (X_n, Y_n)$ of matched pairs, and let $D_i = Y_i - X_i$, $i = 1, \ldots, n$ denote the difference in paired observations; in testing the population median ν_y, of the distribution of Y, define $D_i = Y_i - \nu_y$. Assume that the distribution of the random variable D is continuous. Let θ be the median of its distribution. We consider only the case

$p_0 = .5$ in (4.1) and so equivalently consider the hypotheses

$$H_0: \theta \leq 0 \quad \text{versus} \quad H_1: \theta > 0; \tag{4.5}$$

in comparing two treatments X and Y, the null hypothesis corresponds to the statement that treatment Y is no more effective than treatment X.

To describe the Wilcoxon test, consider the absolute values $|D_1|,\ldots,|D_n|$. Assume that they are all positive; in applications zero differences can be discarded to satisfy this assumption. Then, the Wilcoxon test statistic is computed in the following steps:

(a) Arrange the absolute differences in order of their magnitude from the smallest to the largest and rank them from 1 (the smallest) to n (the largest). If there are ties first assign successive ranks to the absolute values and then assign the average of these ranks to each of the tied values; that is, if n_i absolute differences are tied at the ith smallest value, then assign the rank

$$r_i = n_1 + \cdots + n_{i-1} + (n_i + 1)/2$$

to each of them, where $\sum_{i=1}^{n} n_i = n$.

(b) Assign a plus sign or a minus sign to each rank depending on the sign of the original D_i's.

(c) Compute the Wilcoxon test statistic as the sum of those ranks to which a plus sign was assigned.

Let W be the resulting statistic on which to base the Wilcoxon signed-rank test. Since each rank is more likely to receive a plus sign when $\theta > 0$, large values of W support the alternative hypothesis, thus indicating the appropriate extreme region. The p-value of the Wilcoxon test is then

$$p = Pr(W \geq w) = 1 - F_W(w - 1), \tag{4.6}$$

where w is the observed value of the test statistic.

To find the distribution of W, let V_i denote the Bernoulli random variable indicating whether or not rank r_i is assigned a plus sign. Then, the Wilcoxon statistic can be expressed as

$$W = r_1 V_1 + r_2 V_2 + \cdots + r_n V_n, \tag{4.7}$$

where $r_i = i$ if there are no ties. When $\theta = 0$ (i.e. at the boundary of the hypotheses) the distribution of D is symmetric about 0 and therefore each rank is equally likely to receive a plus sign or a minus sign. In other words, if there is no difference between the two treatments then $V_1, \ldots, V_n$ form a random sample from

$$V = \begin{cases} 1 & \text{with probability } \tfrac{1}{2} \\ 0 & \text{with probability } \tfrac{1}{2} \end{cases} \qquad (4.8)$$

Since the mean and the variance of the Bernoulli random variable V are 1/2 and 1/4, when there are no ties the mean and the variance of W can also be easily computed:

$$\begin{aligned} E(W) &= \sum_{i=1}^{n} iE(V_i) = \frac{1}{2}\sum_{i=1}^{n} i \\ &= \frac{n(n+1)}{4}, \end{aligned} \qquad (4.9)$$

$$\begin{aligned} Var(W) &= \sum_{i=1}^{n} i^2 Var(V_i) = \frac{1}{4}\sum_{i=1}^{n} i^2 \\ &= \frac{n(n+1)(2n+1)}{24}. \end{aligned} \qquad (4.10)$$

Moreover, W is asymptotically normally distributed. When the sample size is large, tests based on this asymptotic approximation are quite accurate. When the sample size is not large, there is no need to resort to an approximate test based on these results, because equations (4.6), (4.7), and (4.8) yield exact tests. The exact permutational distribution of W is obtained by assigning positive or negative signs to the n ranks in all possible 2^n ways. For small-to-moderate sample sizes, many published statistical tables provide the tail probabilities of W as a function of its possible values. Some statistical software packages such as StatXact and TESTIMATE also provide tools to carry out exact tests for typical sample sizes that arise in practical applications.

To illustrate the computation of the exact permutational distribution of W and the p-value, consider the case where the sample size is $n = 4$. Suppose there are no tied ranks so that the ranks are 1, 2, 3, and 4. Then, there are 2^4 possible combinations of signs that the ranks 1, 2, 3, and 4 can receive. Enumerating all 16 possible assignments of signs to the ranks and computing the sum of positive ranks, we obtain Table 4.2.

Table 4.2. Possible signs and values of the Wilcoxon statistic

Rank				Sum of positive ranks
1	2	3	4	W
-	-	-	-	0
+	-	-	-	1
-	+	-	-	2
-	-	+	-	3
+	+	-	-	3
+	-	+	-	4
-	-	-	+	4
+	-	-	+	5
-	+	+	-	5
+	+	+	-	6
-	+	-	+	6
+	+	-	+	7
-	-	+	+	7
+	-	+	+	8
-	+	+	+	9
+	+	+	+	10

When the median of the distribution is 0, each rank has probability 1/2 of being positive and consequently each sign combination in Table 4.2 has probability 1/16. Hence we get the following probability distribution for the Wilcoxon test statistic:

w	0	1	2	3	4	5
$Pr(W = w)$	.0625	.0625	.0625	.125	.125	.125
$Pr(W \leq w)$	.0625	.125	.1875	.3125	.4375	.5625
$Pr(W \geq w)$	1.0	.9375	.875	.8125	.6875	.5625
w	6	7	8	9	10	
$Pr(W = w)$	.125	.125	.0625	.0625	.0625	
$Pr(W \leq w)$	.687	.8125	.875	.9375	1.0	
$Pr(W \geq w)$	.4375	.3125	.1875	.125	.0625	

For example, if $w = 8$ has been observed the p-value given by the Wilcoxon test for the left-sided null hypothesis is $p = Pr(W \geq 8) = .1875$. The computational procedure is further illustrated by Example 4.2.

Permutation tests

Recall that in performing the Wilcoxon test we replaced the observed absolute differences between matched pairs by their rank and zero differences were dropped from the analysis. This made the tabulation of Wilcoxon distribution possible as the sample size is the only parameter that needs to be specified. In general, a nonparametric test can be carried out with any arbitrary score in place of the ranks appearing in (4.7). When the Wilcoxon scores are not desired, often the observed differences between matched pairs are used as the scores. Moreover, in applications involving many zero differences, the idea of dropping these data from the test as required by the Wilcoxon test may not be appropriate (cf. Pratt 1959).

Once the signs are assigned to the desired scores, a test statistic similar to the Wilcoxon statistic is computed using the formula

$$S = w_1 V_1 + w_2 V_2 + \cdots + w_n V_n,$$

where w_i is the score assigned to the ith difference and V_i is a Bernoulli random variable indicating whether or not D_i is positive. In particular, if the test is to be carried out with observed absolute differences, then $w_i = d_i$, the observed value of D_i. The idea of using the data themselves to define scores is due to Fisher (1925, 1935) and the process that makes the exact inference possible is called *Fisher's randomization method*; the procedure can be employed in many problems in nonparametric methods. A test based on S is called a *permutation test* or a *randomization test*. The test can be generally carried out under the concept of *conditional inference*. In this method the permutation distribution of S is obtained by conditioning on the observed magnitudes of the scores. More precisely, all possible values of S are found when each score takes on positive or negative signs, which occurs with probability $1/2$ if the median of the distribution of D is zero. As before, this enumeration of S for all 2^n combinations of signs will give rise to the permutation distribution of S. Then, left-sided hypotheses concerning the median of the distribution of D are tested on the basis of the p-value $p = Pr(S \geq s)$, where s is the observed value of s. A test based on the permutation distribution of S is called a permutation test. The Wilcoxon test is a particular permutation test. The computational procedure is illustrated by the following example.

Example 4.2. Testing the effectiveness of a course (continued)

Consider again the problem of testing the effectiveness of the writing skills course discussed in Example 4.1. Let us retest the desired hypothesis using the Wilcoxon test and the permutation test based on positive differences. Table 4.3 provides details of the necessary computations to carry out the two tests.

Table 4.3. The signed ranks of the scores

Student	x_i	y_i	d_i	Rank($\lvert d_i \rvert$)	Signed Rank
A	78	84	6	8	8
B	65	61	-4	5.5	-5.5
C	67	64	-3	4	-4
D	58	62	4	5.5	5.5
E	52	59	7	9	9
F	85	75	-10	12	-12
G	60	72	12	14	14
H	46	48	2	3	3
I	55	64	11	13	13
J	72	73	1	1.5	1.5
K	80	75	-5	7	-7
L	63	72	9	11	11
M	57	65	8	10	10
N	49	48	-1	1.5	-1.5
O	62	75	13	15	15

The observed value of the Wilcoxon statistic is the sum of the positive ranks in the last column of Table 4.3. Having computed $w = 90$, the p-value is computed as $p = Pr(W \geq 90) = .0457$. On the other hand, the sum of the positive score differences appearing in the third column of Table 4.3 is $s = 73$ and the p-value can be found using the permutation distribution of S. Thee exact p-value of this test is $p = Pr(S \geq 73) = .0453$. Therefore, according to the two permutation tests, the scores from the two exams actually provide sufficient evidence to conclude that the course is effective in improving writing skills. Recall that the sign test failed to reject the null hypothesis. In this application the permutation tests have performed more efficiently to detect significant differences between the two sets of scores.

4.4. The Rank Sum Test and Allied Tests

The previous section demonstrated the usefulness of a nonparametric test based on ranks for comparing two correlated random variables from which paired data are available. The purpose of this section is to develop a similar rank test procedure for comparing two populations when independent random samples are drawn from them; some other related permutation tests will also be discussed briefly. Nonparametric tests based on ranks for comparing two independent distributions were proposed independently by Wilcoxon (1945) and Mann and Whitney (1947). The two tests are equivalent (see Exercise 4.2) and lead to identical results. A form of the common test is known as the rank sum test; sometimes it is also referred to as *Wilcoxon-Mann-Whitney* test.

To describe this test, suppose we have m observations from the first population with distribution function F_x, and n observations from the second population with the distribution function F_y. Assume that either the distributions are identical or one is obtained by shifting the median of the other. Let $X_1, \ldots, X_m$ and $Y_1, \ldots, Y_n$ be the two random samples drawn from the two populations. Consider the null hypothesis that X is stochastically smaller than Y. That is, we are interested in testing

$$H_0: F_x(w) \geq F_y(w) \quad \text{against} \quad H_1: F_x(w) < F_y(w) \text{ , for all } w \tag{4.11}$$

To carry out the test, the observations from the two populations are combined, arranged in order of their magnitudes, and then the ranks $1,2,\ldots,m+n$ are assigned according to the positions of the ordered data. If two or more (say k) observations are tied, then we assign the mean of k consecutive ranks for each of these observations. As in the previous section, one can of course, construct a permutation test by using any desired set of scores, including the *raw data scores*, in place of the ranks. The *Wilcoxon scores* are the ranks of the combined data defined more precisely as

$$w_i = k_1 + \cdots + k_{i-1} + (k_i + 1)/2, \tag{4.12}$$

where k_i is the number of observations tied at the ith smallest distinct value in the combined data set. To outline two other popular scores, let w_i be the score assigned to the ith distinct value of the ordered observations and $K_i = k_1 + \cdots + k_i$ be the sum of the observations tied at first i distinct values of the ordered data. Then, the *normal scores* are defined as

$$w_i = \frac{1}{k_i} \sum_{j=K_{i-1}+1}^{K_i} \Phi^{-1}\left[\frac{j}{m+n+1}\right] \tag{4.13}$$

where $\Phi()$ is the cdf of the standard normal distribution and the *logrank scores* are defined as

$$w_i = \frac{1}{k_i} \sum_{j=K_{i-1}+1}^{K_i} \sum_{l=1}^{j} \left[\frac{1}{m+n-l+1}\right] - 1 . \tag{4.14}$$

The tests in this context are based on the property that if there is no difference between the two populations, then the combined data are identically and independently distributed. As a result the $X_1, \ldots, X_m$ data tends to be dispersed throughout all $m+n$ ordered data. For example, in the case of Wilcoxon ranks with no tied data, the ranks that the data from the first population get will be the same as if they were a random sample of m ranks drawn at random without replacement from the set of ranks $\{ 1,2,\ldots,m+n \}$. Deviations of observed ranks from this distribution will suggest departures from the assumption of identical distributions. The ranks of the first sample concentered among larger values of combined ranks suggest a larger stochastic ordering of the distribution of the first population over the second population.

To describe the nature of the permutation tests, first consider the Wilcoxon test applied to observations with no ties (see (4.26) for a brief discussion of how ties can be handled in a more general setting of comparing a number of populations). As in the case of the signed rank test, it is quite natural to base a test on the sum of the ranks of one population for detecting differences in the two populations. Let S be the sum of the ranks that are assigned to the m observations from the first population. A test based on this statistic is a rank sum test. It follows from the argument given in the previous paragraph that the null hypothesis H_0 should be rejected if the p-value

$$p = Pr(S \geq s) \tag{4.15}$$

is too small, where s is the observed value of the S statistic.

It can be argued or proved (Exercise 4.1) that the mean and the variance of this random variable are

$$E(S) = \frac{m(m + n + 1)}{2} \tag{4.16}$$

and

$$Var(S) = \frac{mn(m + n + 1)}{12} . \tag{4.17}$$

Furthermore, S is asymptotically normally distributed. But the exact distribution of S can also be worked out and therefore there is no need to resort to an approximate test based on these results. There are $\binom{m+n}{m}$ ways that m ranks randomly chosen out of $m + n$ ranks can be assigned to the observations from the first population. The exact probability distribution of S can be found by enumerating all possible assignments and noting that the probability of each assignment is $m!n!/(m + n)!$. For small to moderate values of m and n, many published statistical tables provide the tail probabilities of S. Statistical software packages such as StatXact and TESTIMATE also provide tools to carry out exact tests for typical sample sizes that arise in practical applications. The following example illustrates the procedure of finding the distribution (sometimes called the Wilcoxon-Mann-Whitney distribution) of the S statistic. Consider the case where $m = 2$ and $n = 3$ so that the possible ranks are 1, 2, 3, 4, and 5. Let A and B denote the observations from population 1 and population 2, respectively. Table 4.4 gives $\binom{5}{2} = 10$ all possible ranks that A (or B) can take and the corresponding values of the test statistic.

Table 4.4. Possible ranks and values of the rank sum statistic

Rank					Sum of *A* ranks
1	2	3	4	5	*S*
A	A	B	B	B	3
A	B	A	B	B	4
A	B	B	A	B	5
B	A	A	B	B	5
A	B	B	B	A	6
B	A	A	A	B	6
B	A	B	B	A	7
B	B	A	A	B	7
B	B	A	B	A	8
B	B	B	A	A	9

Since the probability of each combination of ranks that *A* gets is .1, we get the following probability distribution for the *S* statistic:

w	3	4	5	6	7	8	9
$Pr(S = s)$	.1	.1	.2	.2	.2	.1	.1
$Pr(S \leq s)$	.1	.2	.4	.6	.8	.9	1.0
$Pr(S \geq s)$	1.0	.9	.8	.6	.4	.2	.1

For example, if $w = 8$ has been observed, the p-value given by the Wilcoxon test for the left-sided null hypothesis is .2.

Other permutation tests can be also performed by taking a similar approach and appropriate conditioning. To define the general form of the test statistic, given that there are k_i observations tied at the *i*th smallest value of the combined sample, let V_i denote the random variable representing the number of observations belonging to the first population. If the two populations are identical, then given k_i; $i = 1,...r$, m, n, and $m + n$, the conditional distribution of $\mathbf{V} = (V_1, \ldots, V_r)$ is given by the hypergeometric probability distribution

$$Pr(\mathbf{V} = \mathbf{v}) = \frac{\prod_{i=1}^{r} \binom{k_i}{v_i}}{\binom{m+n}{m}}. \tag{4.19}$$

The general form of the test statistic based on a desired set of scores $\{w_1, \ldots, w_r\}$ is the sum of the scores assigned to the first population; that is,

$$S = w_1 V_1 + w_2 V_2 + \cdots + w_r V_r. \tag{4.20}$$

A test based on *S* is called a *linear rank test*.

Since the distribution of *V* given by (4.19) is free of unknown parameters, exact conditional tests of (4.5) can be based on (4.20). The p-value for testing the left-sided null hypothesis in (4.5) is

$$p = Pr(S \geq s) = \sum_{S \geq s} \frac{\prod_{i=1}^{r} \binom{k_i}{V_i}}{\binom{m+n}{m}}, \tag{4.21}$$

where $s = w_1 v_1 + w_2 v_2 + \cdots + w_r v_r$ is the observed value of the test statistic. The computation of the p-value requires the enumeration of every $\mathbf{v}$ such that

$$\sum_{i=1}^{r} v_i = k_i \quad \text{and} \quad \sum_{i=1}^{r} w_i v_i \geq s$$

and then computation of (4.19). Without highly efficient algorithms this is a formidable task; the computation of confidence limits is even more difficult. Recent algorithms developed by Pagano and Tritchler (1983), Vollset, Hirji, and Elashoff (1991), and Mehta, Patel, and Senchaudhuri (1992) have made this possible with typical sample sizes for which asymptotic results are inadequate.

It is worth noting in passing that the test can be also employed to compare two multinomial populations with r ordinal categories and to test the equality of 'success' probabilities of r independent binomial distributions. In the former case, the number of observations falling into the ith category are v_i and $u_i = k_i - v_i$. The test is carried out conditional on the observed sums of observations, $v_i + u_i = k_i, i = 1,...,r$. In the latter case, v_i represents the number of successes obtained in k_i independent and identical Bernoulli trials. The null hypothesis of equal probabilities is tested against the alternative that they follow a trend such as that defined by the *trend scores* $w_i = i - 1$, typically representing the ordered levels of a treatment.

Example 4.3. A comparison of two anesthetics

An experiment is conducted to compare the effect of two anesthetics on plasma epinephrine concentration. The first anesthetic was given to seven randomly selected dogs and the second was given to nine randomly selected dogs. The plasma epinephrine concentrations were measured while the dogs were under the influence of the anesthetics. Table 4.5 shows the results of the experiment.

Table 4.5. Observed plasma epinephrine concentrations

Anesthetic A	.48	.97	.43	.76	1.03	.84	.51		
Anesthetic B	.84	.43	.39	.34	.73	.44	.41	.31	.54

Consider the null hypothesis that the effect of Anesthetic A is no larger than the effect of Anesthetic B. To test this hypothesis by the rank sum procedure, the observations are ordered and the ranks are assigned as follows:

Order	Observation	Rank	Anesthetic
1	.31	1	B
2	.34	2	B
3	.39	3	B
4	.41	4	B
5	.43	5.5	A
6	.43	5.5	B
7	.44	7	B
8	.48	8	A
9	.51	9	A
10	.54	10	B
11	.73	11	B
12	.76	12	A
13	.84	13.5	A
14	.84	13.5	B
15	.97	15	A
16	1.03	16	A

Let us first test the hypothesis by applying the rank sum test. The sum of the ranks assigned to A is $s = 5.5 + 8 + 9 + 12 + 13.5 + 15 + 16 = 79$. If the two underlying random variables are identically distributed, then the random variable S has the Wilcoxon-Mann-Whitney distribution with parameters $m = 7$ and $n = 9$. The observed significance level based on the rank sum test is

$$\begin{aligned} p &= Pr(S \geq 79) \\ &= .021 . \end{aligned}$$

Therefore, the null hypothesis can be rejected at the .05 level. The data from the experiment provide sufficient evidence to conclude that the effect of Anesthetic A on the plasma epinephrine concentration is greater than that of Anesthetic B. Let us now retest the hypothesis by performing a permutation test with the raw data scores that appear in the second column of the table. The p-value of this permutation test can be computed by applying the formula in (4.21). The exact p-value, $p = .028$ leads to the same conclusion.

4.5. Comparing k Populations

Suppose we have independent random samples from k populations with continuous distributions $F_1, F_2, \ldots, F_k$. Let $n_i, i = 1,\ldots,k$ be the sample sizes. Let $X_{ij}, j = 1,\ldots,n_i, i = 1,\ldots,k$, denote the jth observation taken from ith sample. Consider the problem of testing the null hypothesis that the distributions are identical; that is,

$$H_0: F_1(x) = F_2(x) = \cdots = F_k(x) \quad \text{for } all \ x. \tag{4.22}$$

The alternative hypothesis states that at least one of the distributions is different; that is, the populations are not homogeneous. In this situation also we can easily obtain (see Exercise 4.3) a counterpart to the sign test based on the counts of observations from each population that fall above the sample median of the combined data; under H_0, we expect about half the observations from each population to fall above the sample median. As this test does not use the relative magnitudes of the observations the test tends to be inefficient in the sense that it fails to reject the null hypothesis even when the data actually provides sufficient information to do so. So what is desirable is a test such as the rank sum test which recovers some of this information. Kruskal and Wallis (1952) extended the Wilcoxon-Mann-Whitney test for two independent samples to the case of k independent samples. As in the rank sum test, their test, which is commonly known as *Kruskall-Wallis test*, effects the desired comparison by replacing each observation by its rank in the combined data.

Let R_{ij} be the rank assigned to X_{ij} and let R_i be the sum of ranks assigned to the sample from the ith population; that is,

$$R_{ij} = \text{rank of } X_{ij}, \; R_i = \sum_{j=1}^{n_i} R_{ij}, \; j = 1,\ldots, n_i, \; i = 1,\ldots, k. \tag{4.23}$$

Then, the Kruskall-Wallis test statistic is defined as

$$W(\mathbf{X}) = \frac{(N-1)\sum_{i=1}^{k} n_i[R_i/n_i - (N+1)/2]^2}{\sum_{i=1}^{k}\sum_{j=1}^{n_i}[R_{ij} - (N+1)/2]^2}, \tag{4.24}$$

where $N = \sum_{i=1}^{k} n_i$ is the total sample size. It can be shown (Exercise 4.4) that if there are no ties, then the test statistic reduces to

$$W = \frac{12}{N(N+1)} \sum_{i=1}^{k} \frac{R_i^2}{n_i} - 3(N+1). \tag{4.25}$$

For any departure from the null hypothesis this statistic tends to take large values, and consequently, the right tail of the distribution of W can be used to define the appropriate extreme region. Hence, the p-value of the Kruskal-Wallis test is

$$p = Pr(W \geq w) ,$$

where w is the observed value of W. The asymptotic distribution of W is

$$W \sim \chi^2_{k-1} ;$$

that is, W, in the limit, has a chi-squared distribution with $k - 1$ degrees of freedom, and this result is useful in performing an approximate test. As in the case of $k = 2$, when there are no ties, the exact distribution can be found by enumerating all possible $n_1!n_2! \cdots n_k!/N!$ assignments of ranks. Explicit formulation of the generalized version of the problem is an extension of (4.21). Many statistical tables provide exact tail probabilities of W for $k = 3$ and small sample sizes. More extensive tables for a range of typical sample sizes are given by Iman, Quade, and Alexander (1975).

Although the probability of ties is zero with continuous random variables, they do occur quite frequently in practical applications. When ties occur as in previous problems, we assign the average rank to the tied observations by the same convention. When ties occur within the same sample this does not affect the value of the Kruskal-Wallis test statistic. When there are many ties often a correction is made in the definition of W. In the presence of ties the Kruskal-Wallis statistic W is corrected as

$$W' = \frac{W}{1 - \sum_{i=1}^{r} (m_i^3 - m_i)/(N^3 - N)} , \tag{4.26}$$

where r is the number of tied categories and m_i is the number of ties in ith category. If there are no ties, W' reduces to W. It is also possible to treat the current problem in the setting of a contingency table; this treatment is especially useful when there are many ties. Then, a suitable test can be obtained by the method of conditional inference. We shall address this problem in a little more detail in the next section. The Kruskal-Wallis test can be easily performed using the statistical software packages such as StatXact and TESTIMATE.

Example 4.4. Comparing three stimulants

A psychologist wishes to study the effect of two stimulants against a control stimulant, a placebo. She measured reaction times (in seconds) of a number of individuals who were given one of these stimulants. The results of the experiment are shown in Table 4.6.

Table 4.6. Reaction times due to three stimulants

Placebo	2.2	3.7	4.0	4.3	2.5	2.0	5.5
Stimulant I	3.0	3.2	2.8	1.8	2.1	4.2	
Stimulant II	1.5	2.3	3.3	1.0	2.4	1.9	

In this application we have no ties and we have the parameters $k = 3, N = 19$, $n_1 = 7$, $n_2 = 6$, and $n_3 = 6$. Necessary ranks and rank sums to analyze this data via the Kruskal-Wallis procedure are displayed in the following table:

Ranks R_{ij} and rank sums R_i								
Placebo	7	15	16	18	10	5	19	90
Stimulant I	12	13	11	3	6	17		62
Stimulant II	2	8	14	1	9	4		38

The observed value of the test statistic can be computed from (4.25) as

$$W = \frac{12}{380}\left(\frac{8100}{7} + \frac{3844}{6} + \frac{1444}{6}\right) - 60$$

$$= 4.373$$

The asymptotic distribution of W is chi-squared with 2 degrees of freedom and this result can be utilized in this application to obtain an approximation to the distribution of W. The p-value under the hypothesis of no difference in the stimulant effects is

$$\begin{aligned} p &= Pr(W \geq 4.373) \\ &= 1 - F_{\chi^2_2}(4.373) \\ &= .1123 \ . \end{aligned}$$

Therefore, the differences in the stimulants' effects are not quite statistically significant.

4.6. Contingency Tables

A contingency table is an array of numbers in matrix form representing counts or frequencies from a multiple classification. We consider only classifications based on just two criteria (attributes) so that the outcomes can be displayed in a two-way contingency table. For example, in a public opinion poll, the individuals surveyed may be classified according to their political affiliation and their inclination towards a political proposal. As another example, a sample of individuals living in the United States may be classified according to their blood type and their ethnic group. Assume that under each criterion the individuals in the sample are classified according to a finite number of categories. For example, the categories in political affiliation may be Democratic, Republican, and Other. Even when dealing with continuous variables such as height, weight, and blood pressure, often subjects in an experiment are classified into a few categories. Table 4.7 gives an example of a two-way classification of 250 subjects according to two classification criteria, namely weight and blood pressure; row and column totals representing the counts according to each attribute are also shown in the table.

Table 4.7. Distribution of weight and blood pressure

Weight	Low	Normal	High	Row
Underweight	14	22	9	45
Normal	22	71	27	120
Overweight	14	47	24	85
Column total	50	140	60	250

Table 4.7 is a 3×3 contingency table. If the first attribute has r categories (displayed in rows) and the second attribute has c categories (displayed in columns) it is said to be an $r \times c$ contingency table. In this general setting, N individuals or items are classified according to an attribute, A, with categories $A_1, A_2, \ldots, A_r$ and another attribute, B, with categories $B_1, B_2, \ldots, B_c$. Let x_{ij} denote the number of individuals who fall into both A_i and B_j categories under the two criteria. The general form of the $r \times c$ contingency table obtained with this classification is shown in Table 4.8.

Table 4.8. $r \times c$ Contingency table

	Criterion B				
Criterion A	B_1	$\cdots$	B_j	$\cdots$	B_c
A_1	x_{11}	$\cdots$	x_{1j}	$\cdots$	x_{1c}
A_2	x_{21}	$\cdots$	x_{2j}	$\cdots$	x_{2c}
$\cdot$	$\cdot$	$\cdot$	$\cdot$	$\cdot$	$\cdot$
A_i	x_{i1}	$\cdots$	x_{ij}	$\cdots$	x_{ic}
$\cdot$	$\cdot$	$\cdot$	$\cdot$	$\cdot$	$\cdot$
A_r	x_{r1}	$\cdots$	x_{rj}	$\cdots$	x_{rc}

Let

$$N_{i+} = \sum_{j=1}^{c} x_{ij},\ i = 1,2,\ldots,r \tag{4.27}$$

denote the raw totals and let

$$N_{+j} = \sum_{i=1}^{r} x_{ij},\ j = 1,2,\ldots,c \tag{4.28}$$

denote the column totals of the $r \times c$ contingency table. Of course, $\sum_{i=1}^{r} N_{i+} = \sum_{j=1}^{c} N_{+j} = N.$

Let p_{ij} denote the probability that that an item or an individual selected at random from the underlying population is classified as belonging to the ith row and jth column of the contingency table. The marginal probability that the item is classified in the ith row is

$$p_{i+} = p_{i1} + p_{i2} + \cdots + p_{ic},\ i = 1,2\ldots,r \tag{4.29}$$

and the marginal probability that it is classified in the jth column is

$$p_{+j} = p_{1j} + p_{2j} + \cdots + p_{rj},\ j = 1,2\ldots,c. \tag{4.30}$$

4.7. Testing the independence of criteria of classification

First consider the problem of testing the independence of the two criteria. Here N individuals chosen at random from a certain population are classified according to the two criteria. In the current setting, the probabilities satisfy the relation

$$\sum_{i=1}^{r}\sum_{j=1}^{c} p_{ij} = \sum_{i=1}^{r} p_{i+} = \sum_{j=1}^{c} p_{+j} = 1;$$

later we shall study situations where only one set of probabilities is supposed to sum to one. With this notation the null hypothesis of interest can be formulated as

$$H_0: p_{ij} = p_{i+}p_{+j} \quad \text{for all } i = 1,\dots,r \text{ and } j = 1,\dots,c; \tag{4.31}$$

that is, if the criteria of classification are independent, then the joint probability is equal to the product of the marginal probabilities. Equation (4.31) can also be written as

$$H_0: \log(p_{ij}) = \mu_i + \nu_j,$$

where $\mu_i = \log(p_{i+})$ and $\nu_j = \log(p_{+j})$; that is, the hypothesis states that the model for the logarithms of the cell probabilities is a linear function of the logarithms of the marginal probabilities. This is a particular case of analyses that one can perform under the setting of general *loglinear models*. The formulation of $\log(p_{ij})$ as a linear model is especially useful when one need to perform additional inferences on the parameters of the model and in dealing with contingency tables with three or more dimensions. For an in-depth treatment of loglinear models the reader is referred to Bishop, Fienberg, and Holland (1975), Santner and Duffy (1989), and Agresti (1990).

The MLEs (also unbiased estimates) of marginal probabilities are

$$\hat{p}_{i+} = \frac{N_{i+}}{N} \quad \text{and} \quad \hat{p}_{+j} = \frac{N_{+j}}{N}.$$

Perhaps the most popular and widely used test for independence in a contingency table is the *Pearson chi-squared test* based on the expected cell frequencies under the null hypothesis; if H_0 is true, then we expect the frequency

$$e_{ij} = N\hat{p}_{i+}\hat{p}_{+j} = \frac{N_{i+}N_{+j}}{N} \tag{4.32}$$

for ijth cell of the table. The chi-squared test statistic which tends to take larger values for deviations from H_0 is

$$C(\mathbf{x}) = \sum_{i=1}^{r}\sum_{j=1}^{c}\frac{(x_{ij} - e_{ij})^2}{e_{ij}}. \tag{4.33}$$

For example, for the hypothetical data in Table 4.7 we obtain the following expected frequencies by applying (4.32):

Expected frequencies under the independence hypothesis				
Weight	Low	Normal	High	Row total
Underweight	9.0	25.2	10.8	45
Normal	24.0	67.2	28.8	120
Overweight	17.0	47.6	20.4	85
Column total	50	140	60	250

The observed value of the chi-squared statistic is

$$
\begin{aligned}
C(\mathbf{x}) &= \frac{25}{9} + \frac{10.24}{25.2} + \frac{3.24}{10.8} + \frac{4}{24} + \frac{14.44}{67.2} \\
&\quad + \frac{3.24}{28.8} + \frac{9}{17} + \frac{.36}{47.6} + \frac{12.96}{20.4} \\
&= 5.15.
\end{aligned}
$$

The p-value of the test is $p = Pr(C \geq c)$ and conventionally, its computation is facilitated by the fact that asymptotically, C has a chi-squared distribution with $(r - 1)(c - 1)$ degrees of freedom. In the above example the cell frequencies are large so that the approximate distribution of the chi-squared statistic, $C \sim \chi^2_4$, should be good enough for practical purposes. The p-value given by the approximation is $p = 1 - F_C(5.15) = .272$, suggesting that the evidence provided by the hypothetical data in Table 4.7 is weak to doubt the null hypothesis of independence between the blood presure and the weight.

Another widely used application of the chi-squared test is for testing the *goodness-of-fit* of models which specify the probability distribution p_{ij}, $i = 1,...,r$ and $j = 1,...,c$ of a contingency table with or without unknown parameters. The goodness-of-fit tests can be performed with one-way categorical data as well. Descriptions of goodness-of-fit tests and various kinds of applications can be found, for instance, in Mood, Graybill, and Boes (1974) and Agresti (1990, 1992).

Computation of exact p-values

As in the problems discussed in previous sections of this chapter, the exact p-value of even the chi-squared test can be obtained under conditional inference. Here the conditioning is made on the observed marginals of the contingency table so that the conditional distribution [see (4.35)] will be free of nuisance parameters. The idea of conditioning to devise an exact test for any reasonable test statistic is suggested by a test due to Fisher (1925), originally designed for 2×2 contingency tables.

To find the desired conditional distribution, first note that, the joint distribution of $\mathbf{X}$ is simply the multinomial distribution

$$
f_{\mathbf{x}}(\mathbf{X}) = N! \frac{\prod_{i=1}^{r} \prod_{j=1}^{c} p_{ij}^{x_{ij}}}{\prod_{i=1}^{r} \prod_{j=1}^{c} x_{ij}!} \tag{4.34}
$$

It is seen by summing this probability mass function with respect to i and j that the row sums N_{i+} and the column sums N_{+j} are independently distributed with the joint probability distribution

$$f_{\mathbf{N}}(\mathbf{N}) = \frac{N!^2}{\prod_{i=1}^{r} N_{i+}! \prod_{j=1}^{c} N_{+j}!} \prod_{i=1}^{r} p_{i+}^{N_{i+}} \prod_{j=1}^{c} p_{+j}^{N_{+j}}.$$

It is evident from these two equations that given the marginals, $\sum_{i=1}^{r} x_{ij} = N_{+j}$ and $\sum_{j=1}^{c} x_{ij} = N_{i+}$, the conditional probability distribution of cell frequencies is given by

$$\begin{aligned} p(\mathbf{x}) &= Pr(\mathbf{X} = \mathbf{x} \mid \sum_{i=1}^{r} x_{ij} = N_{+j}, \sum_{j=1}^{c} x_{ij} = N_{i+}) \\ &= \frac{1}{N!} \frac{(\prod_{i=1}^{r} N_{i+}!)(\prod_{j=1}^{c} N_{+j}!)}{\prod_{i=1}^{r}\prod_{j=1}^{c} x_{ij}!} \frac{\prod_{i=1}^{r}\prod_{j=1}^{c} p_{ij}^{x_{ij}}}{\prod_{i=1}^{r} p_{i+}^{N_{i+}} \prod_{j=1}^{c} p_{+j}^{N_{+j}}}. \end{aligned}$$

Under the null hypothesis H_0: $p_{ij} = p_{i+}p_{+j}$, we have

$$p(\mathbf{x}) = \frac{1}{N!} \frac{(\prod_{i=1}^{r} N_{i+}!)(\prod_{j=1}^{c} N_{+j}!)}{\prod_{i=1}^{r}\prod_{j=1}^{c} x_{ij}!}. \tag{4.35}$$

In the extended version of ***Fisher's exact test***, the p-value is computed by adding all $p(\mathbf{X})$ that corresponds to $\mathbf{X}$ which are as extreme as or more extreme than the observed $\mathbf{x}$; that is the p-value is computed as

$$p_F = \sum_{p(\mathbf{X}) \le p(\mathbf{x})} p(\mathbf{X}).$$

As the study by Gail and Mantel (1977) illustrates, exact computation of this p-value is a formidable task for moderate and large sample sizes. For this reason, conventionally a chi-squared approximation of Fisher's test is obtained by means of a monotone transformation of $p(\mathbf{X})$ so that the resulting distribution is asymptotically chi-squared with $(r-1)(c-1)$ degrees of freedom. When the frequencies in a contingency table are small, now there is no need to resort to any such approximation, because recently computationally efficient algorithms to compute the exact p-values that arise in contingency tables have been proposed. Mehta and Patel (1980, 1983, and 1986), and Pagano and Halvorsen (1981) developed algorithms that have made the computation of exact p-values in this context feasible even for fairly large sample sizes. Although some of these algorithms were originally designed for performing Fisher's exact test, they can be adapted to many other exact permutation tests.

Using this approach, the computation of exact p-values can be accomplished for other tests of independence in a contingency table as well. In terms of the the probability distribution $p(\mathbf{x})$ given by (4.35), the exact p-value of the chi-squared test is computed as

$$p_C = \sum_{C(\mathbf{X}) \leq C(\mathbf{x})} p(\mathbf{X}); \tag{4.36}$$

that is, the conditional probability of the extreme region defined by the chi-squared test statistic. The method works equally well for any other measure of discrepancy of observed frequencies from expected frequencies that may be used to define a suitable extreme region. Another test of particular interest is the one based on the likelihood ratio test statistic (Exercise 4.5)

$$\lambda(\mathbf{x}) = \frac{\prod_{i=1}^{r} N_{i+}^{N_{i+}} \prod_{j=1}^{c} N_{+j}^{N_{+j}}}{N^N \prod_{i=1}^{r} \prod_{j=1}^{c} x_{ij}^{x_{ij}}}.$$

It is computationally advantageous to express the *likelihood ratio test* in terms of the statistic

$$L(\mathbf{x}) = -2\log\lambda(\mathbf{x}) = 2\sum_{i=1}^{r}\sum_{j=1}^{c} x_{ij}\log\left(\frac{x_{ij}}{N_{i+}N_{+j}/N}\right) \tag{4.37}$$

which is also asymptotically distributed as a chi-squared distribution with $(r-1)(c-1)$ degrees of freedom. Like the previous two tests, this test is also inherently two-sided in the sense that the test statistic tends to take large values for any departure from the hypothesis of independent probabilities under which the likelihood ratio is computed. Therefore, the exact p-value of the likelihood ratio test is computed as

$$p_L = \sum_{L(\mathbf{X}) \geq L(\mathbf{x})} p(\mathbf{X}). \tag{4.38}$$

Most popular statistical software packages including BMDP, IMSL, SAS, and SPSS provide some tools to perform a few nonparametric tests including Fisher's exact test. In contrast, software packages specializing in exact nonparametric inference such as StatXact and TESTIMATE provide a wide variety of tools to perform many exact tests as well as approximate tests which are based on asymptotic results and Monte Carlo sampling methods.

Example 4.5. Testing the independence of heart condition and hypertension

Consider the hypothetical data in the table below describing the health conditions of 200 subjects (adults) that participated in a study. Suppose this contingency table summarizes the findings of the study conducted to find out whether or not hypertension and heart ailment are related.

Heart condition	Hypertension		
	Distortic	Isolated	Never
Ailment	30	56	6
No ailment	14	83	11

The observed value of the chi-squared statistic is 11.33 and that of the likelihood ratio statistic is 11.44. Asymptotically, each of the underlying random variables have a chi-squared distribution with 2 degrees of freedom. The exact p-values of the two tests are $p_C = .0029$ and $p_L = .0037$ and they both suggest the rejection of the independence hypothesis. So, we can conclude that the heart condition and the hypertension of an individual are somehow related.

4.8. Testing the Homogeneity of *r* Populations

The chi-squared test can be adapted to test the homogeneity of r populations as well. Suppose N individuals in the r populations are classified into c categories. The findings of a study of this nature can also be represented by a $r \times c$ contingency table. The Kruskal-Walllis test is also a problem of testing for homogeneity of a number of populations. In that setting, however, there is a natural ordering in the columns when represented as a contingency table, which is especially appealing when there are ties in the observed data. When there is a natural ordering both in the rows and in the columns one can use the *Jonckheere-Terpstra test* (see, for instance, Hollander and Wolfe (1973) for details) for the homogeneity of r populations. Both StatXact and TESTIMATE (version 5.2) provide software tools to perform the Jonckheere-Terpstra test. When there is a natural ordering in columns and/or in rows it is usually more efficient to use a test that takes advantage of that ordering. The chi-squared test is applied when there is no such natural ordering.

In the current version of the problem, we have for the ith population,

$$p_{i+} = p_{i1} + p_{i2} + \cdots + p_{ic} = 1, \quad \text{for all } i = 1,2...,r$$

(in the notation of contingency tables used in this section). Independent observations taken from each population will thus constitute a multinomial distribution. The null hypothesis of interest here is that all r populations have the same probability distribution over the c categories; that is

$$H_0: p_{1j} = p_{2j} = \cdots = p_{rj} = p_j, \quad \text{for all } j = 1,...c\,, \tag{4.39}$$

where p_j is the common probability of the jth category.

Independent sets of independent observations from each multinomial distribution now yield

$$f_{\mathbf{x}}(\mathbf{X}) = \frac{\prod_{i=1}^{r} N_{i+}!}{\prod_{i=1}^{r}\prod_{j=1}^{c} x_{ij}!} \prod_{i=1}^{r}\prod_{j=1}^{c} p_{ij}^{x_{ij}} \tag{4.40}$$

in place of (4.34). Under H_0, (4.40) reduces to

$$f_{\mathbf{x}}(\mathbf{X}) = \frac{\prod_{i=1}^{r} N_{i+}!}{\prod_{i=1}^{r}\prod_{j=1}^{c} x_{ij}!} \prod_{j=1}^{c} p_j^{N_{+j}}$$

Moreover, under the null hypothesis, the joint distribution of the marginal sums N_{+j}, $j = 1,\ldots,c$ is

$$f_{\mathbf{N}_{+.}}(\mathbf{N}_{+.}) = \frac{N!}{\prod_{j=1}^{c} N_{+j}!} \prod_{j=1}^{c} p_j^{N_{+j}} .$$

Consequently, given the marginal sums, the conditional null probability distribution of cell frequencies is free of nuisance parameters. More precisely, equation (4.35) is still valid under the current null hypothesis as well.

The definition of the chi-squared test statistic is the same except that the expected frequencies are computed under (4.39). But under the null hypothesis the MLE of p_j is

$$\hat{p}_j = \frac{N_{+j}}{N}$$

implying that the expected frequency for the ijth cell is

$$e_{ij} = \frac{N_{i+}N_{+j}}{N}. \tag{4.41}$$

This is the same formula we had before, and therefore the test can be carried out exactly as before. Nevertheless, the conclusions we make in the two situations should be presented appropriately. It is also evident that, when represented in the form of a contingency table, the p-value of the Kruskal-Wallis test can be also computed similar to (4.38):

$$p_W = \sum_{W(\mathbf{X}) \geq W(\mathbf{x})} p(\mathbf{X}). \tag{4.42}$$

Example 4.5. Testing the equality of income distributions by ethnic group

Consider the hypothetical data in the following table showing the annual income distribution (in thousands of dollars) of recent college graduates with engineering degrees who were chosen at random from three ethnic groups, all from the same state; 50 engineers are chosen each from Group 1 and Group 2, and 25 engineers are chosen from Group 3:

Income	Under 25	25-35	35-45	More than 45
Group 1	10	18	16	6
Group 2	18	20	10	2
Group 3	14	5	5	1
Column total	44	43	31	9

To investigate if there is discrimination against certain ethnic groups, suppose one first wishes to test whether or not the income distributions of the three populations are the same. The exact p-value of the chi-squared test is .043. Hence, there is sufficient evidence to conclude that the income distributions are significantly different among different ethnic groups and one can now proceed with further investigation of this matter.

Exercises

4.1. Consider the Wilcoxon-Mann-Whitney test statistic S applied to samples of sizes m and n from two populations. Show that

$$E(S) = \frac{m(m + n + 1)}{2}$$

and that

$$Var(S) = \frac{mn(m + n + 1)}{12}.$$

4.2. Let $X_1, \ldots, X_m$ and $Y_1, \ldots, Y_n$ be independent random samples drawn from two populations. Let S be the rank sum test based on these observations. Define

$$U = \sum_{i=1}^{m} \left[\text{number of } Y_j\text{'s} < X_i \right].$$

Show that the U statistic is related to the S statistic as

$$U = S - \frac{m(m + 1)}{2},$$

and thus will lead to an equivalent testing procedure.

4.3. Consider the problem of comparing k independent samples discussed in Section 4.5. Define $\mathbf{X} = (X_1, \ldots, X_k)$, where X_i is the number of observations from the ith sample of size n_i that fall above the sample median. Let $\mathbf{x}_{obs}$ be the observed value of $\mathbf{X}$. Let $N = \sum n_i$ and $T = \sum X_i$. Assume that the median of the combined data is unique so that under the null hypothesis in (4.22), T is equal to $N/2$ or $(N - 1)/2$ according to whether N is even or odd. Show that the null hypothesis can be tested based on the p-value

$$p = \sum_{\mathbf{x} \geq \mathbf{x}_{obs}} p(\mathbf{x}).$$

where

$$p(\mathbf{x}) = Pr(\mathbf{X} = \mathbf{x} \mid \sum X_i = T) = \frac{\prod_{i=1}^{k} \binom{n_i}{x_i}}{\binom{N}{T}}.$$

4.4. Show that if there are no ties in the data, then the Kruskal-Walis test statistic defined by (4.24) reduces to

$$W = \frac{12}{N(N+1)} \sum_{i=1}^{k} \frac{R_i^2}{n_i} - 3(N+1).$$

Hint: $\displaystyle\sum_{i=1}^{k} \sum_{j=1}^{n_i} R_{ij}^2 = \frac{N(N+1)(2N+1)}{6}.$

4.5. Consider the $r \times c$ contingency table considered in Section 4.6. Show that, in the notation of that section, the likelihood ratio test statistic for testing the hypothesis of independence in the contingency table is

$$\lambda(\mathbf{x}) = \frac{\prod_{i=1}^{r} N_{i+}^{N_{i+}} \prod_{j=1}^{c} N_{+j}^{N_{+j}}}{N^N \prod_{i=1}^{r} \prod_{j=1}^{c} x_{ij}^{x_{ij}}}.$$

4.6. Consider the data in Table 4.1. Test each of the following hypotheses by the sign test.

(a) The median score of a student before the course is less than 60.

(b) The median score of a student after the course is greater than 65.

(c) The 75th percentile of scores after the course is greater than 80.

4.7. At one time only 60% of patients with Leukemia survived more than 6 years. Of 42 patients on whom a new drug is tested 27 were alive after 6 years. Test whether the new drug is better than the former treatment.

4.8. The table below gives the pulse rate of fourteen subjects before and after smoking cigarettes. By applying (a) the sign test; then (b) the signed rank test, test the hypothesis that smoking increases the pulse rate.

The pulse rate before and after smoking		
Subject	Before	After
A	78	77
B	65	69
C	66	66
D	70	72
E	62	63
F	73	72
G	68	72
H	63	65
I	65	66
J	68	69
K	70	72
L	73	72
M	67	75
N	69	71

4.9. Suppose that the numbers in the following table represent the mean scores given by judges from two groups of countries to gymnasts who competed on uneven bars at an international competition:

Group 1	9.35	9.64	9.41	9.79	9.29	
Group 2	9.53	9.64	9.78	9.75	9.69	9.61

Using the rank sum test, compare the two groups of countries.

4.10. Consider again the testing problem in Exercise 4.9. Perform permutation tests with each of the following scores:

(a) raw data scores;

(b) normal scores;

(c) logrank scores.

4.11. In telemarketing a new product, a firm offered two types of incentives, A and B, chosen at random. The figures in the following table are the number of items sold in 8 consecutive days; the data are classified according to the incentive offered:

Incentive A	24	14	17	19	32	21	12	19
Incentive B	18	25	35	20	19	25	15	26

Test whether one incentive is more effective than the other in marketing the product.

4.12. Consider the data in Table 4.6. Use the Wilcoxon-Mann-Whitney test to compare (a)Stimulant I and the placebo; and (b) Stimulant II and the placebo.

4.13. A class of nineteen students in Grade 12 of a school was divided into three groups in a random manner. The three groups were given special training in Calculus with three different methods of instructions. At the end of the training, a common test in Calculus was given to the three groups and their scores are recorded in the following table.

Group I	69	51	92	48	76	64	71
Group II	72	39	47	84	61	70	
Group III	44	58	90	63	56	82	

By performing a suitable nonparametric test, examine whether the three instructional methods produce the same or different results. If they are different, which instructional method is most effective?

4.14. To find out whether or not the blood type of an individual is independent of the color of the individual's eyes, 200 individuals are classified according to the two criteria. The results of a hypothetical study are given in the following contingency table:

	Blood type			
Eye color	A	B	AB	O
Blue	21	9	7	19
Brown	36	14	2	11
Other	40	17	8	16

By applying each of the following methods, test whether the two attributes are independent:

(a) Chi-squared test;

(b) Fisher's exact test;

(c) Likelihood ratio test.

4.15. Consider the data in Table 4.7. Conduct exact tests for independence in the contingency table by each of the three methods in Exercise 4.14 and compare the results.

Chapter 5

Generalized p-Values

5.1. Introduction

As should be clear from various applications we have looked at in the previous three chapters, testing hypotheses based on p-values is more general than fixed-level testing. Most of all, p-values can be employed to test parameters of discrete distributions such as binomial and Poisson distributions, thus making exact nonparametric testing possible. Non-randomized fixed-level tests at standard nominal levels do not usually exist for discrete distributions. Although one can find randomized fixed-level tests for testing parameters of discrete distributions, such tests are mainly of theoretical interest; in real world applications, an experimenter would usually resort to an approximate test rather than performing a randomized fixed-level test.

There is another substantial class of problems for which non-randomized fixed-level (conventional) tests based on sufficient statistics are not available. In some situations (e.g. the Behrens-Fisher problem) they do not even exist. As we shall discuss later, however, fixed-level tests with unconventional interpretations may exist. The limited availability of conventional fixed-level tests is a very serious problem. The classical approach to linear models has been adversely affected by the unavailability of exact fixed-level tests for comparing normal populations. For example, in ANOVA (Analysis of Variance) discussed in Chapter 8, most classical procedures are based on the assumption that all populations under comparison have equal variances, an assumption made only for mathematical tractability rather than anything else. As we will see in Chapters 7 through 10, when the assumption of equal variances is violated, classical F-tests and t-tests in linear models can lead to very serious repercussions. For example, in ANOVA, the classical F-test fails to detect significant differences in treatments being compared even when the available data provide sufficient evidence to do so. In many applications including

Biomedical experiments this can be a very costly drawback of conventional tests.

Illustrative examples

When the underlying family of distributions contains two or more unknown parameters, conventional tests are typically available only for special functions of the parameters. This is because in many situations it is not possible or easy to find test statistics having distributions free of nuisance parameters. To illustrate the nature of the problem caused by nuisance parameters and to motivate the need for extending the approach taken in previous chapters, consider a normal population with unknown mean μ and unknown variance σ^2. Let $X_1, \ldots, X_n$ be a random sample of size n from that population. We saw in Chapter 3 that a limited class of functions of μ and σ can be tested on the basis of p-values or conventional fixed-level tests. In some applications the parameter of interest could be a more complicated function of μ and σ^2. For instance, if one is interested in the second moment of the normal random variable about a point other than the mean, say k, then the parameter of interest is

$$E(x-k)^2 = \mu^2 + \sigma^2 - 2k\mu + k^2 . \tag{5.1}$$

Classical tests are not available for this parameter unless $k = \mu$. To give a more interesting example, let X denote the yields per acre of a certain hybrid corn. If the revenue from the crop is a linear function of X and if the *utility function* (see, for instance, DeGroot (1970)) of a farmer who uses this corn is of negative exponential form, then the expected utility of the farmer is of the form

$$\begin{aligned} EU(X) &= E\left[a(1 - e^{-bx})\right] \\ &= a\left[1 - e^{-b\mu + \sigma^2 b^2/2}\right] , \end{aligned}$$

where a and b are two constants which are determined by the desired scale and the famer's risk aversion level. It is now evident that, for the purposes of interval estimation and hypothesis testing, the parameter of interest can be taken to be

$$\theta = \mu + k\sigma^2 , \tag{5.2}$$

where $k = -b/2$. In this type of application it is not easy to find a test variable whose value and distribution depend on the parameters only through the parameter of interest. Here either μ or σ^2 can be considered as the nuisance parameter.

Extended p-values

The purpose of this chapter is to extend the definition of test variables so that, even in applications like those above, one can obtain procedures which do not depend on nuisance parameters. We shall do this extension without affecting the interpretation of the p-value given in Chapter 2. The problems that can be tackled with generalized p-values defined explicitly in Section 5.2 include linear models based on the normal theory.

The possibility of extending the usual definitions of p-values to solve problems such as the Behrens-Fisher problem (see Chapter 7) has been suggested by a number of authors. Kempthorne and Folks (1971) indicated how significance tests can be obtained in the presence of nuisance parameters which can not be tackled by conventional approaches, but did not give explicit definitions. Barnard (1984) and Rice and Gaines (1989) argued that Behrens-Fisher type tests can be interpreted without relying on fiducial arguments and gave formulae for computing exact p-values. Unconventional p-values reported in these articles (yet, in the spirit of the way Fisher treated problems of significance testing) are in fact generalized p-values; that is, they are exact probabilities of well-defined extreme regions of underlying sample spaces. Other unconventional approaches to statistical inference include Fraser's structural inference (see Fraser (1979)) and Dempster's (see Dempster (1967)) upper and lower probability inferences. Unlike these approaches, however, the generalized inferences presented in this book are extensions (not alternatives) of conventional inferences.

Weerahandi (1987) used a generalized p-value for comparing parameters of two regressions with unequal variances and established that it is the exact probability of a well defined unbiased extreme region. Motivated by that application, Tsui and Weerahandi (1989) gave explicit definitions and methods of finding generalized p-values. Thursby (1992), Griffiths and Judge (1992), and Zhou and Mathew (1994) carried out a number of investigations and applications of generalized p-values in regression and in mixed models. Meng (1994) defined a p-value in a Bayesian treatment which, under noninformative priors and natural discrepancy measures, yields numerically equivalent p-values in many applications. A number of other articles have found a variety of other applications in linear models based on normal theory. Some of these important applications will be discussed in Chapters 7 through 10. Sections 5.6 and 5.8 present two interesting applications of generalized p-values. The usefulness of the extension is further illustrated by other examples and exercises in this chapter.

Fixed-level testing

As is the case with all the exact nonparametric methods given in Chapter 4 and the Bayesian procedures, generalized p-values defined in this chapter also do not necessarily yield exact fixed-level tests with conventional repeated sampling properties. As Kempthorne and Folks (1971) argued, a procedure should not be rejected because it does not possess a property that was not sought in the first place. Practitioners who prefer to take the Neyman-Pearson approach and perform tests at a nominal level α would also find generalized p-values extremely useful, because the rule

$$\text{reject } H_0 \text{ if } p < \alpha,$$

provides either tests with size not exceeding the intended level or excellent approximate tests in almost all applications. According to a number of simulation studies reported in the literature, Type I error and power performance of such approximations are usually better than the performance of more complicated approximate tests.

Even when the actual size of tests given by the above rule exceeds the intended level, usually it is possible to construct size-guaranteed tests based on extended p-values, including generalized p-values and *posterior predictive p-values* (defined by Meng(1994)). In other words, it is usually possible, at the cost of a little loss of power, to obtain fixed-level tests to guarantee any desired nominal level. Resorting to such tests is worthwhile only in situations where repeated sampling properties of fixed-level testing can be entertained as practically useful rather than a matter of convention or habit.

Statistical quality control is a good example of a situation where repetition of an experiment naturally arises (with parameters that may have changed from one trial to the next), and fixed-level testing is appealing. On the other hand, in most other applications, including those in biomedical experiments, fixed-level tests are not appealing even when such tests do exist, because the sufficiency of evidence provided by the data concerning the hypotheses depends on the prevailing circumstances and what is being tested. For example, it is not reasonable for one to insist on a test of .05 level both in a situation where there is no treatment available for a certain disease such as AIDS, and in a situation such as the common cold where many treatments are readily available. In such situations it is best for a practitioner to report the p-value as a measure of evidence (in favor of or against the hypotheses being tested) so that experts, decision-makers, and policy-makers can make appropriate use of the evidence provided by the data.

Exactness and Unbiasedness

It should be emphasized that, despite unconventional ways of constructing extreme regions and unconventional structures of test variables that we need to define in the construction, the p-values in the applications discussed in this chapter and in the following chapters are all exact in the following sense: With respect to a specified probability measure, a sample space, and the parameter of interest fixed at the specified value, *p-value is the exact probability of an extreme region* (a well-defined subset of the sample space with the observed sample on its boundary) When desirable and optimum properties are imposed, this subset would become unique up to equivalent p-values. In particular, by an unbiased significance test we mean that *the probability of the extreme region increases for any deviation from the null hypothesis.* These two properties are all that is needed in Fisher's treatment of significance testing. Exactness and unbiasedness in the treatment of the Neyman-Pearson theory do not necessarily hold with p-values and therefore one should not check for such properties or any other expected implication as verification of claimed properties.

The p-values given in Chapter 4 (nonparametric methods) and posterior predictive p-values are also exact only in this sense rather than in conventional frequency interpretations. In all cases, these p-values serve as measures of how well the data support or discredit the underlying null hypotheses. One can of course reject the null hypothesis if the observed p-value is too small, say less than .05. One should not, however, interpret this to mean that in repeated sampling of the same experiment, exactly 5% of the time will a point null hypothesis be rejected when in fact it is true. Although some p-values may have such implications consistent with the Neyman-Pearson decision-theoretic treatment, it is not a requirement in Fisher's treatment of the problem. Barnard (1984), Weerahandi (1993), and Meng (1994) attempt to give other interpretations of extended p-values which do not have uniform distributions leading to conventional interpretations.

5.2. Generalized Test Variables

Like the conditional testing procedures that made exact inference in nonparametric methods possible, the generalized p-values are also in the spirit of Fisher's thinking. His treatment of significance testing is simple and yet general enough so that the definition of a p-value can be extended without deviating from the original treatment. The idea in Fisher's approach to significance testing is to define an extreme region (an unbiased subset of the sample space with the observed sample on its boundary) and then to use its probability under the null hypothesis to perform the test. The only other requirement is that one should be able to compute the probability of the extreme region without knowledge of the underlying nuisance parameters. It is possible to find such subsets and derive tests without the notion of generalized test variables introduced by Tsui and Weerahandi (1989). Nevertheless, test variables do facilitate such derivations and so the purpose of this section is to define them clearly with some illustrative examples.

The test variables that we defined in Chapter 2 can attain only some of the subsets of the sample space thus making problems such as the Behrens-Fisher problem unsolvable even when extreme regions with the required properties do exist. So, we only need to extend the definition of test variables so that such extreme regions become attainable. This can be accomplished simply by letting test variables depend on every quantity that an extreme region may possibly depend on. Although we refer (to be consistent with the literature) to p-values based on such tests as generalized p-values, except for the way we specify the extreme region, there is no other difference between classical p-values and generalized p-values in their interpretation in significance testing.

To illustrate the underlying problem more clearly and to formulate the generalized version of test variables, consider an observable random vector $\mathbf{X}$ with the cumulative distribution function $F(\mathbf{x};\zeta)$, where $\zeta = (\theta,\boldsymbol{\delta})$ is a vector of unknown parameters, θ being the parameter of interest, and $\boldsymbol{\delta}$ is a vector of nuisance parameters. Let Ξ be the sample space of possible values of $\mathbf{X}$ and Θ be the parameter space of θ. As before, an observation from $\mathbf{X}$ is denoted by $\mathbf{x}$, where $\mathbf{x} \in \Xi$.

Recall that an observed significance level for hypothesis testing is defined on the basis of a data-based extreme region, a subset of the sample space with $\mathbf{x}$ on its boundary. In significance testing with p-values, it is properties such as the unbiasedness (as opposed to the repeated sampling considerations in the Neyman-Pearson treatment of fixed-level testing) that qualify such a subset as an extreme region whose probability can be used as a measure of how the observed data supports or discredits the null hypothesis. The drawback of test statistics and test variables defined in Chapter 2 is that when nuisance parameters are present they may not be able to attain all subsets of the sample space with desirable properties. For example, this is the case if we use generalized likelihood ratio test statistics to construct procedures for testing the parameters in (5.1) or (5.2). As illustrated by Figure 5.1, an extreme region with observed $\mathbf{x}$ on its boundary can depend on ζ as well as on $\mathbf{x}$; for the purpose of comparison the structure of a critical region (on which a fixed-level test can be based) is shown in Figure 5.2. An unbiased extreme region may exist even when a critical region of size α does not exist. There is no reason why every extreme region $C(\mathbf{x};\theta,\delta)$ should have a representation of the form $\{\mathbf{X} \mid T(\mathbf{X};\theta) \geq T(\mathbf{x};\theta)\}$. The boundary of the extreme region can be any complicated function of all the underlying quantities $\mathbf{x}$, θ, and δ. This restrictive nature is the motivation for extending Definition 2.1. Therefore, we need to allow a test variable to depend all these quantities in any form. In any case, the probability of the extreme region should be required to be independent of ζ.

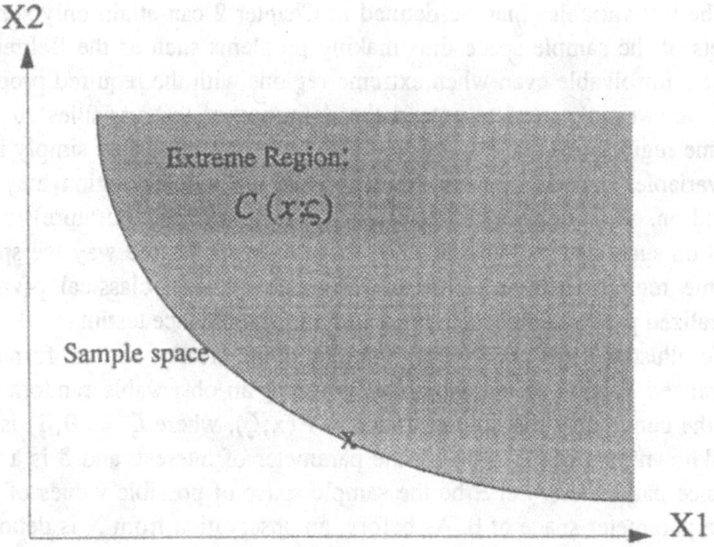

Figure 5.1. Generalized extreme region for significance testing.

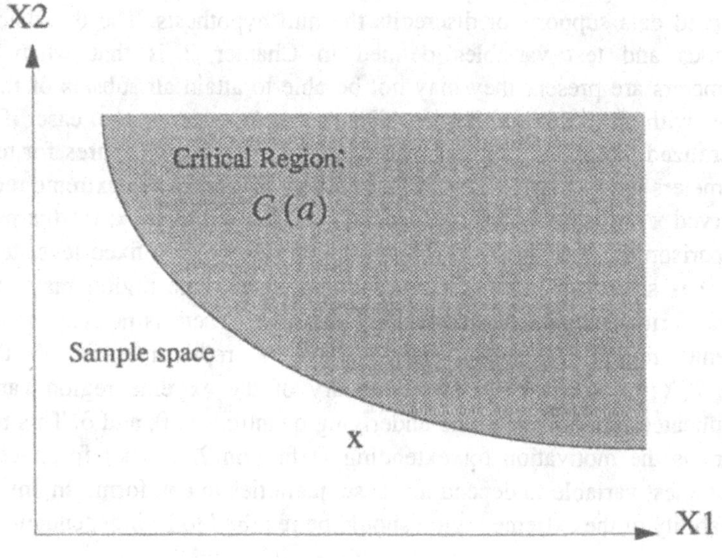

Figure 5.2. Critical region for fixed-level testing.

In order to define a subset which can be regarded as an extreme region, we first need to establish a stochastic ordering of the sample space according to the possible values of θ, the parameter of interest. In Chapter 2, this was accomplished by means of test statistics and test variables. Recall that a test variable is a function of $\mathbf{X}$ and θ only and is a monotonic function of θ (Property 2). In particular, a test variable is free of $\boldsymbol{\delta}$, the nuisance parameters. Note that it was Property 2 and not other properties that was vital in obtaining a stochastic ordering of the sample space. The role of the other properties was to ensure that the probability of the resulting extreme region can be computed without any knowledge of the nuisance parameters. By relaxing some of the requirements in Definition 2.1, we obtain the following extension.

Definition 5.1. A random variable of the form $T = T(\mathbf{X};\mathbf{x},\boldsymbol{\zeta})$ is said to be a *generalized test variable* if it has the following three properties:

Property 1. $t_{obs} = t(\mathbf{x};\mathbf{x},\boldsymbol{\zeta})$ does not depend on unknown parameters.

Property 2. When θ is specified, T has a probability distribution that is free of nuisance parameters.

Property 3: For fixed $\mathbf{x}$ and $\boldsymbol{\delta}$, $Pr(T \le t;\theta)$ is a monotonic function of θ for any given t.

Without loss of generality, we can consider Property 1 to be redundant, because if the property is not satisfied, then we can define a generalized test variable $\tilde{T}$ as $\tilde{T} = T(\mathbf{X};\mathbf{x},\boldsymbol{\zeta}) - T(\mathbf{x};\mathbf{x},\boldsymbol{\zeta})$ and impose Properties 2 and 3 on $\tilde{T}$. Property 2 is imposed to ensure that p-values based on generalized test variables are computable when θ is specified. This property is related to the notions of similarity and unbiasedness of a test defined in Section 2.6. Finally, Property 3 ensures that the sample space can be stochastically ordered on the basis of the generalized test variable. In some applications this property is somewhat too restrictive, and therefore, in Section 5.7 the property will be relaxed so as to handle such applications. If $Pr(T > t)$ is a nondecreasing function of θ, then T is said to be *stochastically increasing* in θ.

It should be noted that any probability statement written in terms of T and t_{obs} is nothing but the probability of an event (with any interpretation that applies to the probability distribution function of $\mathbf{X}$) defined relative to the observed sample point $\mathbf{x}$. In particular, statements such as $\{\mathbf{X} \mid T(\mathbf{X};\mathbf{x},\boldsymbol{\zeta}) \ge T(\mathbf{x};\mathbf{x},\boldsymbol{\zeta})\}$ define a subset of the sample space Ξ with the observed value $\mathbf{x}$ on its boundary. In defining generalized p-values we need to resort to this type of statement, because, as we will see in applications undertaken in this chapter and in the following chapters, many unbiased extreme regions leading to practical solutions cannot be attained without having $\mathbf{x}$ on both sides of the inequality. In dealing with test variables, extreme care must be taken to make sure that the role of $\mathbf{X}$ is to find the probability of the above extreme region and the role of $\mathbf{x}$ is only to keep track of the observed sample point on the boundary of that region. The roles are not substitutive and the extended structure of the test variable is just a devise to facilitate the computation of the above extreme region.

As in the case of test statistics, without loss of information about the parameter θ, the search for generalized test variables can be confined to functions of complete sufficient statistics. In other words, we will consider test variables which depend on **X** and **x** only through a vector of sufficient statistics **S(X)** and its observed value **s(x)**. The following example illustrates how generalized test variables having desirable properties can be obtained in simple problems.

Example 5.1. The second moment of the normal distribution

Let $X_1, \ldots, X_n$ be a random sample from a normal population with mean μ and variance σ^2. Suppose the second moment of the distribution,

$$\theta = E(X^2) = \mu^2 + \sigma^2$$

is the parameter of interest. A test statistic having properties 1 and 2 of Definition 2.1 is not available for the problem of testing hypotheses about θ. Therefore, we shall find a generalized test variable T based on the sufficient statistics

$$\bar{X} \quad \text{and} \quad S^2 = \frac{\sum_{i=1}^{n} (X_i - \bar{X})^2}{n},$$

namely, the sample mean and the sample variance. We shall show that the subset

$$C_{\mathbf{x}}(\mu,\sigma) = \left\{ \mathbf{X} \mid \left(\bar{x} - \frac{s(\bar{X} - \mu)}{S} \right)^2 + \frac{s^2\sigma^2}{S^2} \geq \mu^2 + \sigma^2 \right\}$$

is a subset of the sample space with its probability free of nuisance parameters. Obviously, this is a well-defined subset having observed **x** on its boundary but it is not attainable using a test statistic or a test variable which does not depend on $\bar{x}$ and s^2.

Since generalized test variables may depend on all the unknown parameters, we can base the construction of a generalized test variable on the random quantities

$$Z = \sqrt{n}\,\frac{\bar{X} - \mu}{\sigma} \sim N(0, 1), \quad \text{and} \quad U = \frac{nS^2}{\sigma^2} \sim \chi^2_{n-1} \qquad (5.3)$$

having standard distributions with no unknown parameters. In view of the identity

$$\theta = (\bar{X} - Z\sigma/\sqrt{n})^2 + \sigma^2$$
$$= (\bar{X} - Z\frac{S}{\sqrt{U}})^2 + \frac{nS^2}{U},$$

we define a potential generalized test variable $T = T(\bar{X},S)$ as

$$T = (\bar{x} - s\frac{Z}{\sqrt{U}})^2 + \frac{ns^2}{U} - \theta, \tag{5.4}$$

where $\bar{x}$ and s^2 are the observed values of the sample mean and the sample variance, respectively. Notice that when T is expressed in terms of $\bar{x}$, $\bar{X}$, s, and S, the right hand side of the inequality that defines the extreme region $C_{\mathbf{x}}$ is nothing but $T + \theta$. It is clear from the definitions of the random variables Z and U in (5.3) that the observed value of T is $t_{obs} = T(\bar{x},s) = 0$. Moreover, the distributions of Z and U given by (5.3) imply that the distribution of T depends on the parameters of the normal distribution only through θ, the parameter of interest. Hence, T satisfies Properties 1 and 2 of Definition 5.1. Property 3 immediately follows from the fact that θ is the location parameter of the distribution of T. In other words, T is stochastically decreasing in θ. To see this more clearly, let

$$R = (\bar{x} - s\frac{Z}{\sqrt{U}})^2 + \frac{ns^2}{U}.$$

Obviously, the distribution of R, say $F_R(r)$, is free of unknown parameters. The cumulative distribution of T can be expressed as

$$Pr(T \le t) = Pr(R \le t + \theta) = F_R(t + \theta). \tag{5.5}$$

It is now clear from equation (5.5) that the cdf of T is an increasing function of θ, thus implying Property 3. Therefore, T defined by (5.4) is indeed a generalized test variable for θ.

5.3. Definition of Generalized p-Values

Consider again the problem of testing one-sided hypotheses of the form

$$H_0: \theta \le \theta_0, \quad H_1: \theta > \theta_0, \tag{5.6}$$

where θ_0 is a prespecified value of the parameter θ. Recall that in the usual setting for significance testing of hypotheses on the basis of a test statistic $T(\mathbf{X})$, a data-based extreme region $C_{\mathbf{x}}$ is of the form

$$C_{\mathbf{x}} = \{\mathbf{X} \in \Xi : T(\mathbf{X}) \ge T(\mathbf{x})\},$$

where smaller values of T indicate stronger evidence against H_0. To extend this definition, let $T = T(\mathbf{X};\mathbf{x},\zeta)$ be a generalized test variable for θ. Assume without loss of generality that T is stochastically increasing in θ. Then, T has the desired property that its larger values support larger values of θ. Therefore, we can generalize the definition of the extreme region so that the desirable property is preserved, and in turn, the p-value can be defined as in Section 2.3.

Definition 5.2. Let $T = T(\mathbf{X};\ \mathbf{x},\zeta)$ be a stochastically increasing (in the parameter of interest) test variable. Then, the subset of the sample space defined by

$$C_{\mathbf{x}}(\zeta) = \{\mathbf{X} \in \Xi : T(\mathbf{X};\ \mathbf{x},\zeta) \geq T(\mathbf{x};\ \mathbf{x},\zeta)\}$$

is said to be a *generalized extreme region* for testing H_0 against H_1.

Definition 5.3. A subset $C_{\mathbf{x}}(\zeta)$ of the sample space is said to be a *continuous generalized extreme region* if there exists a continuous generalized test variable of the form $H(\mathbf{X};\ \mathbf{x},\zeta)$ such that

$$C_{\mathbf{x}}(\zeta) = \{\mathbf{X} : H(\mathbf{X};\ \mathbf{x},\zeta) \geq H(\mathbf{x};\ \mathbf{x},\zeta)\}\ .$$

Definition 5.4. If $C_{\mathbf{x}}$ is a generalized extreme region, then

$$p = \underset{\theta \leq \theta_0}{Sup}\ Pr(\mathbf{X} \in C_{\mathbf{x}}(\zeta) \mid \theta)$$

is said to be its *generalized p-value* for testing H_0.

The generalized p-value will be sometimes referred to as the 'generalized significance value', or merely as the 'p-value' ('observed level of significance') when this causes no confusion. If $T = T(\mathbf{X};\ \mathbf{x},\zeta)$ is a generalized test variable which is stochastically increasing in θ, then the generalized p-value can be computed conveniently as

$$p = Pr(T \geq t_{obs} \mid \theta = \theta_0)\ , \tag{5.7}$$

where $t_{obs} = T(\mathbf{x};\ \mathbf{x},\zeta_0)$ and $\zeta_0 = (\theta_0,\boldsymbol{\delta})$. As in Section 2.3 we refer to

$$\pi(\mathbf{x};\theta) = Pr(T(\mathbf{X};\ \mathbf{x},\zeta) \geq t_{obs} \mid \theta) \tag{5.8a}$$

and

$$\pi_0(\mathbf{x};\theta) = Pr(T(\mathbf{X};\ \mathbf{x},\zeta_0) \geq t_{obs} \mid \theta) \tag{5.8b}$$

as *generalized power functions* of the generalized test variables $T(\mathbf{X};\ \mathbf{x},\zeta)$ and $T(\mathbf{X},\ \mathbf{x},\zeta_0)$, respectively, where $\zeta_0 = (\theta_0,\boldsymbol{\delta})$.

Alternative uses of these two power functions were discussed in Section 2.3. They play the same roles in the present context also. In particular, while $\pi_0(\theta)$ can be employed in comparisons of alternative extreme regions, $\pi(\theta)$ is useful in stating general propositions without any reference to specific hypotheses. It should be emphasized, however, that since the optimality of most powerful tests in fixed-level testing is readily incorporated in the definition of test variables based on sufficient statistics and in the definition of generalized p-values, the power function $\pi_0(\theta)$ does not play a vital role in significance testing.

If T is stochastically decreasing in θ, the observed level of significance for testing H_0 is given by $p = Pr(T \le t_{obs} \mid \theta = \theta_0)$. Moreover, if the null hypothesis H_0 is right-sided, then the generalized p-value for testing H_0 can be computed using the formula $p = Pr(T \le t_{obs} \mid \theta = \theta_0)$, or the formula $p = Pr(T \ge t_{obs} \mid \theta = \theta_0)$ depending on whether T is stochastically increasing or decreasing in θ.

In each of these cases, the generalized p-value is the exact probability of an unbiased extreme region with the observed sample point on its boundary. It serves to measure how well the data supports or contradicts the null hypothesis; the smaller the p-value, the greater the evidence against the null hypothesis. In other words, smaller values of the observed level of significance favor the alternative hypothesis and its larger values favor the null hypothesis.

Example 5.2. The second moment of the normal distribution (continued)

Let $X_1, \ldots, X_n$ be a random sample from a normal population with mean μ and variance σ^2. Let $\bar{X}$ and S^2 be the maximum likelihood estimators of μ and σ^2, respectively. Consider the problem of testing the hypotheses

$$H_0\colon \theta \le \theta_0, \quad H_1\colon \theta > \theta_0, \tag{5.9}$$

where $\theta = \mu^2 + \sigma^2$ is the second central moment of the distribution. It was seen in Example 5.1 that

$$T = \left(\bar{x} - s\frac{Z}{\sqrt{U}} \right)^2 + \frac{ns^2}{U} - \theta \,,$$

is a generalized test variable for θ, where

$$Z = \frac{\bar{X} - \mu}{\sigma/\sqrt{n}} \sim N(0, 1)\,, \quad \text{and } U = \frac{n\,S^2}{\sigma^2} \sim \chi^2_{n-1}\,.$$

Since T was found to be stochastically decreasing in θ and the observed value of T is 0, the generalized p-value for testing the hypotheses in (5.9) is

$$\begin{aligned} p &= Pr(\, T \le t_{obs} \mid \theta = \theta_0\,) \\ &= Pr\left[(\bar{x} - s\frac{Z}{\sqrt{U}})^2 + \frac{ns^2}{U} \le \theta_0 \right]. \end{aligned} \tag{5.10}$$

The p-value in (5.10) can be computed by numerical integration with respect to Z and U, which are independent random variables with known probability density functions. The probability of the inequality in (5.10) can also be evaluated by the Monte Carlo method. This method can be easily implemented by generating a large number of random numbers from the standard normal distribution and from the chi-squared distribution with $n-1$ degrees of freedom, and then finding the fraction of random pairs for which the inequality is satisfied.

Testing point null hypotheses

Next consider point null hypotheses and composite alternative hypotheses of the form

$$H_0: \theta = \theta_0, \qquad H_1: \theta \neq \theta_0 ,$$

where θ_0 is a particular value of the parameter that has been specified. Suppose $T = T(\mathbf{X}; \mathbf{x}, \zeta)$ is a generalized test variable. Let τ be the sample space of possible values that T can take. In the case of point null hypotheses, there are alternative ways of defining p-values that measure how well data supports the hypotheses. We use a natural extension of Definition 5.3 when there exists a function of T which tends to take larger values for greater discrepancies between θ and θ_0.

Let T be a generalized test variable for testing θ and let $C_t(\zeta)$ be a subset of τ. We define a generalized p-value based on $C_t(\zeta)$ for testing H_0 if it has the following property:

Property 4. Given any fixed t and δ, the probability, $Pr(T \in C_t(\zeta))$ is a nondecreasing function of (i) $\theta - \theta_0$ when $\theta \geq \theta_0$, and (ii) $\theta_0 - \theta$ when $\theta < \theta_0$.

As pointed out in Section 2.3, in some applications this property is somewhat too restrictive. In such situations a property parallel to Property 3' defined in that section can be used in place of Property 4. This problem will be further addressed in Section 5.7. Once a generalized test variable with Properties 1, 2, and 3 has been found, the methods discussed in Section 2.3 can be used to find a subset of the sample space having Property 4.

If there exists a set $C_t(\zeta)$ with Property 4, then that set is considered an extreme region for testing the hypotheses under consideration. The p-value for testing H_0 against H_1 based on the extreme region $C_{t_{obs}}$ is defined as

$$p = Pr(T \in C_{t_{obs}}(\zeta) \mid \theta = \theta_0),$$

where $t_{obs} = T(\mathbf{x}; \mathbf{x}, \zeta)$ is the observed value of the T. The corresponding data-based power function is

$$\pi(\mathbf{x}; \theta) = Pr(T \in C_{t_{obs}}(\zeta) \mid \theta).$$

Applications of these definitions can be found in Chapters 7 through 10. The procedure of finding an extreme region having Property 4 is illustrated by Example 5.3.

5.4. Frequency Interpretations and Generalized Fixed-Level Tests

Except for a few applications, there is no common agreement about the practical use of repeated sampling properties. If the same experiment can be repeated, one should rather combine the data to perform a more powerful test. Some statisticians including Bayesians advocate the use of procedures having desirable properties with respect to the current sample rather than other possible samples that could have been observed, but were not. Insisting on the repeated sampling property with the same experiment can sometimes lead to procedures with highly undesirable features (see, for instance, Pratt (1961), Kiefer (1977) and the discussions of that paper); we shall address some of these difficulties in Chapter 6 and establish more appealing procedures. Fixed-level testing and interval estimation in random-effects models and mixed models provide a class of applications suffering from such undesirable features.

In any case, many practitioners would like to have repeated sampling properties when they do not jeopardize more desirable properties, and therefore one would resort to extended p-values only when exact tests having such repeated sampling properties are not available, when the available tests are not powerful, or when they have undesirable properties. For example, in linear models that we will deal with in Chapters 7 through 10, depending on whether or not the assumption of equal error variances is reasonable, tests based on generalized p-values tend to be less powerful or more powerful than conventional F-tests and t-tests.

As in the case of p-values in nonparametric methods, generalized p-values also have unconventional frequency interpretations implied by the definition. First of all, if the experiment is repeated with new samples, the proportion of samples that fall into the current extreme region is of course equal to p, provided that $\theta = \theta_0$. To describe this more clearly and to describe related material, consider for convenience, the problem of testing the hypotheses

$$H_0\colon \theta \le \theta_0 \quad \text{versus} \quad H_1\colon \theta > \theta_0,$$

on the basis of the random observation $\mathbf{x}$ obtained from the distribution $F(\mathbf{x} \mid \zeta)$, where $\zeta = (\theta, \delta)$ is a vector of unknown parameters, θ is the parameter of interest, and δ is a vector of nuisance parameters. Recall that a generalized test variable was defined to be a function of the form $T = t(\mathbf{X};\, \mathbf{x}_{obs}, \zeta)$ having the three properties spelled out by Definition 5.1. Assume that T is stochastically increasing in θ, the parameter being tested. If the current observed value of the generalized test variable is $t_{obs} = t(\mathbf{x}_{obs};\, \mathbf{x}_{obs}, \zeta)$, then the generalized p-value is $p(t_{obs}) = Pr(\, T \ge t_{obs} \mid \theta = \theta_{obs})$. Consider the function

$$p(t) = Pr(\, T(\mathbf{X};\, \mathbf{x}_{obs}, \zeta) \ge t \mid \theta = \theta_0\,).$$

When T is a continuous random variable, it follows from the probability integral transform that $F_T(T)$, and hence $p(T)$, has a uniform distribution over the interval $[0,1]$; that is

$$p(T) \sim U(0,1) . \tag{5.11}$$

when $\theta = \theta_0$. In view of this result, a decision-maker may reject the null hypothesis if $p < \alpha$, where α is the nominal level of testing.

Definition 5.4. Let $p(t_{obs})$ be a generalized p-value based on a continuous generalized test variable $T = t(\mathbf{X}; \mathbf{x}_{obs}, \zeta)$. Let H_0: $\theta \in \Theta_0$ be the null hypothesis being tested against the alternative H_1: $\theta \in \Theta_1$. Then, the rule defined as

$$\text{reject } H_0 \text{ if } p(t_{obs}) < \alpha \tag{5.12}$$

is said to be a *generalized fixed-level test* of level α.

Since the class of test statistics is a particular case of test variables, this is indeed a generalization of conventional fixed-level tests. Generalized fixed-level tests should be of interest to decision theorists who insist on conventional fixed-level tests as well as to others. This is because, according to various simulation studies reported in the literature, in many applications the generalized fixed-level tests (constructed by imposing various optimal and desirable properties) often provide level guaranteed tests or excellent approximate tests (conventional), which outperform simple and more complicated approximations, both in power and in size. Therefore, even if one has philosophical differences, the advantages of generalized fixed-level tests should not be ignored as they tend to have superior power and size performance in conventional approaches to hypothesis testing as well.

What kind of repeated sampling properties does a generalized fixed-level test have? If the same original value of $\mathbf{x}_{obs}$ is used in the computation of $t_{obs} = t(\mathbf{x}; \mathbf{x}_{obs}, \zeta)$ with new $\mathbf{x}$ values, (5.11) implies that when $\theta = \theta_0$, H_0 will be rejected a fraction α of the time, in the long run. Moreover, the proportion of times that H_0 is rejected when it is true is less than or equal to α. Although this is fine for the purpose of a hypothetical treatment, in reality, one can not practically compute the p-value except when $x = x_{obs}$ without knowledge of the values of nuisance parameters. If one computes t_{obs} as $t(x; x, \zeta)$ for the observed x in each trial, there is no difficulty computing the p-value in each trial, but (5.11) and hence the desired repeated sampling property are no longer valid. On the other hand, regardless of whether or not there is such a property, even this experiment cannot be entertained as practically useful, as opposed to hypothetical. This is because, if the same experiment can be repeated, the confidence statement is no longer valid and the parameter of interest will become known exactly.

To overcome this dilemma, Weerahandi (1993) argued that one should insist on the repeated sampling property only with independent experiments. Suppose in a series of independent experiments one performs generalized fixed-level tests at level α. The question then is whether or not the proportion of times that H_0 is rejected by generalized fixed-level tests of level α is less than or equal to α. Simulated experiments seem to support this result under certain conditions. In particular, the experiments should be independent in every respect, including the directions of hypotheses. However, a formal proof or a set of sufficient conditions for this to be valid are not currently available. In a related problem, Meng (1994) also attempted to establish a repeated sampling property of tests based on posterior predictive p-values. To do this he provided an interesting theorem which also describes why fixed-level tests based on extended p-values always provide good approximations. Since it is often possible to obtain numerically equivalent tests by the generalized p-value approach, the results in Meng(1994) have implications on generalized fixed-level tests as well.

In any case, classical repeated sampling property can be considered practically useful in a few applications such as the statistical quality control. According to many simulation studies of repeating the same experiments in important applications, the actual size of fixed-level tests based on (5.12) do not exceed the intended level. Griffiths and Judge (1992), Thursby (1992), Weerahandi and Johnson (1992), and Zhou and Mathew (1994), Ananda and Weerahandi (1996, 1997), and Weerahandi and Amaratunga (1996) carried out various simulation studies to understand the power and size performance of generalized fixed-level tests in sampling from a single sample space, and also repeated the studies for various typical values of underlying nuisance parameters. According to their findings, not only did the generalized tests provide excellent tests having (not exceeding) the intended level, but also they performed as well as, or often substantially better than other approximate tests which are not exact in any sense.

5.5. Invariance

The purpose of this section is to extend the results of Section 2.5 to the case of generalized p-values. When a problem of finding a test statistic has been reduced by the principle of sufficiency, even with generalized p-values, we can exploit symmetries in the underlying distributions and symmetries in the testing problem to reduce further the number of statistics on which a generalized test variable can be based.

Suppose the underlying family of distributions $F(\mathbf{X};\ \zeta)$ parameterized by a vector of unknown parameters ζ is invariant under a group G of transformations on the sample space Ξ. Let θ be the parameter being tested and let $\overline{G}$ be the induced group of transformations on the parameter space Θ of θ. Consider again the general form of the hypothesis testing problem,

$$H_0: \theta \in \Theta_0 \quad H_1: \theta \in \Theta_1,$$

where Θ_0 and Θ_1 are subsets of Θ. In Section 2.5 we defined the hypothesis testing problem to be invariant under the group G if each hypothesis is not affected by the transformation of the distribution; that is if

$$\overline{g}(\Theta_0) = \Theta_0 \quad \text{and} \quad \overline{g}(\Theta_1) = \Theta_1$$

for all $\overline{g} \in \overline{G}$. In many applications this may not be the case and yet it may possible to transform the parameters so that the transformed problem is invariant. Unlike with classical p-values, in the present context this transformation is allowed to be a function of **x**, the observed value of the random vector **X**. This procedure is illustrated by Example 5.3 and by the application in Section 5.6.

The invariance of tests based on generalized p-values have the same motivation as ordinary p-values and can be defined in a similar manner. Since generalized p-values also serve to measure the evidence in favor of or against the hypotheses, it is still natural to require that the p-value as a function of **x** is not affected by the group of transformations, provided that the testing problem is invariant. Further, this property should hold regardless what θ_0 is. Therefore, the property can be imposed on the data-based power function.

Definition 5.5. A data-based power function $\pi(\mathbf{x};\theta)$ based on a generalized extreme region $C_{\mathbf{x}}(\zeta)$ is said to be invariant under G if $\pi(g(\mathbf{x});\theta) = \pi(\mathbf{x};\theta)$ for all $\mathbf{x} \in \Xi$, $\theta \in \Theta$, and $g \in G$, where $\pi(\mathbf{x};\theta) = Pr(\mathbf{X} \in C_x(\zeta) | \theta)$.

When the testing problem is invariant under a certain group of transformations, Theorem 5.1 suggests how to reduce the number of statistics on which to base a test variable. The theorem applies equally well to a number of random variables having the structure of a generalized test variable. This method of deriving a generalized test variable will be referred to as the *method of invariance*.

Theorem 5.1. Suppose that the testing problem is invariant under a group, G, of transformations on the sample space Ξ. Let $\mathbf{M}(\mathbf{X})$ be a set of maximal invariants with respect to that group. Suppose that $\mathbf{T} = \mathbf{T}(\mathbf{X};\, \mathbf{x},\zeta)$ is a continuous random vector. If the observed value $\mathbf{t}(\mathbf{x}) = \mathbf{T}(\mathbf{X};\, \mathbf{x},\zeta)$ and the distribution of **T** depend on the data **x** only through the observed maximal invariant $\mathbf{m} = \mathbf{M}(\mathbf{x})$, then the data-based power function of any invariant test can be obtained using **T** and **m** only.

Proof. Although the result is generally valid for a vector of random quantities, for convenience, consider the particular case where T is a single random variable. Let $\pi(\mathbf{x};\theta)$ be an invariant power function of an extreme region. Then, it follows from Theorem 1.2 that $\pi(\mathbf{x};\theta)$ depends on $\mathbf{x}$ only through $\mathbf{m} = \mathbf{M}(\mathbf{x})$. Therefore, the power function can be expressed as

$$\pi(\mathbf{x};\ \theta) = G(\mathbf{m};\ \theta),$$

where G is a real-valued function. Now it remains to show that $G(\mathbf{m};\theta)$ is attainable by the generalized test variable T. Since T is absolutely continuous, the random variable $W = F(T;\mathbf{m},\theta)$ has a uniform distribution over the interval $[0,1]$, where $F(t;\mathbf{m},\zeta)$ is the cdf of the random variable T. Let w be the observed value of W. Since the observed value of T depends on the data only through $\mathbf{m}$, w also depends on $\mathbf{x}$ only through $\mathbf{m}$. To this end, letting $g = G(\mathbf{m},\theta)$ for a given value of θ, consider the particular extreme region defined as

$$C_{\mathbf{m}}(\zeta) = \begin{cases} w \le W \le w + g & g \le .5,\ w \le .5 \\ w - g \le W \le w & g \le .5,\ w > .5 \\ 0 \le W \le g & g \ge .5,\ w \le .5 \\ 1 - g \le W \le 1 & g \ge .5,\ w > .5 \end{cases} .$$

Observe that $Pr(W \in C_{\mathbf{m}}(\zeta)) = g$, because W is a uniform random variable. Moreover, by construction the observed value w of W belongs to the extreme region $C_{\mathbf{m}}(\zeta)$. Hence, the generalized extreme region $\{\mathbf{X}\colon F(T;\mathbf{m},\zeta) \in C_{\mathbf{m}}(\zeta)\}$ yields the invariant power function $\pi(\mathbf{x};\ \zeta)$, thus completing the proof.

Example 5.3. Two-parameter exponential distribution (continued)

Consider again the the two-parameter exponential distribution with unknown parameters α and β, and with probability density function

$$f_X(x) = \frac{1}{\beta}\, e^{-(x-\alpha)/\beta} \qquad \text{for} \quad x > \alpha\ ,\quad \beta > 0$$

from which a random sample $X_1, \ldots, X_n$ is available, where α and β are unknown parameters. In Example 2.8 we addressed the problem of testing hypotheses about β. Now consider the problem of testing hypotheses about the parameter α. In order to illustrate the procedure of deriving invariant extreme regions in the case of point null hypotheses, consider the following hypotheses:

$$H_0\colon \alpha = \alpha_0 \qquad H_1\colon \alpha \ne \alpha_0\ . \tag{5.13}$$

According to Example 2.8 and Exercise 2.6, $U = min(X_i)$ and $V = \bar{X} - min(X_i)$ are independent sufficient statistics which are distributed as

$$U - \alpha \sim G(1, \frac{\beta}{n}) \text{ and } V \sim G(n-1, \frac{\beta}{n}) . \quad (5.14)$$

We shall now find, by the method of invariance, a generalized test variable based on U and V. Notice that the family of distributions parameterized by (α, β) is invariant under the scale transformation

$$(U,V) \to (cU, cV) ,$$

where c specifies the particular change of scale of the original random variable X. The induced transformation on the parameters is

$$(\alpha, \beta) \to (c\alpha, c\beta) .$$

However, the testing problem is not invariant in the usual sense unless $\alpha_0 = 0$. In the present situation this problem can be tackled by considering the equivalent testing problem in which the parameter of interest is $\tilde{\alpha} = \alpha - \alpha_0$ and the underlying observable random variable is $\tilde{X} = X - \alpha_0$. This method of defining an invariant testing problem is possible only in a very limited class of applications. In the context of testing hypotheses with generalized p-values, we shall now demonstrate a much more general method of obtaining an invariant testing problem.

Let u and v be observed values of U and V be respectively. Define a scale invariant parameter θ as $\theta = \alpha/u$; other choices such as α/v or $\alpha/(u+v)$ in place of θ will perform equally well and will have no bearing on the p-value. In terms of this parameter and $\theta_0 = \alpha_0/u$, the hypotheses in (5.13) can be expressed in an equivalent form as

$$H_0: \theta = \theta_0, \quad H_1: \theta \neq \theta_0 .$$

With this reparametrization the testing problem is invariant regardless of the value α_0. Consider the potential generalized test variable

$$T((U,V);(u,v),\theta) = \frac{U}{V} + \theta u \left[\frac{1}{v} - \frac{1}{V} \right] . \quad (5.15)$$

To show that this is indeed a generalized test variable, define two independent random variables with distributions free of unknown parameters as

$$Y_1 = \frac{2n(U-\alpha)}{\beta} \sim \chi^2_2 \text{ ,and } Y_2 = \frac{2nV}{\beta} \sim \chi^2_{2(n-1)} .$$

In terms of these random variables, the random variable defined in (5.15) can be rewritten as

$$T = \frac{Y_1}{Y_2} + \theta \frac{u}{v} .$$

It is evident from this equation that (i) the distribution of T is free of unknown parameters; (ii) T is stochastically increasing in θ; and (iii) the distribution of T

depends on the data only through u/v. Moreover, the observed value of T is also u/v. Hence, T is a generalized test variable with both the distribution and the observed value depending on the data only through u/v. Since u/v is a maximal invariant, according to Theorem 5.1, any scale invariant test can be obtained from T. Having found a generalized test variable which can produce any scale invariant extreme region, we can now express (5.15) as

$$T = \frac{W}{n-1} + \frac{\alpha}{v} ,$$

where W has an F-distribution with 2 and $2(n-1)$ degrees of freedom. Since

$$W = (n-1)(T - \frac{\alpha}{v}) \sim F_{2,2(n-1)} ,$$

T is an observable random variable with observed value u/v, and $v > 0$, it now immediately follows from Corollary 2.1 that $Pr(\tilde{w} \leq W \leq w \mid \alpha = \alpha_0)$ is a decreasing function of $|\alpha - \alpha_0|$, where $\tilde{w}$ is chosen such that $f_W(\tilde{w}) = f_W(w)$, f_W being the pdf of W. Therefore, the appropriate extreme region for testing (5.13) is the complement of the set $\{X: \tilde{w}_0 \leq W \leq w_0 \mid \alpha = \alpha_0\}$ and the generalized p-value is

$$p = 1 - Pr(\tilde{w}_0 \leq W \leq w_0 \mid \alpha = \alpha_0) , \tag{5.16}$$

where $w_0 = (n-1)(u - \alpha_0)/v$.

5.6. Comparing the Means of Two Exponential Distributions

The purpose of this section is to undertake an important application for which conventional methods in fixed-level testing and significance testing do not provide exact solutions. The application concerns an important problem in lifetime testing, where it is desired to compare the mean lifetimes of two brands of an equipment or a certain component of the equipment (see, for instance, Lawless (1982)). Consider the problem when the underlying lifetime distributions are exponential and the observed data are possibly right-censored.

Suppose that $X_1, \ldots, X_M$ and $Y_1, \ldots, Y_N$ form two sets of independent and identically distributed lifetime random variables having exponential distributions with means μ_x and μ_y respectively. Suppose the two sets of lifetimes are censored after observing, respectively, the first m and n failures. Let the observed lifetimes be $X_{(1)}, \ldots, X_{(m)}$ and $Y_{(1)}, \ldots, Y_{(n)}$. The problem is to test hypotheses about the difference in the two means, namely $\theta = \mu_x - \mu_y$, on the basis of the available censored data. To establish a testing procedure based on generalized p-values, consider hypotheses of the form

$$H_0: \mu_x - \mu_y \leq \theta_0 , \quad H_1: \mu_x - \mu_y > \theta_0 . \tag{5.17}$$

Assume, without loss of generality, that $\theta_0 \geq 0$. In Example 2.9 we solved this problem when $\theta_0 = 0$, $m = M$, and $n = N$. Although it was not necessary to assume that $m = M$ and $n = N$, the approach taken in Example 2.9 does not work when $\theta_0 \neq 0$.

It can be shown (Exercise 5.1) that $U = \sum_{i=1}^{m} X_{(i)} + (M-m)X_{(m)}$ and $V = \sum_{i=1}^{n} Y_{(i)} + (N-n)Y_{(n)}$ are sufficient statistics for μ_x and μ_y respectively. Furthermore, these random variables are independently distributed as

$$U \sim G(m, \mu_x) \quad \text{and} \quad V \sim G(n, \mu_y). \tag{5.18}$$

Let u and v be the observed values of U and V, respectively.

In order to find a generalized test variable, notice first of all that the family of joint distributions of these random variables is invariant under the group of common scale transformations

$$(U,V) \rightarrow (kU, kV) \quad \text{and} \quad (\mu_x, \mu_y) \rightarrow (k\mu_x, k\mu_y),$$

where $k > 0$. The induced transformation on the parameter of interest is $\theta \rightarrow k\theta$. Therefore, the testing problem is invariant in the usual sense only when $\theta_0 = 0$. This is not an inherent problem of the exponential distribution, but rather it is a shortcoming of the classical formulation of the testing problem. The point is that, in specifying the hypotheses, one should be allowed to choose any desirable scale. For instance, the null hypothesis may be $H_0\colon \theta \leq 2$ or $H_0\colon \theta \leq 24$ according to whether the lifetimes are measured in years or in months. In few applications such as that in Example 5.3, one can resolve this difficulty by transforming the original problem so that the resulting testing problem will have $\theta_0 = 0$. In many applications including the current one this is not possible. Therefore, we have to take the approach suggested in Section 5.5.

Consider the scale invariant parameter $\beta = \theta/v$ (any other scale invariant parameter such as $\theta/(u+v)$ will work equally well, because the reparameterization will have no bearing on the resulting generalized p-value). The problem of testing hypotheses in (5.18) is equivalent to testing

$$H_0\colon \beta \leq \beta_0 \quad \text{versus} \quad H_1\colon \beta > \beta_0,$$

where $\beta_0 = \theta_0/v$. With this representation of the hypotheses, the testing problem is invariant for all $\theta_0 \geq 0$.

Consider the potential test variable

$$\begin{aligned} T(U,V;\, u,v,\delta) &= \frac{u}{v}\frac{\mu_x}{U} - \frac{\mu_y}{V} - \beta \\ &= \frac{u}{v}\frac{2}{W_1} - \frac{2}{W_2} - \beta, \end{aligned} \tag{5.19}$$

where W_1 and W_2 are chi-squared random variables defined as

$$W_1 = \frac{2U}{\mu_x} \sim \chi^2_{2m} \quad \text{and} \quad W_2 = \frac{2V}{\mu_y} \sim \chi^2_{2n} ,$$

and $\delta = (\beta, \mu_x, \mu_y)$. We have defined T such that its observed value is $t_{obs} = 0$ and the distribution of T does not involve any unknown parameters. It can also be seen from (5.19) that the distribution of T depends on the data only through u/v, a maximal invariant for scale transformations. Therefore, by Theorem 5.1 any scale invariant test concerning β can be obtained using T. From the cdf

$$\begin{aligned} F_T(t) &= Pr(T \leq t) \\ &= Pr\left[\frac{u}{v}\frac{2}{W_1} - \frac{2}{W_2} \leq t + \beta \right], \end{aligned}$$

it is clear that T is stochastically decreasing in θ as well as in β. Therefore, T is indeed a generalized test variable.

We can now find the appropriate extreme region and its p-value in terms of the original parameters. Since the cdf of T is increasing in θ, the appropriate extreme region defined by T is

$$C_{u,v}(\delta) = \{(U,V) \mid u\frac{\mu_x}{U} - v\frac{\mu_y}{V} \leq \theta\}.$$

The generalized p-value is the probability of this set when $\theta = \theta_0$. This probability can be conveniently computed using the representation of T given by (5.19). Hence, the hypotheses in (5.17) can be tested on the basis of the generalized p-value

$$\begin{aligned} p &= Pr\left[\frac{u}{W_1} - \frac{v}{W_2} \leq \frac{\theta_0}{2} \right] \\ &= Pr\left[\frac{u}{(\theta_0/2 + v/W_2)} \leq W_1 \right] && (5.20) \\ &= 1 - E\left[F_{\chi(2m)}\left[u(\theta_0/2 + v/W_2)^{-1} \right] \right] , && (5.21) \end{aligned}$$

where $F_{\chi(2m)}(\ )$ is the cdf of the chi-squared distribution with $2m$ degrees of freedom and the expectation in (5.21) is taken with respect to the random variable $W_2 \sim \chi^2_{2n}$. The expectation can be computed by numerical integration or by means of a large number of random numbers generated from the chi-squared distribution with $2n$ degrees of freedom.

Observe that when $\theta_0 = 0$, equation (5.20) reduces to

$$p = Pr\left(\frac{u}{v} \le \frac{W_1}{W_2}\right)$$

$$= 1 - F_{2m,2n}\left(\frac{n}{m}\frac{u}{v}\right) .$$

This is the same result that we obtained in Example 2.9 (when $m = M$ and $n = N$). Hence the p-value given by (5.21) is indeed an extension of the classical F-test for comparing exponential populations.

Example 5.4. Comparing the lifetimes of two brands of light bulbs

A manufacturer of light bulbs produces two brands of light bulbs, say A and B. In justifying a higher price for brand A, the manufacturer claims that, on average, brand A lasts at least one year longer than brand B. An accelerated testing of 15 light bulbs of brand A and 10 light bulbs of brand B are transformed into data expected to be valid under normal household operating conditions. The transformed data (in years) are shown in Table 5.1; each datum marked with a star corresponds to a censored lifetime.

Table 5.1 Observed life times

(a) Brand A

3.327*	0.828	3.327*	3.327	2.139
0.613	1.492	0.231	0.281	0.310
0.792	2.479	1.473	0.395	3.404

(b) Brand B

2.828	1.476	1.681	0.934	0.905
3.327*	1.002	0.283	1.294	0.304

Let us examine whether or not this data supports the manufactures claim. The underlying problem is to test the hypotheses

$$H_0: \mu_A - \mu_B \ge 1 \quad \text{versus} \quad H_1: \mu_A - \mu_B < 1 ,$$

where μ_A and μ_B are the mean lifetimes of brand A and brand B, respectively.

Verifying the distributional assumptions

Let us first examine whether each data set can be assumed to have come from an exponential distribution so that the results of this section can be employed to perform a test. A simple graphical test can be based on the observed (empirical) and expected *survival function*, which is defined in the literature on lifetime distributions as $S(t) = Pr(T \geq t)$, where T denotes a lifetime random variable. Clearly, the survival function of an exponential distribution with mean μ is

$$S(t) = e^{-t/\mu}.$$

The formula for the empirical survival function (an estimate in the case of censored data) $\hat{S}(t)$ can be found, for instance, in Lawless (1982). If the exponential distribution is reasonable, a plot of $\hat{H}(t) = -\log\hat{S}(t)$ against t should be roughly linear and pass through the origin, where $\hat{H}(t)$ is an estimate of the *cumulative hazard function* $H(t) = -\log S(t)$, in the terminology of the literature. This plot can be rather jagged and so that it is often preferable to plot just the time points $(t_{(1)}, \cdots, t_{(N)})$, the order statistic of observed failure times, where N is the sample size in the case of uncensored data. In this case, a plot of the points $[t_{(i)}, -\log\{1-(i-.5)/N\}]$ should be roughly linear (see, for instance Crowder, Kimber, Smith, and Sweeting (1991)). If the data are type II censored after observing n failures, the plot is still valid but will include points only for the first n ordered observations. Figure 5.3 and Figure 5.4 show the plots of the data for Brand A and Brand B, respectively. Both plots look reasonably linear and pass through the origin suggesting that we can proceed with the assumption of exponentially distributed lifetimes for each brand.

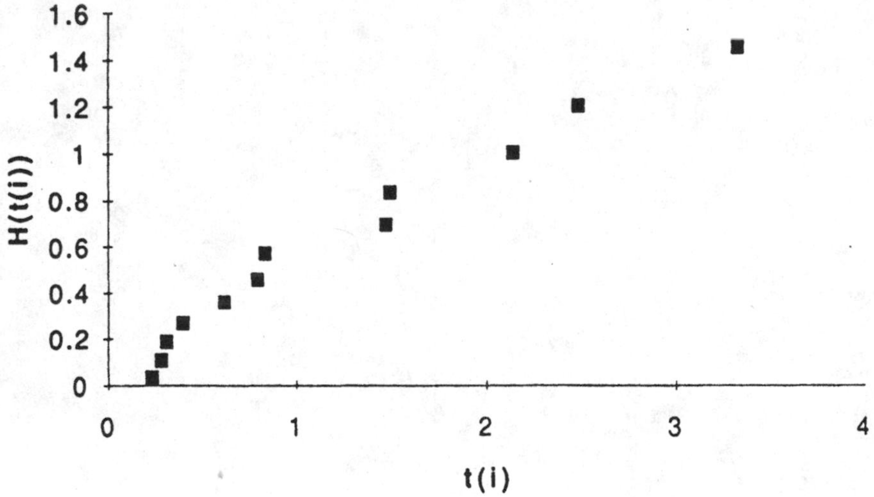

Figure 5.3. Plot for Brand A light bulbs.

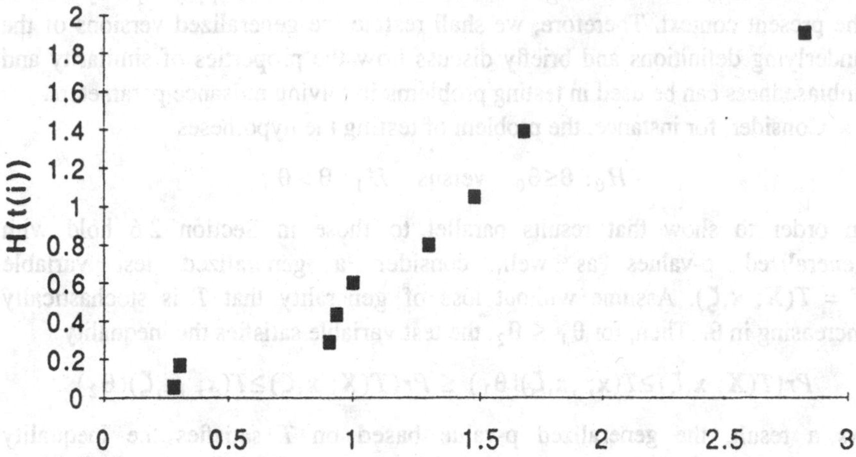

Figure 5.4. Plot for Brand B light bulbs.

The observed values of sufficient statistics for the parameters are the sums of the two sets of the data in Table 5.1. They are $u = 24.418$ and $v = 14.034$ in the notation of this section. Further, in this example $M = 15$, $N = 10$, $m = 13$, and $n = 9$. The appropriate p-value for testing the above hypotheses is

$$p = E\left\{ F_{\chi(26)}\left[u(1/2 + v/W)^{-1} \right] \right\} = .027 , \qquad (5.21)$$

where $F_{\chi(26)}()$ is the cdf of the chi-squared distribution with 26 degrees of freedom and the expectation is taken with respect to the random variable W, which has a chi-squared distribution with 18 degrees of freedom. This p-value strongly discredits the manufacturer's claim that the mean lifetime of brand A light bulbs exceeds that of brand B light bulbs at least by one year.

5.7. Unbiasedness and Similarity

In Section 2.6 we studied the notions of unbiasedness and similarity in the context of significance testing. The notions are useful and apply equally well in the present context. Therefore, we shall restate the generalized versions of the underlying definitions and briefly discuss how the properties of similarity and unbiasedness can be used in testing problems involving nuisance parameters.

Consider, for instance, the problem of testing the hypotheses

$$H_0: \theta \le \theta_0 \quad \text{versus} \quad H_1: \theta > \theta_0 \,.$$

In order to show that results parallel to those in Section 2.6 hold with generalized p-values as well, consider a generalized test variable $T = T(\mathbf{X};\, \mathbf{x}, \zeta)$. Assume without loss of generality that T is stochastically increasing in θ. Then, for $\theta_1 \le \theta_2$, the test variable satisfies the inequality

$$Pr(T(\mathbf{X};\, \mathbf{x},\zeta) \ge T(\mathbf{x};\, ,\mathbf{x},\zeta) \mid \theta_1) \le Pr(T(\mathbf{X};\, \mathbf{x},\zeta) \ge T(\mathbf{x};\, ,\mathbf{x},\zeta) \mid \theta_2) \,.$$

As a result, the generalized p-value based on T satisfies the inequality $p \le \pi(\mathbf{x};\, \theta)$ for all $\theta > \theta_0$, where $\pi(\mathbf{x};\, \theta) = Pr(T(\mathbf{X};\, \mathbf{x},\zeta) \ge T(\mathbf{x};\, \mathbf{x},\zeta) \mid \theta)$ is the power function and $p = \pi(\mathbf{x};\theta_0)$ is the generalized p-value. This property is a consequence of Property 3 (Property 4 in the case of point null hypotheses) of Definition 5.1 and is related to the property of p-unbiasedness explicitly defined below. On the other hand Properties 1 and 2 of a generalized test variable ensure that the generalized p-value is independent of nuisance parameters. We call this property p-similarity, or just similarity.

It is now clear that, just as is with classical test variables, we have readily incorporated the notions of unbiasedness and similarity in our definitions of generalized test variables. In testing situations where the hypotheses involve more than one parameter of interest and in a few applications with a single parameter, the properties that we imposed on generalized test variables are somewhat too restrictive or difficult to verify. If that is the case we can directly impose the properties of unbiasedness and similarity rather than Properties 1 to 4 of a generalized extreme region. It is the property of unbiasedness which characterizes the generalized p-value as a suitable quantity for measuring the statistical evidence in favor of or against the null hypothesis. The property of similarity merely requires that this measure be computable without any knowledge of nuisance parameters.

Consider the problem of testing the hypotheses

$$H_0: \theta \in \Theta_0 \quad \text{versus} \quad H_1: \theta \in \Theta_1 \tag{5.22}$$

on the basis of the generalized test variable $T(\mathbf{X};\, \mathbf{x},\zeta)$. Let $\boldsymbol{\theta}_0 = \overline{\Theta}_0 \cap \overline{\Theta}_1$ be the boundary (possibly vector valued) of the hypotheses. Let $C_{\mathbf{x}}(\zeta)$ be the generalized extreme region given by T and let

$$\pi(\mathbf{x};\ \boldsymbol{\theta}) = Pr(\ \mathbf{X} \in C_{\mathbf{x}}(\zeta) \mid \boldsymbol{\theta}\ ) \tag{5.23}$$

be its power function.

Definition 5.6. A test based on a generalized extreme region $C_{\mathbf{x}}(\zeta)$ is said to be *p-unbiased* if

$$\pi(\mathbf{x};\ \boldsymbol{\theta}) \geq p(\mathbf{x}) \quad \text{for } all\ \ \boldsymbol{\theta} \in \Theta_1\ , \tag{5.24}$$

where

$$p(\mathbf{x}) = \pi(\mathbf{x};\ \boldsymbol{\theta}_0)$$

is defined as the generalized p-value of the extreme region.

Definition 5.7. A test based on a generalized extreme region $C_{\mathbf{x}}(\zeta)$ is said to be *p-similar* (on the boundary) if, given any $\mathbf{x} \in \Xi$,

$$\pi(\mathbf{x};\ \boldsymbol{\theta}_0) = p \tag{5.25}$$

does not depend on the nuisance parameters $\boldsymbol{\delta}$, where $p = p(\mathbf{x})$ is the generalized p-value of the extreme region.

Even if a random variable of the form $T(\mathbf{X};\ \mathbf{x},\zeta)$ does not quite have Properties 1 through 3, from now on we shall call it a generalized test variable if extreme regions based on T as given by Definition 5.2 are p-similar and p-unbiased. It can be easily seen that an unbiased test based on a generalized power function continuous in $\boldsymbol{\theta}$ is similar (on the boundary). It is easier to derive tests by imposing the latter rather than the former. After a potential test variable is derived so that tests based on that variable are p-similar, it is usually a minor task to define a generalized extreme region so that the resulting p-value is p-unbiased as well. This method of finding test variables will be called the *method of similarity*. Although this process somewhat simplifies the problem of deriving unbiased tests, it is not as easy as the the method of invariance. While the method of invariance and the method of similarity produce equivalent test variables in some applications, the two methods compliment each other in other applications. In many applications the problem of finding a generalized extreme region leading to an optimum testing procedure with desirable properties can be solved in the following steps:

1. Find a set of minimal sufficient statistics for the unknown parameters of the underlying distribution.

2. Find a suitable group of transformations under which the distribution and the testing problem are both invariant.

3. Reduce the number of statistics by the method of invariance so that resulting p-values will be invariant under the transformation.

4. By the method of similarity, find a generalized test variable based on the

statistics found in Step 3.

5. Obtain an unbiased extreme region based on the test variable found in Step 4, by exploiting the properties of its distribution (i.e. its stochastic monotonicity etc.)

In some applications , p-similar tests can be obtained using Theorem 2.3. In the context of generalized p-values the following theorem may be more useful in deriving similar tests.

Theorem 5.2. Let $\mathbf{T} = \mathbf{T}(\mathbf{X};\ \mathbf{x},\zeta)$ be a continuous (in all arguments of T) random vector and $S = S(\mathbf{X};\ \mathbf{x},\zeta)$ be a continuous random variables based on an observable random vector **X**. If

(i) the observed value of S depends on at least one nuisance parameter δ;

(ii) the distribution of S is free of unknown parameters and **x**;

(iii) there exists δ_0 such that $S(\mathbf{x};\mathbf{x},\ \zeta)$ does not depend on **x** when $\delta = \delta_0$ (or as $\delta \to \delta_0$); and

(iv) the observed value and the distribution of **T** is free of δ (at least when θ is specified),

then the power function (the p-value when θ is specified) of any measurable generalized extreme region similar in δ (when θ has been specified) based on $\mathbf{R} = (\mathbf{T},S)$ can be obtained using **T** and its observed value alone.

Proof. Let $C_r(\zeta)$ be any measurable similar region and let

$$\pi(\mathbf{t}(\mathbf{x}),s(\mathbf{x};\delta),\delta|\theta) = Pr(\ \mathbf{R}\in C_r(\zeta)\ |\ \theta\ )$$

be its power function. Since the extreme region is similar with respect to δ we have

$$\pi(\mathbf{t}(\mathbf{x}),s(\mathbf{x};\delta),\delta\ |\ \theta) = \pi(\mathbf{t}(\mathbf{x}),s(\mathbf{x};\delta_0),\delta_0\ |\ \theta)\ .$$

But $s(\mathbf{x};\delta_0)$ is independent of **x**. Therefore, the power function of $C_r(\zeta)$ can be obtained using the random variables $(\mathbf{T},S)$ and the observed value, **t**, of **T**. Moreover, the distribution function of S does not depend on **x**. Since **T** is continuous, it now follows (as in Theorem 5.1) from the probability integral transform that the power function can be obtained using the random variable **T** and its observed value, **t**, alone.

The procedure of finding invariant and unbiased tests is illustrated by the application in the next section. More applications can be found in Chapters 6 through 10 in which the test variables are derived by invoking the notion of similarity as well as the principle of invariance

5.8. Comparing the Means of an Exponential Distribution and a Normal Distribution

Let $X_1, \ldots, X_m$ be a random sample from an exponential population with mean μ_x and let $Y_1, \ldots, Y_n$ be a random sample from a normal distribution with mean μ_y and variance σ^2. Suppose it is of interest to compare the means of the two populations. This is an example of a testing problem which cannot be solved by conventional methods even when the testing problem is invariant. The procedure outlined in Section 5.7 can solve many hypotheses concerning μ_x and μ_y. To illustrate the procedure, consider the hypotheses

$$H_0\colon \mu_x - \mu_y \le \delta_0 \,, \quad H_1\colon \mu_x - \mu_y > \delta_0 \,, \tag{5.26}$$

where $\delta_0 \ge 0$. This problem has two nuisance parameters and therefore we shall invoke both the invariance property and the similarity property to derive a generalized test variable.

It is easily seen that the statistics $\sum X_i$, $\bar{Y}$, and S^2 are sufficient for the parameters of the two distributions, where $S^2 = \sum (Y_j - \bar{Y})^2/(n-1)$. By sufficiency we can confine our search of a test variable to the class of variables which are functions of these three random quantities. Their distributions are

$$\sum X_i \sim G(m,\mu_x) \;,\; \bar{Y} \sim N(\mu_y, \frac{\sigma^2}{n}) \;,\; \text{and}\; S^2 \sim G(\frac{n-1}{2}, \frac{2\sigma^2}{n-1}) \,,$$

which are all mutually independent. Observe that the family of joint distributions of these random variables parameterized by (μ_x, μ_y, σ^2) is invariant under the group of common scale changes

$$(\sum X_i, \bar{Y}, S^2) \to (k\sum X_i, k\bar{Y}, k^2 S^2) \,,$$

where k is a positive constant. The induced transformation on the parameters is

$$(\mu_x, \mu_y, \sigma^2) \to (k\mu_x, k\mu_y, k^2\sigma^2) \,.$$

However, the testing problem is not invariant in its current form. Nevertheless, the equivalent problem of testing

$$H_0\colon \bar{\mu}_x - \bar{\mu}_y \le \theta_0 \quad \text{versus} \quad H_1\colon \bar{\mu}_x - \bar{\mu}_y > \theta_0 \tag{5.27}$$

is scale invariant, where $\bar{\mu}_x = \mu_x/s$, $\bar{\mu}_y = \mu_y/s$, and $\theta_0 = \delta_0/s$ are all scale invariant. Consider the random quantities

$$T_1 = \sum \frac{X_i}{s} \sim G(m, \bar{\mu}_x) \;\text{ and }\; T_2 = \sqrt{n}\frac{(\bar{Y} - s\bar{\mu}_y)}{S} \sim t_{n-1} \,.$$

Noting that the observed values of these two random variables depend on the data only through the maximal invariants $\sum x_i/s$ and $\bar{y}/s$, we can conclude from Theorem 5.1 that any scale invariant test can be constructed using T_1 and T_2.

To this end, consider the potential generalized test variable based on T_1 and T_2 (and their observed values and parameters) defined as

$$T = \sqrt{n}\left[\frac{\bar{\mu}_x \sum x_i}{\sum X_i} + \frac{(\bar{Y} - s\bar{\mu}_y)}{S} - \theta\right] \quad (5.28)$$

$$= \sqrt{n}\,\bar{\mu}_x t_1/T_1 + T_2 - \sqrt{n}\,\theta\,,$$

where $\theta = \bar{\mu}_x - \bar{\mu}_y$ is the equivalent parameter of interest. Notice that the observed value of T, namely $t_{obs} = \sqrt{n}\,(\bar{y}/s)$, and the distribution T do not depend on any nuisance parameters. Obviously, any test based on T_1 and T_2 can be constructed using T and T_2 as well. Since (a) the observed value of T_2 depends on the nuisance parameter $\bar{\mu}_y$; (b) the distribution of T_2 is free of parameters and the data; and (c) the observed value of T_2 does not depend on the data when $\bar{\mu}_y = \sqrt{n}\,\bar{y}/s$, by Theorem 5.2, we can conclude that all p-similar tests based on T and T_2 can be obtained using T alone.

In order to find a p-unbiased test for (5.26), observe that the cdf of T,

$$F_T(t) = Pr(\,T \le t\,) \quad (5.29)$$

$$= Pr\left[\frac{2\sum x_i}{sW} + \frac{T_2}{\sqrt{n}} \le \frac{t}{\sqrt{n}} + \theta\right],$$

is an increasing function of θ, where

$$W = \frac{2\sum X_i}{\mu_x} \sim \chi^2_{2m} \quad \text{and} \quad T_2 \sim t_{n-1}\,, \quad (5.30)$$

are independent random variables. Therefore, the left tail of the distribution of T yields unbiased extreme regions. Hence, the generalized p-value for testing (5.26) as well as (5.27) is

$$p(\textstyle\sum x_i, \bar{y}, s) = Pr\left[\frac{2\sum x_i}{sW} + \frac{T_2}{\sqrt{n}} \le \frac{t_{obs}}{\sqrt{n}} + \theta_0\right],$$

$$= Pr(\,T_2 \le \sqrt{n}\,(\frac{\bar{y}}{s} - \frac{2\sum x_i}{sW} + \theta_0)\,)$$

$$= E\left[F_{t(n-1)}\left[\frac{\sqrt{n}}{s}(\delta_0 + \bar{y} - 2\textstyle\sum x_i/W)\right]\right], \quad (5.31)$$

where the expectation in (5.31) is taken with respect to the chi-squared random variable W and $F_{t(n-1)}$ is the cdf of Student's t distribution with $n-1$ degrees of freedom.

It is of interest to note that when $\delta_0 = 0$, the testing problem defined by (5.26) is scale invariant in its original form. Recall that in the case of comparing two exponential means, classical methods provided exact solutions when the testing problem is scale-invariant. In contrast, in the present situation classical methods do not provide an exact solution to at least this special case.

Exercises

5.1. Let $X_1, \ldots, X_M$ and $Y_1, \ldots, Y_N$ be two sets of independent and identically distributed exponential random variables with means μ_x and μ_y, respectively. Suppose the random variables are censored as described in Section 5.6. Let $X_{(1)}, \ldots, X_{(m)}$ and $Y_{(1)}, \ldots, Y_{(n)}$ be the observed random variables before censoring. Show that

$$U = \sum_{i=1}^{m} X_{(i)} + (M-m)X_{(m)} \quad \text{and} \quad V = \sum_{j=1}^{n} Y_{(j)} + (N-n)Y_{(n)}$$

are sufficient statistics for (μ_x, μ_y). Are these sufficient statistics complete? Show that U and V are independent gamma random variables.

5.2. Let $X_1, \ldots, X_m$ and $Y_1, \ldots, Y_n$ be two independent sets of exponential random variables with means μ_x and μ_y, respectively. For given constants a, b, and c, establish the procedure based on generalized p-values for testing the hypotheses

$$H_0\colon a\mu_x + b\mu_y \geq c \quad \text{versus} \quad H_1\colon a\mu_x + b\mu_y < c\,.$$

Also find the posterior probability of H_0 under the improper prior $\pi(\mu_x, \mu_y) = 1$.

5.3. Let $X_1, \ldots, X_n$ be a random sample from the uniform distribution with density

$$f_X(x) = \frac{1}{\beta - \alpha} \quad \text{for } \alpha \leq x \leq \beta\,.$$

Let $X_{(1)}, \ldots, X_{(n)}$ be the order statistic. Suppose that $\theta = \alpha/\beta$ is the parameter of interest. Consider the problem of testing the hypotheses

$$H_0\colon \theta \leq \theta_0 \quad \text{versus} \quad H_1\colon \theta > \theta_0\,,$$

where θ_0 is a prespecified value of the parameter.

(a) Show by the method of invariance that invariant tests can be obtained using the generalized test variable

$$T = \alpha^{-1}\,[(X_{(n)} - \alpha)x_{(1)} - (X_{(1)} - \alpha)x_{(n)}]\,.$$

(b) Derive the above test variable by the method of similarity and find the p-unbiased test.

(c) Find the posterior probability of H_0 under the improper prior $\pi(\alpha, \beta) = 1$.

5.4. Let $X_1, \ldots, X_m$ and $Y_1, \ldots, Y_n$ be two sets of independently and identically distributed random variables with means μ_x and μ_y respectively. Suppose that each random variable in the former set is exponentially distributed and those of the latter set are normally distributed with variance σ^2. For given constants a, b, and c, establish by the method of invariance the procedure based on generalized p-values for testing the hypotheses

$$H_0\colon a\mu_x + b\mu_y \geq c \quad \text{versus} \quad H_1\colon a\mu_x + b\mu_y < c \,.$$

Also find the posterior probability of H_0 under the improper prior $\pi(\mu_x, \mu_y, \sigma) = \sigma^{-1}$.

5.5. Let $X_1, \ldots, X_n$ be a random sample from a normal population with mean μ and variance σ^2. Let $\bar{X}$ and S^2 be the sample mean and the sample variance. Show that extreme regions similar in σ can be obtained using the random quantities

$$T_1 = \frac{\bar{X} - \mu}{S} + \frac{\mu}{s} \quad \text{and} \quad T_2 = \frac{S^2}{\sigma^2} \,.$$

Hence deduce by applying Theorem 5.2 that p-similar extreme regions for testing hypotheses about μ can be constructed using T_1 alone. Find the p-value for testing

$$H_0\colon \mu \leq 0 \quad \text{versus} \quad H_1\colon \mu > 0 \,.$$

5.6. Consider again the hypothesis testing problem in Exercise 5.4. By a method similar to that in Exercise 5.5, find the p-value for testing the hypotheses by the method of p-unbiasedness.

5.7. Let $X_1, \ldots, X_n$ be a random sample from a normal population with mean μ and variance σ^2. Let $p_k = Pr(X \leq k)$, an important parameter in reliability testing. Consider the problem of testing

$$H_0\colon p_k \geq .99 \quad \text{against} \quad H_1\colon p_k < .99 \,.$$

(a) Derive a generalized test variable for testing the hypotheses and find the generalized p-value (Hint: Express the hypotheses in terms of the parameter $\theta = (k - \mu)/\sigma$).

(b) Find the Bayes test under the noninformative prior.

5.8. let $Y_1, \ldots, Y_n$ be a random sample from a distribution with density function

$$f(y) = \frac{\lambda\mu}{\lambda - \mu} \frac{1}{y^2} \quad \text{for} \quad \mu \leq y \leq \lambda \,,$$

where λ and μ are positive parameters.

(a) Find the complete sufficient statistics for (λ, μ).

(b) Show that $X = 1/Y$ has a uniform distribution.

(c) Hence, by the method of invariance derive a generalized test variable for

testing hypotheses about the range of the distribution of Y, namely $\theta = \lambda - \mu$.
(d) Derive the test variable by the method of unbiasedness.

5.9. Let $X_1, \ldots, X_m$ and $Y_1, \ldots, Y_n$ be two independent sets of independent exponential random variables with parameters α and β. Using Theorem 5.2, derive a test statistic for testing hypotheses about the ratio of the two parameters. Show that the same test statistic can be employed for testing hypotheses concerning the probability, $Pr(X < Y)$.

5.10. Let $X_1, \ldots, X_n$ be a random sample from the two-parameter exponential distribution with parameters α and β. Let μ be the mean of the distribution. Consider the problem of testing

$$H_0: \mu = \mu_0 \quad \text{against} \quad H_1: \mu \neq \mu_0 .$$

(a) Use the method of invariance to obtain the p-value for testing the hypotheses.

(b) Derive the same test by the method of similarity.

(c) Find the form of the noninformative prior.

(d) Find the posterior probability of H_0 and the posterior odds ratio.

5.11. Let $X_1, \ldots, X_n$ be a random sample from a normal population with mean μ and variance σ^2. Let $\bar{X}$ and S^2 be the sample mean and the sample variance. Establish procedures for testing hypotheses about the parameter $\theta = \mu + k\sigma^2$.

5.12. Consider again the lifetime data in Table 5.1. Suppose the data that were censored at the experiment later became available. Suppose the actual lifetimes of two censored data on Brand A are 4.162 and 3.894 and that of Brand B is 4.261.

(a) Find the MVUE of the difference in mean lifetimes of the two brands of light bulbs.

(b) Retest the manufacturer's claim with the new information.

(c) Test the hypothesis that Brand A light bulbs last, on average, 14 months longer than Brand B light bulbs.

5.13. In a clinical trial for studying the ability to maintain remission in acute leukemia patients, two groups of patients were given either a placebo or a drug being tested. The experiment was terminated after observing data from 22 patients. The table below shows, for patients in each group, the lengths of remission in days; the starred data corresponds to the censored observations.

Observed Remission Periods

(a) Placebo

37	10	86	141	7	100
52	117	3	10	76	16

(b) Drug

324*	324	16	9	305	134
165	324*	8	14	8	20

Examine whether it is reasonable to assume that each data set follows an exponential distribution. Test the hypothesis that the drug prolongs the remission time by at least 8 weeks, on average.

Chapter 6
Generalized Confidence Intervals

6.1. Introduction

The purpose of this chapter is to develop a counterpart of generalized p-values in interval estimation. Even those practitioners who insist on conventional confidence intervals will find the generalization useful to obtain excellent approximate interval estimates for problems such the interval estimation in mixed models. According to simulation studies, such approximations have outperformed more complicated approximations reported in the literature. The conventional definition of a confidence interval will be generalized so that problems such as that of constructing exact interval estimates for the difference in two exponential means can be solved. As in the case of hypothesis testing, exact confidence intervals (classical) for statistical problems involving nuisance parameters are available only in special cases. For example, exact confidence intervals for the second moment of a normal distribution are not available. The Behrens-Fisher problem is probably the most well known problem in which exact confidence intervals based on minimal sufficient statistics do not exist. In Chapter 7 this problem will be discussed in detail and the problem will be solved by the generalized approach.

The current inference problem is more serious than that of hypothesis testing. Even when the problem does not involve nuisance parameters and there is no difficulty in writing down exact probability statements yielding confidence intervals, in some applications conventional confidence intervals lead to results which contradict the very meaning of 'confidence'. Probably Pratt (1961) provided the first example of a uniformly most accurate confidence interval having highly undesirable properties. In Section 6.8 we shall study Pratt's example in detail and then resolve the paradox by resorting to the generalization of interval estimation developed in this chapter. Similar problems also arise in linear models based on the normal theory. As we will see in Chapter 9, this is especially true in the case of variance components in random effects models where, with some data sets, conventional methods yield negative confidence bounds for variance components. These are also consequences of the inherent limitations of the conventional definition of confidence intervals.

Some frequently quoted advantages of the Bayesian treatment are the absence of existence problems in interval estimation, and the fact that interval estimates obtained by the Bayesian approach do not violate the meaning of confidence. But, one does not necessarily have to be a Bayesian, treat a constant parameter as a random variable, or have a prior distribution on the parameter to resolve such drawbacks of confidence intervals. The only thing exact about an interval estimate such as the one discussed by Pratt (1961) is the exact probability statement on which the interval is based. The purpose of this chapter is to enhance the class of available procedures by insisting on exact probability statements rather than on repeated sampling properties. Note that the confidence intervals given by nonparametric methods, such as those presented in Section 3.2, are also exact in this sense. This will enable us to solve many problems that cannot be solved by conventional methods and also to obtain more desirable properties of resulting intervals. As in the Bayesian treatment, the idea is to do the best with the observed data rather than on all possible samples that could have been observed but were not. In many applications, the idea of repeating the same experiment is self contradictory (as the interval is no longer valid when additional samples are taken) and is of little practical use. Nevertheless, in a few of applications (e.g. statistical quality control in which the parameter of interest may vary around the intended value of the parameter), the problem of repeating related experiments arises. Therefore, in Section 6.3, the issues related to repeated sampling properties are addressed briefly.

In Section 6.2, we first enhance the class of interval estimates by considering procedures based on exact probability statements with or without repeated sampling properties and then later we examine their implications in repeated experiments. The enhanced class of interval estimates will be referred to as generalized confidence intervals as defined precisely in the next section.

6.2. Generalized Definitions

To describe the underlying problem in more detail, consider a population with cumulative distribution function $F(\mathbf{x} \mid \zeta)$, where $\mathbf{X}$ is a random vector, $\zeta = (\theta, \delta)$ is a vector of unknown parameters, θ is a parameter of interest, and δ is a vector of nuisance parameters. We are interested in finding interval estimates of θ based on observed values of $\mathbf{X}$. The problem is to construct generalized confidence intervals of the form $[A(\mathbf{x}), B(\mathbf{x})] \subset \Theta$, where $A(\mathbf{x})$ and $B(\mathbf{x})$ are functions of $\mathbf{x}$, the observed data.

The unavailability of exact confidence intervals in many situations can be attributed to the conventional approach to constructing interval estimates. In order to understand the restrictive nature of conventional confidence intervals, let us first examine the conventional definition. To fix ideas, consider a random sample $\mathbf{X} = (X_1, \ldots, X_n)$ taken from a distribution depending on a parameter θ. Let $A(\mathbf{X})$ and $B(\mathbf{X})$ be two statistics satisfying the equation

$$Pr[\ A(\mathbf{X}) \leq \theta \leq B(\mathbf{X})\ ] = \gamma\,, \tag{6.1}$$

where γ is a prespecified constant between 0 and 1. Let $a = A(\mathbf{x})$ and $b = B(\mathbf{x})$ be the observed values of the two statistics. In the terminology used in Section 1.8, $[a,b]$ is called a *confidence interval* for θ with the *confidence coefficient* γ. For example, if $\gamma = .95$, the interval $[a,b]$ obtained in this manner is a 95% confidence interval. Various repeated sampling properties of interval estimates obtained in this manner are mere implications of the probability statement (6.1) rather than something deliberately required in the construction of the interval.

Although this approach to constructing interval estimates is conceptually simple and easy to implement, its drawback is that in most applications involving nuisance parameters it is not easy or impossible to find $A(\mathbf{X})$ and $B(\mathbf{X})$ so as to satisfy (6.1) for all possible values of the nuisance parameters. The idea in generalized confidence intervals is to make this possible by making probability statements relative to the observed sample, as done in Bayesian and nonparametric methods. In other words, we allow the functions $A(\,)$ and $B(\,)$ to depend not only the observable random vector $\mathbf{X}$ but also on the observed data $\mathbf{x}_{obs}$.

In Section 1.8 we found confidence intervals of parameters of continuous distribution using pivotal quantities. When there are no nuisance parameters the method is fairly general. Unfortunately, the situation is quite different when nuisance parameters are present. Even with distributions with just two parameters one cannot usually find pivotals except for very special functions of the parameters. For example, in the case of normal distribution with mean μ and variance σ^2, while there is no difficulty finding a pivotal quantity for any linear combination of μ and σ, there is no pivotal quantity available for functions such as $\mu^2 + \sigma^2$. In Chapter 5 we saw how nonstandard functions of parameters can arise naturally in practical applications. Now we shall extend the conventional definition of pivotal quantities so that we can construct interval estimates in this type of application as well. The extension is the counterpart of generalized test variables defined in Chapter 5.

Definition 6.1. Let $R = r(\mathbf{X};\mathbf{x},\zeta)$ be a function of **X** and possibly **x**, ζ as well. The random quantity R is said to be a *generalized pivotal quantity* if it has the following two properties:

Property A: R has a probability distribution that is free of unknown parameters.

Property B: r_{obs} defined as $r_{obs} = r(\mathbf{x};\mathbf{x},\zeta)$ (this will be referred to as the *observed pivotal*) does not depend on nuisance parameters, δ.

As in the case of conventional pivotal quantities, Property A is imposed to ensure that a subset of the sample space ρ of possible values of R can be found at any desirable confidence coefficient γ with no knowledge of parameters. This property is related to the notion of *similarity* that we will study in Section 6.6. Property B is imposed to ensure that probability statements based on a generalized pivotal quantity will produce confidence regions involving observed data **x** only.

As in the Bayesian treatment, we can now attempt to obtain interval estimates on θ relative to the observed sample with no special regard to samples that could have been observed but were not. Unlike the Bayesian treatment, however, we would like to write the probability statement based on the random vector **X**. Given a generalized pivotal $R = r(\mathbf{X};\mathbf{x},\zeta)$ and a desired confidence coefficient γ, consider a subset C_γ of the sample space ρ of R which satisfies

$$Pr(R \in C_\gamma) = \gamma . \tag{6.2}$$

This method of specifying a subset of the sample space will be referred to as the *generalized pivotal quantity method.* Compared to the pivotal quantity method discussed in Section 1.8, this is a more general way of obtaining a subset $C(\mathbf{x};\theta)$ of the sample space such that

$$Pr(\mathbf{X} \in C(\mathbf{x};\theta)) = \gamma . \tag{6.3}$$

Unlike the subsets obtained in Section 1.8, these depend not only on γ and θ, but also on the observed sample point **x**. If R is a continuous random variable we may be able to find C_γ directly from its distribution or by applying the probability integral transform (see Section 1.8). Recalling that the underlying probability statement is the only exact statement of a conventional confidence interval as well, we can now construct an interval for θ as before by replacing **X** appearing in the probability statement by **x**:

Definition 6.2. If the subset C_γ of the sample space ρ of R satisfies equation (6.2), then the subset Θ_c of the parameter space given by

$$\Theta_c(r) = \{\theta \in \Theta \mid r(\mathbf{x};\mathbf{x},\zeta) \in C_\gamma\} \tag{6.4}$$

is said to be a $100\gamma\%$ *generalized confidence interval* for θ.

As in the case with conventional confidence intervals, when there exists a generalized confidence interval with $100\gamma\%$ confidence, there is usually a class of $100\gamma\%$ generalized confidence intervals. Depending on the application, a particular one-sided interval, the shortest confidence interval, or some other interval might be preferable.

Although this generalization is not based on repeated sampling considerations, in Section 6.3 we will examine its implications in repeated experiments. In the following chapters we do not always make the distinction between a generalized confidence interval and a confidence interval, because in all applications we undertake in those chapters, the former is merely an extension of the latter and a generalized confidence interval reduces to a classical confidence interval in the special cases in which the latter exists. Moreover in those applications, the generalized confidence intervals are also numerically equivalent to Bayesian confidence intervals under certain noninformative priors.

Example 6.1. The coefficient of variation of a normal distribution

Let $X_1, \ldots, X_n$ be a random sample from a normal population with mean μ and variance σ^2. Suppose that the coefficient of variation $\eta = \sigma/\mu$ is the parameter of interest. In the context of interval estimation we can equivalently consider the problem of constructing confidence intervals for the parameter

$$\theta = \mu/\sigma = \eta^{-1},$$

the mean of the distribution expressed in σ units. Let us find a generalized pivotal quantity, R, based on the sufficient statistics

$$\bar{X} \quad \text{and} \quad S^2 = \frac{\sum_{i=1}^{n} (X_i - \bar{X})^2}{n},$$

namely, the sample mean and the sample variance, respectively. Since a generalized pivotal can be a function of all unknown parameters, we can base the construction of R on the random quantities

$$Z = \frac{\bar{X} - \mu}{\sigma/\sqrt{n}} \sim N(0, 1), \text{ and } U = \frac{nS^2}{\sigma^2} \sim \chi^2_{n-1}, \tag{6.5}$$

whose distributions are free of unknown parameters. Consider the identity

$$\begin{aligned}\theta &\equiv \frac{\bar{X}}{\sigma} - \frac{Z}{\sqrt{n}} \\ &\equiv \frac{\bar{X}}{S}\left(\frac{U}{n}\right)^{1/2} - \frac{Z}{\sqrt{n}}.\end{aligned}$$

In view of this identity we can define a potential generalized pivotal

$R = r(\mathbf{X};\mathbf{x},\mu,\sigma)$ as

$$R = \frac{\bar{x}}{s}\frac{S}{\sigma} - \frac{\bar{X}-\mu}{\sigma} = \frac{\bar{x}}{s}\left(\frac{U}{n}\right)^{1/2} - \frac{Z}{\sqrt{n}}, \tag{6.6}$$

where $\bar{x}$ and s^2 are the observed values of the sample mean and the sample variance, respectively. Clearly the observed value of R is θ, thus satisfying Property B of Definition 6.1. Moreover, the distributions of Z and U given by (6.5) imply that the distribution of R does not depend on unknown parameters, as required by Definition 6.1. Therefore, R is indeed a generalized pivotal quantity. In the following sections we shall study various conditions under which this pivotal quantity is unique (up to equivalent intervals).

Now consider the problem of constructing lower confidence bounds for θ. Since $r_{obs} = \theta$, the probability statement

$$\begin{aligned}\gamma &= Pr(\ R \geq c\ ) \\ &= Pr(\ \frac{\bar{x}}{s}\sqrt{U} \geq Z + c\sqrt{n}\ )\end{aligned} \tag{6.7}$$

will lead to a left-sided $100\gamma\%$ confidence interval. But

$$T = \frac{Z + c\sqrt{n}}{\sqrt{U/(n-1)}} \sim t_{n-1}(c\sqrt{n})\ ;$$

that is, the random variable T has a noncentral t-distribution with $n-1$ degrees of freedom and the noncentrality parameter $c\sqrt{n}$. Therefore, (6.7) can be expressed as

$$G_{n-1,c\sqrt{n}}(\sqrt{n-1}\,\frac{\bar{x}}{s}) = \gamma\,, \tag{6.8}$$

where $G_{n-1,c\sqrt{n}}(.)$ is the cumulative distribution function of the noncentral t-distribution with $n-1$ degrees of freedom and the noncentrality parameter $c\sqrt{n}$. It is now evident that

$$\theta \geq c_\gamma(\frac{\bar{x}}{s}) \quad \text{and} \quad \eta \leq c_\gamma(\frac{\bar{x}}{s})^{-1}$$

are $100\gamma\%$ interval estimates for θ and η, respectively, where $c_\gamma(\bar{x}/s)$ is the value of c that satisfies Equation (6.8) for a specified value of γ. It is of interest to note that, in spite of the fact that these interval estimates are based on the generalized pivotal quantity R, they do satisfy conventional repeated sampling properties. This can be seen by noting that the random variable $V = G_{n-1,\eta\sqrt{n}}(\sqrt{n-1}\,\frac{\bar{X}}{S})$ has a uniform distribution over $[0,1]$. In fact, V is a conventional pivotal quantity.

6.3. Frequency Interpretations and Repeated Sampling Properties

As with Bayesian confidence intervals, generalized confidence intervals are not based on conventional repeated sampling considerations, but rather on exact probability statements. As a matter of fact, in most applications generalized confidence intervals are numerically equivalent to Bayesian confidence intervals under certain noninformative priors. Frequency properties of both procedures are not yet fully understood. Nevertheless, as with nonparametric confidence intervals, they do have some unconventional repeated sampling properties and also provide excellent approximations and bounds satisfying other properties. The class of conventional confidence intervals is of course a subset of all generalized confidence intervals and some members outside this class as well as the members within the class satisfy conventional repeated sampling properties. The definition of conventional confidence intervals given by (6.1) has various implications in repeated experiments. Although the following property is not of direct practical use it does have some practically useful implications.

Property 1: Consider a particular case involving interval estimation of a parameter θ. If the same experiment is repeated a large number of times to obtain new sets of observations, **x**, of the same sample size n, then the confidence intervals given by the probability statement (6.1) will correctly include the true value of the parameter θ 95% of the time.

This property is useful when one needs to study the performance of approximate confidence intervals by simulation. Except for this theoretical implication, however, it cannot be entertained as a practically useful repeated sampling property, because if one can indeed obtain repeated samples from the same experiment, the claimed confidence level will no longer be valid. Moreover, if the sampling is repeated a large number of times, then the parameter will become known exactly so that statistical inference on the parameter is no longer an issue. Nevertheless, Property 1 has other useful implications. In particular, the following property is practically useful and does not contradict the confidence level of individual intervals:

Property 2: After a large number of situations of setting 95% confidence intervals for certain parameters of independent statistical problems, an experimenter will have correctly included the true values of the parameters in the corresponding intervals 95% of the time.

It is desirable to have at least this repeated sampling property when available procedures which can guarantee the property do not jeopardize more desirable features. Insisting on the repeated sampling property with the same experiment may leave a practitioner with only a limited class of procedures with undesirable features (see, Pratt (1961), Kiefer (1977) and the discussions of that paper). Interval estimation in random effects models and mixed models provides a class of applications suffering from such undesirable features.

There have been several attempts to study various repeated sampling properties of generalized procedures. The simulation studies carried out by Griffiths and Judge (1992), Thursby (1992), Weerahandi and Johnson (1992), and Zhou and Mathew (1994), and the theoretical results in Weerahandi (1993), and Meng (1994) have implications in interval estimation as well as in hypothesis testing. It follows from the findings of these studies that, in linear models based on the normal theory, the generalized confidence intervals approximately satisfy even Property 1, or rather guarantees the intended confidence level, and often beat other approximations available in the literature, as far as the accuracy of the approximation and the length of intervals are concerned. Examples 6.3 and 6.6, and findings in Ananda and Weerahandi (1996) provide some evidence of similar performance of generalized confidence intervals outside the normal theory as well. Therefore, practitioners who prefer to make inferences without deviating from the classical philosophy of statistical inference can utilize generalized confidence intervals to obtain good approximate or level-guaranteed confidence intervals (classical).

Generalized confidence intervals have some unconventional frequency interpretations implied by the definition. To describe these, note first of all that the two $\mathbf{x}$ terms appearing in (6.4) play different roles. Let $\mathbf{x}_{obs}$ be the actual observed value of $\mathbf{x}$. Let us now imagine that the experiment is repeated with new samples, $\mathbf{x}$. Since resampling from the same experiment, in reality, would change the confidence statement, the purpose of this hypothetical experiment is only to understand its frequency properties. With each new sample the interval for θ given by (6.4) is recomputed relative to the real observed sample; that is, the interval is recomputed using $r(\mathbf{x}; \mathbf{x}_{obs}, \zeta)$. Then, it follows from the law of large numbers applied to Bernoulli trials defined using the the random variable $R = r(\mathbf{X}; \mathbf{x}_{obs}, \zeta)$, that $100\gamma\%$ of the intervals will contain θ, in the long run. Although this frequency property is theoretically valid, of course it is not practically useful or feasible.

Do the generalized confidence intervals applied to continuous families of distributions satisfy at least Property 2? The above mentioned simulation studies imply that with typical values of nuisance parameters, the true coverage of generalized confidence intervals in repeating the same experiment is at least as large as the intended coverage. Furthermore, if the experiments are independent in every respect including the directions of one-sided confidence intervals, the overall coverage seems to be at least as large as the intended level. However, a set of sufficient conditions for this property to be valid or a formal proof is not currently available. Motivated by simulation results and the above theoretical property, Weerahandi (1993) attempted to provide a proof of the desired result, but due to incomplete sufficient conditions, the assertion was later replaced by a weaker result.

6.4. Invariance in Interval Estimation

As in the case of generalized p-values, the derivation of generalized pivotals can be facilitated by invoking the notions of invariance and similarity. Further, the uniqueness of an interval estimate based on a readily available pivotal quantity such as the one used in Example 6.1 can also be established by imposing these desirable properties. In achieving these goals one can first reduce the underlying statistical problem by exploiting the concept of *sufficiency*. When the set of minimal sufficient statistics consists of more than one statistic, perhaps the notion of *invariance* and *similarity* (Property A) can be invoked to further reduce the number of statistics on which a pivotal quantity can be based. The purpose of this section is to study how a problem of interval estimation can be reduced by the method of invariance; the method of similarity will be studied in the following section.

Invariance in interval estimation is usually defined in the literature by means of an appropriate loss function, or by invoking elements of hypothesis testing (see, for instance, Ferguson (1967) and Lehmann (1986)). But in the context of generalized confidence intervals we can usually avoid elements of statistical decision theory and the relation between the problems of setting confidence intervals and hypothesis testing. This is accomplished by first transforming the problem of interval estimation to an equivalent one involving a parameter which is not affected by the underlying groups of transformations; the procedure will be illustrated by the application in Section 6.4.

Definition 6.3. Suppose that the family of distributions, $F(\mathbf{X};\,\theta)$ of $\mathbf{X}$, is invariant under the group, G, of transformations on the sample space Ξ with the induced transformation on the parameter $\theta \rightarrow \theta$ (i.e. θ is unaffected). A generalized confidence region $\Theta(\mathbf{x})$ is said to be invariant under G if $\Theta(g(\mathbf{x}))=\Theta(\mathbf{x})$ for all $\mathbf{x}\in\Xi$ and $g\in G$.

On one hand the principle of invariance provides a simple method of reducing the number of data dependent quantities on which a generalized pivotal can be based. On the other hand, as discussed in Section 2.5, when the underlying inference problem is invariant under a certain data transformation, it is highly desirable to insist on solutions with the invariance property. Otherwise two investigators working on the same statistical problem with the same data measured in different units may come to different conclusions. In the present context, Theorem 6.1 enables us to search for invariant confidence regions within the class of generalized pivotal quantities which depend on the data only through maximal invariants. This is the counterpart of Theorem 5.1 concerning generalized p-values. The proof of the theorem is similar to that of Theorem 5.1. The theorem is illustrated by Example 6.2; the application in Section 6.4 further demonstrates the usefulness of Theorem 6.1.

Theorem 6.1. Let $F(\mathbf{X};\,\theta)$ be a family of distributions which is invariant under the group G of transformations on the sample space Ξ with the induced transformation on the parameter $\theta \to \theta$. Suppose that $\mathbf{R} = r(\mathbf{X};\mathbf{x},\zeta)$ is an absolutely continuous random vector with the observed value $r_{obs} = r(\mathbf{x};\mathbf{x},\zeta)$. If the distribution of $\mathbf{R}$ and the observed value r_{obs} depend on $\mathbf{x}$ only through a set of maximal invariants $\mathbf{m}(\mathbf{x})$, then any invariant generalized confidence region of θ based on $\mathbf{X}$ can be constructed using $\mathbf{R}$ alone.

Example 6.2. The coefficient of variation of a normal distribution (continued)

Consider again the problem of constructing confidence intervals for the coefficient of variation $\eta = \sigma/\mu$, where μ and σ are respectively the mean and the standard deviation of the normal distribution. By sufficiency, the statistical problem can be based on the sample mean $\bar{X}$ and the sample variance S^2. To further reduce the statistical problem, observe from Equation (6.5) that the underlying family of distributions parameterized by (μ,σ) is invariant under the common scale transformations

$$(\bar{X},S) \to (k\bar{X},kS) \text{ and } (\mu,\sigma) \to (k\mu,k\sigma) ,$$

where k is a positive constant. Obviously, the parameter of interest is unaffected by any change of scale, and therefore, the statistical problem is invariant.

Now consider the generalized pivotal

$$R = \frac{\bar{x}}{s}\frac{S}{\sigma} - \frac{\bar{X}-\mu}{\sigma}$$

used in Example 6.1. It is evident from the representation of R given by (6.6) that the distribution of this random variable depends on the data only through $\frac{\bar{x}}{s}$, a maximal invariant. Furthermore, at the observed values of the statistics, R does not depend on the data. Therefore, it immediately follows from Theorem 6.1 that any scale invariant generalized confidence region of η can be constructed from R alone. In particular, the one-sided confidence interval for η found in Example 6.1 is unique within the class of invariant interval estimates based on $\bar{X}$ and S.

6.5. Interval Estimation of the Difference Between Two Exponential Means

In Chapter 5 we considered the problem of testing hypotheses about the difference in two exponential means. Let us now consider the problem of setting interval estimates for the same parameter. Inferences on the parameter are to be performed based on two samples taken from the two exponential populations. Although the exponential distribution is one of the simplest distributions, with just a single parameter, exact confidence intervals based on sufficient statistics are not available for this important problem which arises in many practical applications. Samaranayake and Bain (1988) provided an interval with a view towards meeting or exceeding the intended confidence coefficient. Since the exponential distribution is highly nonnormal, approximations based on asymptotic results are rather poor when the sizes of samples taken from the two populations are small. In this section we shall see that there is no difficulty in constructing generalized confidence regions based on exact probability statements.

Let $X_1, \ldots, X_m$ and $Y_1, \ldots, Y_n$ be two sets of independent and identically distributed exponential random variables with means μ_x and μ_y, respectively. Although censored data in the case of lifetime distributions can be handled as in Section 5.5, let us assume for convenience that the data are not censored. It follows from Example 1.1 that the statistics $U = \sum_{i=1}^{m} X_i$ and $V = \sum_{i=1}^{n} Y_i$ are sufficient for (μ_x, μ_y). Furthermore, these random variables are independently distributed as $U \sim G(m, \mu_x)$ and $V \sim G(n, \mu_y)$. Let u and v be the observed values of U and V, respectively. The problem is to construct interval estimates for $\theta = \mu_x - \mu_y$ based on these two quantities. Samaranayake and Bain (1988) considered this problem and gave an approximate solution.

Having reduced the problem by sufficiency let us now find a single quantity on which generalized confidence intervals can be based. Observe that the family of joint distributions of these random variables is invariant under the scale transformations

$$(U,V) \to (k\,U, k\,V) \quad \text{and} \quad (\mu_x, \mu_y) \to (k\,\mu_x, k\,\mu_y)\,,$$

where k is a positive constant. The induced transformation on the parameter of interest is $\theta \to k\theta$. Unlike the inference problem encountered in Example 6.2, this problem is not invariant. Therefore, first consider the problem of setting confidence intervals for the scale invariant parameter $\theta_u = \theta/u$. It should be emphasized that this kind of reparameterization defined by means of an observed value of a random variable has no effect on the resulting interval estimates of θ.

According to Theorem 6.1, any scale invariant interval estimator of θ_u is a function of a maximal invariant, say the ratio of the sufficient statistics $M(U,V) = V/U$. To facilitate the definition of a generalized pivotal quantity based on this maximal invariant, consider the chi-square random variables W_1 and W_2 given by

$$W_1 = \frac{2U}{\mu_x} \sim \chi^2_{2m} \quad \text{and} \quad W_2 = \frac{2V}{\mu_y} \sim \chi^2_{2n} \,. \tag{6.9}$$

Define a generalized pivotal quantity based on these random variables and the observed value of the maximal invariant M as,

$$\begin{aligned} R &= \frac{u}{v}\frac{\mu_x}{U} - \frac{\mu_y}{V} \\ &= \frac{u}{v}\frac{2}{W_1} - \frac{2}{W_2}. \end{aligned} \tag{6.10}$$

The observed value of R is $r_{obs} = \frac{u}{v}\theta_u$ and the distribution of R depends on the data only through u/v. Therefore, according to Theorem 6.1, R can generate all scale invariant confidence regions based on the sufficient statistics. Moreover, R is a generalized pivotal quantity, because the distribution of R does not involve any unknown parameters and its observed value is free of nuisance parameters.

Having found a single random quantity on which confidence regions can be based we can now proceed as in Example 6.1. To illustrate the procedure, let us find lower confidence bounds of θ. These can be obtained using probability statements of the form

$$\begin{aligned} Pr(R \geq c) &= Pr\left\{\frac{u}{v}\frac{2}{W_1} \geq c + \frac{2}{W_2}\right\} \\ &= Pr\left\{W_1 \leq \frac{u}{v}\frac{2W_2}{2 + cW_2}\right\}. \end{aligned}$$

If $c \geq 0$, then this probability statement implies that the right-sided $100\gamma\%$ generalized confidence interval for θ_u is $[c_\gamma(u/v) \,,\, \infty)$, where $c_\gamma = (v/u)\,c_\gamma(u/v)$ is a constant chosen such that

$$E\left[F_{W_1}\left(\frac{2W_2}{2v/u + c_\gamma W_2}\right)\right] = \gamma \,,$$

where the expectation is taken with respect to W_2 whose distribution is given by (6.9) and F_{W_1} is the cdf of chi-square distribution with $2m$ degrees of freedom. If $c < 0$, the appropriate constant can be found in a similar manner.

Hence, the right-sided $100\gamma\%$ generalized confidence interval of the parameter θ is $[\, a_\gamma(u,v) \,,\, \infty)$, where

$$E\left[F_{W_1}\left(\frac{2uW_2}{2v + a_\gamma W_2}\right)\right] = \gamma \quad \text{if } a_\gamma \geq 0 \text{ and} \tag{6.11}$$

$$E\left[F_{W_2}\left(\frac{2vW_1}{2u - a_\gamma W_1}\right)\right] = 1-\gamma \quad \text{if } a_\gamma < 0\ .$$

Other confidence regions for θ can be found in a similar manner or directly on the basis of the generalized pivotal quantity for θ implied by (6.10), namely,

$$T = \frac{2u}{W_1} - \frac{2v}{W_2} \tag{6.12}$$

with the observed value $t_{obs} = \theta$.

For given values of m and n it is more convenient to compute and tabulate confidence bounds for a scale-invariant parameter such as $\theta_v = \theta/v$ rather than for θ. For example, the lower confidence for θ_v given by (6.11), say $k(u,v) = a_\gamma(u,v)/v$ depends on the data only through u/v. For four values of p, Figure 6.1 gives the $100p\%$ lower confidence bounds for θ_v when $m=10$ and $n=10$. In other words the $100p\%$ lower confidence bound for θ is $vk(u,v)$. Of course, $vk(u,v)$ also represents $100(1-p)\%$ upper confidence bounds for θ. Figure 6.2 provides confidence bounds for θ_v when $m=20$ and $n=20$. In particular, 95% equal tail confidence intervals for θ can be found using these figures.

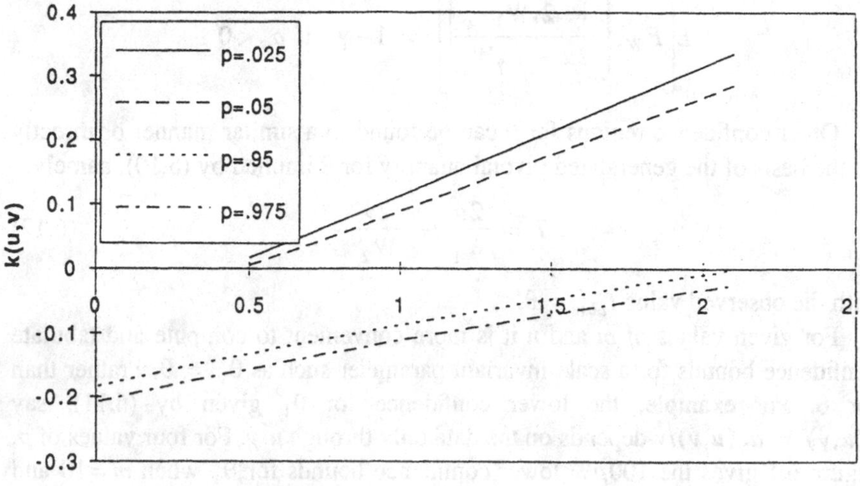

Figure 6.1. Confidence bounds for θ/ν when $m = 10$, $n = 10$.

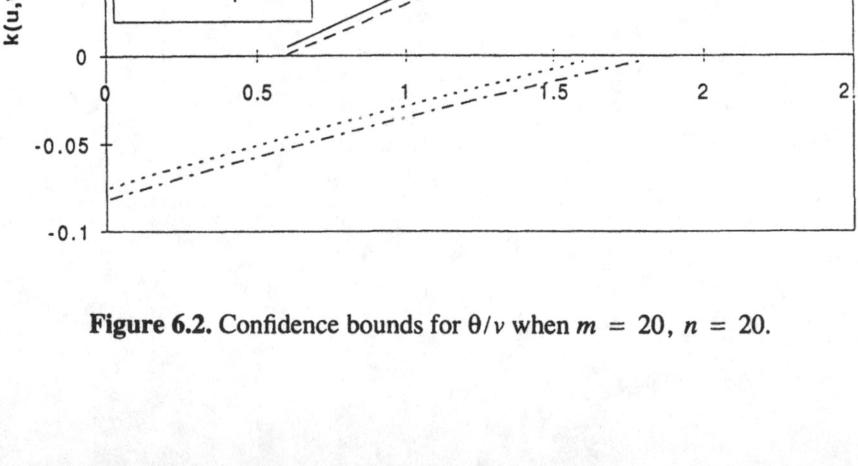

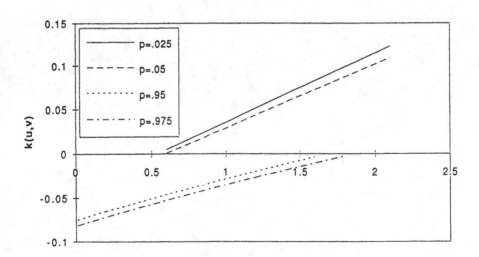

Figure 6.2. Confidence bounds for θ/ν when $m = 20$, $n = 20$.

Example 6.3. Comparing two cancer treatments

A study is carried out to compare the effects of two chemotherapy treatments in prolonging the survival times of advanced lung cancer patients; see Prentice (1973) for a detailed description of this kind of study and for an actual data set. Suppose the data in Table 6.1 are survival times (in days) of 24 patients who were assigned to a new treatment and an old treatment; the starred numbers denote censored survival times obtained after observing the survival times of 10 and 11 individuals in the two respective groups.

Table 6.1 Lung cancer survival times

(b) New treatment					
723*	723	657	563	132	187
46	295	234	42	723*	124
(a) Old treatment					
125	447	95	564	692*	88
121	48	332	692	81	225

Assuming that the survival times are exponentially distributed, as was the case in the study of Prentice (1973), let us find an equal-tail 95% confidence interval for the difference in mean survival times between the new treatment and the old treatment.

Let μ_{new} and μ_{old} be the mean survival times due to the new treatment and the old treatment, respectively. Since the observations are censored, inferences about these parameters can be based on the sufficient statistics U and V defined in Section 5.5. Since U and V have gamma distributions with shape parameters $m=10$ and $n=11$, respectively, the results of this section still apply. The observed values of the sufficient statistics are $u=4449$ and $v=3510$. The maximum likelihood estimate of the difference in survival times is $\mu_{new}-\mu_{old} = u/10-v/11 = 126$ days. Further, in this case the solution of Equation (6.11) is

$$k_\gamma = \begin{cases} 611 & \text{if } \gamma=.025 \\ -233 & \text{if } \gamma=.975 \end{cases}.$$

Therefore, the equal-tail 95% generalized confidence interval for the difference in the mean survival times is $[-233 \le \mu_{new}-\mu_{old} \le 611]$. In contrast, if U and V are assumed to be approximately normally distributed, we get the approximate 95% confidence interval $[-207 \le \mu_{new}-\mu_{old} \le 459]$. The approximation seriously underestimates the length of the confidence interval. This type of approximation may lead to serious repercussions in hypothesis testing, in particular.

6.6. Similarity in Interval Estimation

In this section, we briefly discuss the method of reducing a problem of interval estimation by invoking the notion of similarity. The method is similar to that in hypothesis testing that we studied in Sections 2.6 and in 5.6. On one hand, Property A of a generalized pivotal quantity is related to the notion of similarity, while on the other hand, Property B can be considered as redundant. This is because if a quantity $R = r(\mathbf{X};\mathbf{x},\zeta)$ under consideration does not satisfy Property B, then we can define a potential pivotal as $R\prime = R - r(\mathbf{x};\mathbf{x},\zeta)$ and impose Property A on $R\prime$. In deriving a single random quantity having Property A, we need to invoke the notion of similarity in the presence of a number of random variables on which the interval estimates are to be based. Since Property B is redundant, we can, without loss of generality, confine our attention to subsets of the sample space for which interval estimates computed using (6.4) are free of nuisance parameters whenever the probability of the subset does not depend on the same parameters.

Definition 6.4. A subset $C = C(\mathbf{x};\zeta)$ of the sample space Ξ and the interval estimate $\Theta_c = \{\theta \in \Theta \mid \mathbf{x} \in C\}$ are said to be similar (in $\boldsymbol{\delta}$) if $Pr(\mathbf{X} \in C) = p(\mathbf{x};\ \zeta)$ does not depend on δ, the nuisance parameters.

Although we invoke the notion of similarity as a means of reducing the underlying statistical problem, it should be noted that without this property a confidence interval is practically useless since it cannot provide a workable solution with a given set of data. The derivation of pivots which can generate similar interval estimates is a nontrivial task. Theorem 6.2 is extremely useful in reducing the number of quantities on which a generalized pivotal quantity can be based. It is also useful in proving that a readily available generalized pivot is unique (up to equivalent interval estimates) in generating interval estimates which can be computed without any knowledge of the nuisance parameters. The proof of the theorem is similar to that of Theorem 5.2 and is left to the reader as an exercise. The application of the theorem is illustrated in Example 6.4. The usefulness of this theorem will become further evident from the Behrens-Fisher problem and related problems discussed in Chapter 7.

Theorem 6.2. Let $\mathbf{R} = \mathbf{r}(\mathbf{X};\mathbf{x},\zeta)$ and $S = s(\mathbf{X};\mathbf{x},\zeta)$ be continuous random quantities based on an observable random vector $\mathbf{X}$, where $\zeta = (\theta,\boldsymbol{\delta})$ are the unknown parameters of the distribution of $\mathbf{X}$. If (i) the observed value of S depends on at least one nuisance parameter δ, (ii) the distribution of S are free of unknown parameters and $\mathbf{x}$; (iii) there exists δ_0 such that $S(\mathbf{x};\mathbf{x},\ \zeta)$ does not depend on $\mathbf{x}$ when $\delta = \delta_0$ (or as $\delta \to \delta_0$); and (iv) the observed value and the distribution of $\mathbf{R}$ is free of $\boldsymbol{\delta}$, then any (measurable) subset of the sample space Ξ of $\mathbf{X}$ leading to an interval estimate similar in δ based on $\mathbf{T} = (\mathbf{R},S)$ can be obtained using $\mathbf{R}$ (and its observed value) alone.

Example 6.4. Interval estimation of the difference in two exponential means (continued)

In section 6.4, we derived the generalized pivotal quantity defined by (6.12), namely $T = 2u/W_1 - 2v/W_2$, by the method of invariance, where

$$W_1 = \frac{2U}{\mu_x} \sim \chi^2_{2m} \quad \text{and} \quad W_2 = \frac{2V}{\mu_y} \sim \chi^2_{2n}$$

and U and V are the sums of two sets of data obtained from the two exponential populations with respective means μ_x and μ_y, and u and v are the observed values of the random variables U and V, respectively. The parameter of interest is $\theta = \mu_x - \mu_y$. We shall now derive the same pivot by invoking the notion of similarity; that is by merely insisting that a data dependent (through u and v) quantity is acceptable for setting interval estimates about θ only if it can be computed without any knowledge about the nuisance parameters.

To achieve this goal note first of all that any confidence region (measurable) constructed for θ based on U and V can also be based on the two quantities T and W_1. As we saw in Section 6.4, T is a generalized pivotal quantity. Therefore, it remains to be shown that in the presence of T, W_1 cannot contribute to interval estimation without affecting similarity. It is easily seen that (i) the observed value, $w_1 = 2u/\mu_x$ of W_1 depends on the nuisance parameter μ_x; (ii) the distribution of W_1 is free of the parameters and the data; and (iii) when $\mu_x = u$, w_1 does not depend on the data. Hence, according to Theorem 6.2, all confidence regions free of nuisance parameters can be obtained using T alone.

6.7 Generalized Confidence Intervals Based on p-Values

Consider again the problem of interval estimation of a parameter θ. When the form of a p-value of a one-sided test is readily available, generalized confidence intervals for θ can be deduced directly from its power function. Let $p(t;\theta_0)$ be a p-value constructed for testing a one-sided null hypothesis. Assume that the p-value is based on a generalized test variable whose cdf is stochastically increasing in θ. When the form of the p-value is known in terms of θ_0, the power function $\pi(\theta)$ of the test is obtained simply by replacing θ_0 by θ.

Since $\pi(\theta; T(X; x_{obs}, \zeta))$ is a generalized pivotal quantity having a uniform distribution (with x_{obs} fixed), generalized confidence intervals can be easily found. Regardless of whether the underlying hypothesis is left-sided or right-sided, an exact $100(\gamma_2 - \gamma_1)\,\%$ generalized confidence region of θ can be found by solving the equation

$$[\ \gamma_1 \le \pi(\theta) \le \gamma_2\ ] \tag{6.13}$$

for θ, where γ_1 and γ_2 are numbers (specified appropriately) between 0 and 1. Noting that $\pi(t;\theta)$ is a monotonic function of θ, the confidence interval given by

(6.13) can be obtained by solving its inequalities for θ. In particular, one-sided confidence intervals for θ can be constructed by setting either $\gamma_1 = 0$ or $\gamma_2 = 1$ appropriately. Other confidence regions for θ can be obtained by means of appropriately chosen subsets of $[0,1]$ to be used in (6.13).

Example 6.5. The difference in two exponential means (continued)

In section 5.5 we derived the form of the p-values for testing hypotheses about the difference in two exponential means, θ. It follows from (5.21) that the corresponding power function is

$$\pi(\theta) = \begin{cases} 1 - E\{F_{W_1}(u(\theta/2+v/W_2)^{-1})\} & \text{if } \theta \geq 0 \\ E\{F_{W_2}(v(-\theta/2+u/W_1)^{-1})\} & \text{if } \theta < 0 \end{cases}, \quad (6.14)$$

where W_1 and W_2 are the chi-square random variables with distributions given by (6.9), F_{W_1} is the cdf of W_1, and the expectation appearing in (6.14) is taken with respect to W_2. In order to construct right-sided confidence intervals for θ, consider the inequalities $1-\gamma \leq \pi(\theta) \leq 1$. Solving these inequalities for θ will yield a $100\gamma\%$ generalized confidence interval for θ. Suppose that the constant k_γ is chosen so as to satisfy the equations

$$E\left\{F_{W_1}(u(k_\gamma/2+v/W_2)^{-1})\right\} = \gamma\,, \quad \text{if } k_\gamma \geq 0 \text{ and} \quad (6.15)$$

$$E\left\{F_{W_2}(v(-k_\gamma/2+u/W_1)^{-1})\right\} = 1-\gamma\,, \quad \text{if } k_\gamma < 0\,. \quad (6.16)$$

Now it is evident that $[\, k_\gamma\,,\ \infty\,)$ is the desired $100\gamma\%$ generalized confidence interval for θ. Notice that this is the same as the confidence interval derived in Section 6.4 by means of a generalized pivotal quantity.

Example 6.6. Comparing the lifetimes of two brands of light bulbs (continued)

Consider again the problem of testing the difference in mean lifetimes of two brands of light bulbs discussed in Example 5.4. The problem is to test the null hypothesis $H_0\colon \mu_A - \mu_B \geq 1$ representing the manufacturer's claim against the alternative hypothesis $H_1\colon \mu_A - \mu_B < 1$ on the basis of the data given in Table 5.1. Now suppose that the hypotheses are to be tested at the nominal level .05. Recall that the observed level of significance for testing the hypotheses was $p_{obs} = .253$. Therefore, the manufacturer's claim that on average brand A light bulbs last at least one year longer than brand B cannot be rejected. The equal-tail 95% generalized confidence interval of $\delta = \mu_A - \mu_B$ computed by applying (6.15) is

$$-1.482 \leq \delta \leq 2.417 .$$

Observe that the boundary of the hypotheses is well within the confidence interval, further supporting the null hypothesis. An approximate 95% confidence interval for δ given by asymptotic results is $[-1.0069 \leq \delta \leq 2.007]$, a considerable underestimation of the length of the confidence interval.

6.8. Resolving an Undesirable Feature of Confidence Intervals

Generalized confidence intervals can be useful even in those situations where conventional confidence intervals do exist, but have certain unappealing properties. When a set of minimal sufficient statistics is not complete, confidence intervals with even certain optimum properties may also posses features which are counter to the meaning of 'confidence', as pointed out by Pratt (1961). To describe this in a little detail and to discuss how we can remedy the situation by means of the elements of generalized interval estimation, consider the problem discussed by Pratt (1961). This concerns the problem of constructing interval estimates for the parameter of the uniform distribution $U(\theta - \frac{1}{2}, \theta + \frac{1}{2})$. The inferences about the parameter θ are to be conducted on the basis of a random sample $X_1, \ldots, X_n$ taken from the distribution.

Let $x_{(1)}$ and $x_{(n)}$ be the smallest and the largest of the observations. Let $X_{(1)}$ and $X_{(n)}$ be the corresponding random variables. Then, it can be shown that $[\max(x_{(1)} + d_\gamma, x_{(n)}) - 1/2, \min(x_{(1)}, x_{(n)} - d_\gamma) + 1/2]$ is the uniformly most accurate invariant (see Lehmann(1986) for its definition) confidence interval for θ, where d_γ is the solution of the equation

$$\gamma = \begin{cases} 1 - 2d_\gamma^n & \text{if } \gamma > 1 - \frac{1}{2}^{n-1} \\ 1 - 2d_\gamma^n + (2d_\gamma - 1)^n & \text{otherwise} \end{cases} . \tag{6.17}$$

The unappealing feature of this interval is that, although it is supposed to be an interval with confidence coefficient γ, when $\delta_x = (x_{(n)} - x_{(1)}) \geq d_\gamma$, the interval would certainly include θ, thus leading to a paradox. Moreover, even when $\delta_x < d_\gamma$, the more confident we feel that the confidence interval contains θ the closer the value of δ_x is to d_γ. But, regardless of the value of δ_x, the conventional definition states the same confidence coefficient, thus violating what we commonly mean by 'confidence'.

The problem with this uniformly most accurate invariant confidence interval is that it does not utilize all the information in the data concerning the parameter. The failure to utilize all the information in the data can be attributed to the incompleteness of $(X_{(1)}, X_{(n)})$ for θ. Consequently, there is a class of generalized pivotal quantities which can produce the interval estimate in question.

We shall now derive a procedure having intuitively appealing features, and shed some light on the problem by deriving a pivot which can generate all invariant interval estimates including the above one. This will let us study the root cause of the problem leading to the paradox. For other solutions which may be appropriate under various conditions, the reader is referred to Kiefer (1977)

Although the underlying problem is not location invariant (while the family of distributions is invariant) in the usual sense, in the context of generalized confidence intervals it can be transformed into an equivalent invariant problem, say by the reparameterization $\beta = \theta - x_{(1)}$. Then, all invariant interval estimators can be obtained on the basis of the generalized pivotal quantity $R = X_{(n)} + X_{(1)} - 2x_{(1)} - 2\beta = (X_{(n)} - \theta) + (X_{(1)} - \theta)$, because (i) the distribution of R is free of data and θ; and (ii) the observed value of $R = r_{obs} = x_{(n)} - x_{(1)} - 2\beta$ depends on the data only through the maximal invariant $\delta_x = x_{(n)} - x_{(1)}$. In particular, the uniformly most accurate invariant interval is given by the probability statement

$$Pr(\max(\delta_x, 2d_\gamma - \delta_x) - 1 \leq R \leq \min(\delta_x - 2d_\gamma, -\delta_x) + 1)) \ ;$$

that is by the probability statement

$$-\Delta_x - (1 - \delta_x) \leq R \leq \Delta_x + (1 - \delta_x) \ ,$$

where $\Delta_x = \min(2(\delta_x - d_\gamma), 0)$. Notice that this probability is not really γ, but rather depends on the magnitude of the maximal invariant δ_x, because the distribution of R is free of $x_{(1)}$ and $x_{(n)}$. It is is of interest to note that although R is a generalized pivot for the invariant problem, it is an ordinary pivot for the original problem.

Of course, there is no difficulty obtaining any interval estimate without this drawback by means of an appropriately chosen probability statement of the form $Pr(a \leq R \leq b) = \gamma$. It can be shown (Exercise 6.12) that the probability density function of R is

$$f_R(r) = \begin{cases} n(1+r)^{n-1}/2 & \text{if } -1 \leq r \leq 0 \\ n(1-r)^{n-1}/2 & \text{if } 0 \leq r \leq 1 \end{cases} . \tag{6.18}$$

In particular, the shortest $100\gamma\%$ confidence interval can be found using the probability statement $Pr(-k_\gamma \leq R \leq k_\gamma)$, where

$$\int_{k_\gamma}^{1} f_R(r) = \frac{1-\gamma}{2} \ .$$

It is now clear that the shortest $100\gamma\%$ confidence interval for θ is

$$[(x_{(1)} + x_{(n)})/2 - k_\gamma/2, (x_{(1)} + x_{(n)})/2 + k_\gamma/2] \ , \tag{6.19}$$

where $k_\gamma = 1 - (1-\gamma)^{1/n}$. Interestingly, this confidence interval is symmetric about the Pitman estimator of θ and has a constant length.

6.9. Bayesian and Conditional Confidence Intervals

Another method which can provide a satisfactory solution (see, for instance, Barnard (1976)) to the inference problem addressed in the preceding section is the method of conditional inference (see Kiefer (1977), and Lehmann (1986)). Pratt (1961) pointed out how the problem discussed in Section 6.8 can be resolved by conditioning on δ_X and yet criticized the use of conditional inferences in this context. In this approach the paradox is resolved by constructing confidence intervals based on one random variable conditional on another random variable which is an ancillary statistic for the parameter of interest. An *ancillary statistic* is an observable statistic whose distribution does not depend on the parameter. For further details and methods of finding ancillary statistics the reader is referred to Basu (1959), Basu (1977), and Lehmann (1986). In Chapter 9, we will undertake a class of applications in which there is a need to take the advantage of both the conditional confidence intervals and the generalized confidence intervals.

To illustrate the method of constructing conditional confidence intervals, consider again the problem of conducting inferences about the parameter of $U(\theta-½,\theta+½)$ based on the statistics $X_{(1)}$ and $X_{(n)}$. The statistic $X_{(n)}-X_{(1)}$ is an ancillary for θ as a result of the joint distribution of $X_{(1)}-\theta$ and $X_{(n)}-\theta$ being free of θ. The distribution of $U = X_{(n)}-X_{(1)}$ is

$$f_U(u) = n(n-1)u^{n-2}(1-u) \quad \text{for } 0 \le u \le 1 .$$

The conditional distribution of $V = X_{(1)}$ given $U=u$ is (see Exercise 6.12) uniform over the interval $[\theta-½,\theta+½-u]$. Therefore, conditional on the ancillary, $V-\theta$ is a pivotal quantity for θ. All $100\gamma\%$ confidence intervals based on this pivotal has the constant length $\gamma(1-\delta_x)$, where $\delta_x=u$ is the observed value of the ancillary $U=X_{(n)}-X_{(1)}$. In particular, the $100\gamma\%$ equal-tail conditional confidence interval for θ is

$$\left[x_{(1)} + \frac{\delta_x}{2} - \frac{\gamma(1-\delta_x)}{2} \le \theta \le x_{(1)} + \frac{\delta_x}{2} + \frac{\gamma(1-\delta_x)}{2} \right] . \quad (6.20)$$

As expected, the length of this interval depends on the range of the data on which the inference was conditioned. The expected length of this interval estimator is $\gamma(1-(n-1)/(n+1)) = 2\gamma/(n+1)$; this is not uniformly smaller or larger than the length of the generalized confidence interval given by (6.19).

Let us now turn to the Bayesian solution to the problem. Bayesian solutions do not suffer from the drawback of conventional confidence intervals that we studied in Section 6.8. Consider the problem of constructing credible intervals for θ under the noninformative prior $\pi(\theta) = 1$. The joint probability density function of the sufficient statistics is

$$f(x_{(1)},x_{(n)} \mid \theta) = n(n-1)(x_{(n)}-x_{(1)})^{n-2} \quad \text{if } \theta-\tfrac{1}{2} \le x_{(1)} < x_{(n)} \le \theta+\tfrac{1}{2}\,.$$

Since the posterior density of θ is proportional to the joint density of $(\theta, X_{(1)}, X_{(n)})$, it immediately follows from this equation that

$$\theta \mid x_{(1)},x_{(n)} \sim U(\, x_{(n)}-\tfrac{1}{2}\,,\; x_{(1)}+\tfrac{1}{2}\,)\,.$$

It is now evident from this posterior distribution that the Bayesian confidence intervals are numerically the same as the conditional confidence intervals derived above.

Exercises

6.1. Let $R = r(\mathbf{X};\mathbf{x},\zeta)$ be a generalized pivotal quantity and let $S = s(\mathbf{X};\mathbf{x},\zeta)$ be continuous random variable, where $\zeta = (\theta,\delta)$ are unknown parameters. If S satisfies the conditions of Theorem 6.2, show that any (measurable) confidence region for θ based on (R,S) can be constructed using R alone, provided that it does not depend on δ.

6.2. Let $X_1, \ldots, X_m$ and $Y_1, \ldots, Y_n$ be two independent sets of exponential random variables with means μ_x and μ_y, respectively. For given constants a and b, derive a generalized pivotal quantity for $\theta = a\mu_x + b\mu_y$ by

(a) the method of invariance;

(b) the method of similarity.

Find the form of a one-sided confidence interval based on the generalized pivotal quantity. Find the form of the fixed-level tests for testing hypotheses of the form

$$H_0\colon \theta \le \theta_0 \quad \text{versus} \quad H_1\colon \theta > \theta_0 .$$

6.3. Let $X_1, \ldots, X_m$ and $Y_1, \ldots, Y_n$ be two sets of independently and identically distributed random variables with means μ_x and μ_y respectively. Suppose that each random variable in the former set is exponentially distributed and those of the latter set are normally distributed with variance σ^2. Let $\mu = \mu_y - \mu_x$ be the parameter of interest. Derive a generalized test variable and find a two-sided generalized confidence interval for μ. Find the form of fixed-level tests for testing left-sided null hypotheses.

6.4. Let $X_1, \ldots, X_n$ be a random sample from the uniform distribution with density

$$f_X(x) = \frac{1}{\beta-\alpha} \quad \text{for } \alpha \le x \le \beta .$$

Let $X_{(1)}, \ldots, X_{(n)}$ be the order statistic. Suppose that $\theta = (\beta-\alpha)/(\beta+\alpha)$ is the parameter of interest.

(a) Show that

$$R = \frac{1-\theta}{\theta}\left[\frac{(X_{(n)}-\alpha)}{\alpha}x_{(1)} - \frac{(X_{(1)}-\alpha)}{\alpha}x_{(n)}\right]$$

is a generalized pivotal quantity.

(b) Derive this pivotal quantity by the method of similarity and based on that pivot, find the $100\gamma\%$ left-sided generalized confidence interval for θ.

(c) Find the left-sided Bayesian confidence interval for θ under the diffuse prior

$\pi(\alpha,\beta) = 1.$

6.5. Let $X_1,\ldots, X_n$ be a random sample from a normal population with mean μ and variance σ^2. Let $\bar{X}$ and S^2 be the sample mean and the sample variance. By applying Theorem 6.2 find generalized pivotal quantities for constructing confidence intervals for each of the following parameters:

(a) the mean, μ;

(b) the variance, σ^2;

(c) the second moment, $\mu^2 + \sigma^2$;

(d) the moment generating function, $M(t) = exp(\mu t + \sigma^2 t^2/2)$;

(e) the cumulative probability distribution function, $F(x) = Pr(X \le x)$.

6.6. Let $X_1,\ldots, X_m$ and $Y_1,\ldots, Y_n$ be independent random samples from two lifetime distributions. Assuming that the lifetimes are exponentially distributed with means μ_x and μ_y respectively, find a generalized pivotal quantity for the probability $p = Pr(X > Y)$

(a) by the method of similarity;

(b) by the method of invariance.

Show that interval estimators given by these pivots have Property 1 as well as Property 2 of Definition 6.1.

6.7. Let $R = r(M;\, m_{obs}, \zeta)$ be an absolutely continuous generalized pivotal quantity for a parameter θ based on $M = m(\mathbf{X})$, where $m_{obs} = m(\mathbf{x})$ is the observed value of M. If M is a maximal invariant under a group of transformations of $\mathbf{X}$ and if θ is not affected by the induced transformation on the vector ζ of parameters, then show that any invariant generalized confidence interval of θ based on $\mathbf{X}$ can be obtained using R as well.

6.8. Consider the p-value given by formula (5.10) for testing hypotheses about θ, the second moment of a normal distribution. Based on the form of that p-value, find

(a) a $100\gamma\%$ lower confidence bound for θ;

(b) a $100\gamma\%$ upper confidence bound for θ.

6.9. Consider the p-value given by formula (5.31) for testing hypotheses about θ, the difference between the mean of a normal distribution and the mean of an exponential distribution. Based on the form of that p-value, find

(a) a left-sided $100\gamma\%$ confidence interval for θ;

(b) a two-sided $100\gamma\%$ confidence interval for θ.

6.10. let $Y_1,\ldots, Y_n$ be a random sample from a distribution with density function

$$f(y) = \frac{\lambda\mu}{\lambda-\mu}\frac{1}{y^2} \quad \text{for } \mu \le y \le \lambda ,$$

where λ and μ are positive parameters. Construct one-sided and two-sided interval estimates for the range of the distribution and establish their uniqueness by invoking the notion of similarity.

(a) Examine whether the interval estimates can be obtained by exploiting the location invariance of the problem.

(b) Examine whether the interval estimates can be obtained by the method of scale invariance.

6.11. Let $X_1, \ldots, X_n$ be a random sample from a uniform distribution over the interval $[-\frac{1}{2}, \frac{1}{2}]$. Let $R = (X_{(1)} + X_{(1)})/2$ be the sample mid-range. Show that the probability density function of R is

$$f_R(r) = \begin{cases} n(1+2r)^{n-1} & \text{if } -\frac{1}{2} \le r \le 0 \\ n(1-2r)^{n-1} & \text{if } 0 \le r \le \frac{1}{2} \end{cases} .$$

Find the mean and the variance of R. Also find the distribution of $X_{(n)} - X_{(1)}$ and the conditional distribution of $X_{(1)}$, given $X_{(n)} - X_{(1)} = x_{(n)} - x_{(1)}$.

6.12. Let $X_1, \ldots, X_n$ be independently and uniformly distributed according to the distribution $U(-\theta, \theta)$. Find each of the following intervals for θ and compare them:

(a) the shortest and invariant 95% generalized confidence interval;

(b) the uniformly most accurate invariant 95% confidence interval;

(c) the shortest 95% confidence interval conditional on an appropriate ancillary statistic;

(d) the shortest 95% Bayesian interval under the noninformative prior.

6.13. Let $X_1, \ldots, X_n$ be independently and normally distributed according to the distribution $N(\theta, k\theta^2)$, where k is a known constant. Find the form of each of the following intervals for θ:

(a) the invariant and left-sided 95% generalized confidence interval;

(b) the left-sided 95% confidence interval conditional on the ancillary,

$$U = \bar{X}/\sqrt{\sum X_i^2/n} \; ;$$

(c) the left-sided 95% credible interval under the noninformative prior.

6.14. Consider again the data on survival times of advanced lung cancer patients given in Table 6.1.

(a) Construct a 99% lower confidence bound for the difference in the two mean survival times.

(b) Construct a 99% upper confidence bound for the difference in mean survival times.

(c) Test the hypothesis that, on average, the new treatment prolongs the survival time of a patient six months longer than that due to the old treatment.

6.15. A life test with steadily increasing stress is carried out to compare two brands of aircraft components, A and B. The following table shows the survival times (in hours) of twelve components from each brand:

Observed survival times

(a) Brand A

2.16	1.64	0.25	0.33	0.04	4.31
0.89	0.06	0.02	1.56	2.63	2.67

(b) Brand B

1.18	4.90	5.08	0.11	3.32	0.51
0.14	0.24	0.13	0.64	3.52	0.26

Examine whether the data are exponentially distributed. Construct a 95% equal-tail confidence interval, a 95% lower confidence bound, and a 95% upper confidence bound for the difference in mean survival times of Brand A and Brand B components.

Chapter 7

Comparing Two Normal Populations

7.1. Introduction

In Chapter 3 we considered problems of making inferences about the parameters of a single normal distribution. In this chapter, we will deal with various problems involving parameters of two normal populations. Statistical inferences concerning the parameters of two normal distributions arise in the comparison of two treatments, brands, market segments, services, etc. Experimenters in a variety of fields including biomedical research, agricultural experiments, quality engineering, and socioeconomic investigations often need to perform tests and construct confidence intervals for the difference in two normal means or variances. For example, this is the case when an experimenter compares the yield of a new hybrid corn with a standard kind.

Suppose a set of observations are available from each of the two normal populations. Let (μ_x, μ_y) and (σ_x^2, σ_y^2) be the two sets of means and variances of the two populations. Suppose $X_1, \ldots, X_m$ and $Y_1, \ldots, Y_n$ are the two sets of observations taken from these populations. Assume that $\mathbf{X} = (X_1, \ldots, X_m)$ and $\mathbf{Y} = (Y_1, \ldots, Y_n)$ are all independent random variables. Let $\bar{X}$, $\bar{Y}$, S_x^2, and S_y^2 be the maximum likelihood estimators (see Chapter 3) of μ_x, μ_y, σ_x^2, and σ_y^2, respectively. Since these four quantities are complete sufficient statistics for the parameters of the two distributions, all inferences about the parameters can be based on them. The four random variables are independent. The two sample means are distributed as normal and the two sample variances are distributed as gamma. More precisely, their distributions are given by

$$\bar{X} \sim N\left(\mu_x, \frac{\sigma_x^2}{m}\right), \quad \bar{Y} \sim N\left(\mu_y, \frac{\sigma_y^2}{n}\right), \tag{7.1}$$

and

$$\frac{mS_x^2}{\sigma_x^2} \sim \chi^2_{m-1}, \quad \frac{nS_y^2}{\sigma_y^2} \sim \chi^2_{n-1}. \tag{7.2}$$

Let $(\bar{x}, \bar{y}, s_x^2, s_y^2)$ be the observed values of $(\bar{X}, \bar{Y}, S_x^2, S_y^2)$, respectively. In the following sections we will consider various parameters of interest and will derive procedures based on these four statistics to perform inferences about the parameters.

7.2. Comparing the Means when the Variances are Equal

The purpose of this section and the next is to develop procedures for testing hypotheses about the difference in the means of two normal populations and to construct confidence intervals for that parameter. In many situations experimenters need to deal with the problem of comparing the means of two populations when the value of neither mean is known for certain. Often the populations that they encounter are normally distributed to a reasonable degree of approximation that is adequate for practical purposes.

Assume that the two populations have a common variance, say σ^2; that is $\sigma_x^2 = \sigma_y^2 = \sigma^2$. Usually this is an assumption made for mathematical tractability of the problem and for the simplicity of the solution. Sometimes this assumption is justified by the manner in which the experiment is conducted. In section 7.4 we will study how to test whether this assumption is reasonable. In the next section we will drop this assumption and still develop exact statistical procedures.

When $X_1, \ldots, X_m$ and $Y_1, \ldots, Y_n$ are random samples from the two normal distributions, $N(\mu_x, \sigma^2)$ and $N(\mu_y, \sigma^2)$, inferences about the parameters can be based on the sufficient statistics, $\bar{X}$, $\bar{Y}$, and

$$S^2 = \frac{\sum_{i=1}^{m}(X_i - \bar{X})^2 + \sum_{j=1}^{n}(Y_i - \bar{Y})^2}{m+n},$$

$$= \frac{mS_x^2 + nS_y^2}{m+n}.$$

S^2 is sometimes called the pooled sample variance. As can be easily seen, these are also the maximum likelihood estimators of μ_x, μ_y, and σ^2, respectively. Furthermore, the three random variables are independently distributed. While the distribution of the two sample means are given by (7.1), equation (7.2) implies that

$$\frac{(m+n)S^2}{\sigma^2} \sim \chi^2_{m+n-2} \,. \tag{7.3}$$

Let $\theta = \mu_x - \mu_y$ be the parameter of interest.

First consider the problem of testing the hypotheses

$$H_0\colon \theta \le 0 \quad \text{versus} \quad H_1\colon \theta > 0 \,.$$

In order to derive a testing procedure by the principle of invariance, observe that the family of joint distributions of $\overline{X}$, $\overline{Y}$, and S^2 is both location and scale invariant. On the other hand the testing problem is also invariant with respect to either transformation. By the location invariance, we can confine our search for invariant testing procedures to those based on the two statistics $\overline{X}-\overline{Y}$ and S^2. By the scale invariance we can further reduce the problem to tests based on the statistic

$$T = \frac{\overline{X}-\overline{Y}}{S} \,. \tag{7.4}$$

In other words, as stated by Theorem 2.2, any affine invariant testing procedure can be obtained using T alone.

Since the distribution of $\overline{X}-\overline{Y}$ can be standardized as

$$\frac{\overline{X}-\overline{Y}}{\sigma(1/m + 1/n)^{1/2}} \sim N(\delta,\, 1)\,,$$

it follows from (7.3) that the distribution of T is given by

$$\frac{(mn(m+n-2))^{1/2}}{m+n} T \sim t_{m+n-2}(\delta) \;; \tag{7.5}$$

that is, the noncentral t distribution with $m+n-2$ degrees of freedom and the noncentrality parameter $\delta = \theta/(\sigma\sqrt{1/m+1/n})$. In particular, when $\theta = 0$, this distribution reduces to the Student's t distribution with $m+n-2$ degrees of freedom. Now it is clear that

1. when $\theta=0$, the distribution of T is free of nuisance parameters; and
2. the cumulative distribution function of T is stochastically increasing in θ.

Therefore, T is a test statistic for testing the hypotheses under consideration. The larger values of the observed value $t_{obs} = (\overline{x}-\overline{y})/s$ of T discredit H_0; that is the right tail of the distribution of T is the appropriate extreme region on which the p-value for testing H_0 can be based. The p-value is

$$\begin{aligned} p(t_{obs}) &= Pr(T \ge (\overline{x}-\overline{y})/s \mid \theta=0) \\ &= 1 - G_{m+n-2}\left[\frac{(\overline{x}-\overline{y})}{s}\frac{(mn(m+n-2))^{1/2}}{m+n}\right], \end{aligned}$$

where G_{m+n-2} is the cumulative distribution function of the Student's t

distribution with $m+n-2$ degrees of freedom.

General one-sided and two-sided tests

Now consider general left-sided hypotheses of the form

$$H_0: \theta \le \theta_0 \quad \text{versus} \quad H_1: \theta > \theta_0 . \tag{7.6}$$

By change of variables $X\prime = X - \theta_0$ and $Y\prime = Y$, it can then be deduced from foregoing results that the hypotheses in (7.6) can be tested on the basis of the p-value $p = \pi(\theta_0)$, where

$$\pi(\theta) = 1 - G_{m+n-2}\left(\frac{(\bar{x}-\bar{y}-\theta)}{s} \frac{(mn(m+n-2))^{½}}{m+n} \right) \tag{7.7}$$

is the power function that yields interval estimates for θ. The hypotheses in (7.6) can be tested on the basis of this p-value. In particular, if a fixed-level test at the .05 level is desired, H_0 can be rejected if $p \le .05$.

The power function appropriate for power comparisons is

$$\pi_0(\delta) = 1 - G_{m+n-2,\delta}\left(\frac{(\bar{x}-\bar{y}-\theta_0)}{s} \frac{(mn(m+n-2))^{½}}{m+n} \right), \tag{7.8}$$

where $G_{m+n-2,\delta}$ is the cumulative distribution function of the noncentral t distribution with $m+n-2$ degrees of freedom and the noncentrality parameter δ.

It is clear that if the null hypothesis is right-sided, it can be tested using the p-value $p = 1 - \pi(\theta_0)$. The p-values or fixed-level tests for testing point null hypotheses of the form $H_0: \theta = \theta_0$ can also be obtained from the same test statistic employed above. This is an application of Theorem 2.1 as illustrated in Section 3.2. In view of the parallel of the results in Section 3.2 and those of this section it is evident that the p-value for testing $H_0: \theta = \theta_0$ is

$$p = 2G_{m+n-2}\left(-\frac{|\bar{x}-\bar{y}-\theta_0|}{s} \frac{(mn(m+n-2))^{½}}{m+n} \right). \tag{7.9}$$

Confidence intervals

Interval estimates of the difference in means can be obtained from probability statements involving the test statistic T, the pivotal quantity $(\bar{X}-\bar{Y}-\theta)/S$, or they can be deduced from foregoing results. For example, it can be deduced from Equation 7.7 that the left-sided $100\gamma\%$ confidence interval based on the invariant statistic T is

$$\theta \leq (\bar{x}-\bar{y}) + s\frac{(m+n)}{\sqrt{mn(m+n-2)}}t_{\gamma}, \tag{7.10}$$

where t_γ is the γth quantile of the Student's t distribution with $m+n-2$ degrees of freedom. The symmetric and the shortest $100\gamma\%$ confidence interval for θ is

$$(\bar{x}-\bar{y})-\frac{s(m+n)}{\sqrt{mn(m+n-2)}}t_{\frac{1+\gamma}{2}} \leq \theta \leq (\bar{x}-\bar{y})+\frac{s(m+n)}{\sqrt{mn(m+n-2)}}t_{\frac{1+\gamma}{2}} \tag{7.11}$$

Example 7.1. Comparing the efficiency of factory workers

Two sewers, A and B, of a garment factory were timed to complete a certain job in sewing shirts. The following data shows the amounts of time (in minutes) taken by each worker to do the same job on 10 shirts.

Table 7.1. Job completion times in minutes

Sewer A	3.71 3.35 3.78 4.85 4.05 3.55 3.04 2.62 3.65 3.24
Sewer B	3.95 3.70 3.73 4.29 3.58 3.74 3.54 4.65 1.53 4.23

Is there a statistically significant difference between the mean job completion times by the two workers? To answer this question with the help of the results in this section let us assume that the job completion times are normally distributed with a common variance.

The sample means and the sample variances are $\bar{x}_A = 3.584$, $\bar{x}_B = 3.694$, $s_A^2 = .3278$, and $s_B^2 = .6345$. Let μ_A and μ_B be the true means of the job completion times. The point estimate of the mean difference $\delta = \mu_B - \mu_A$ is $\bar{x}_B - \bar{x}_A = .110$. The pooled sample standard deviation is $s = .6936$.

Consider the null hypothesis that there is no difference between the means μ_A and μ_B; that is the null hypothesis $H_0\colon \mu_A = \mu_B$. According to (7.9) the p-value for testing this null hypothesis is

$$p = 2G_{18}\left(-\frac{.11}{.6936}\times 2.121\right) = 0.7405.$$

Under H_0 this is a typical value for the p-value and so there is no reason to doubt the null hypothesis. In other words, the data do not indicate any statistically significant difference between the performances of the two workers. The 97.5th percentile of the t distribution with 18 degrees of freedom is 2.101, and therefore, the shortest 95% confidence interval for δ given by (7.11) is

$$.11 - \frac{.6936}{2.121}\times 2.101 \leq \delta \leq .11 + \frac{.6936}{2.121}\times 2.101;$$

that is $[-.577, .797]$.

7.3. Solving the Behrens-Fisher Problem

Now consider the typical case where the variances of the two normal populations are not quite equal. We are still interested in exact inferences about the difference in the two means, $\theta = \mu_x - \mu_y$. The inferences are to be based on the set of complete sufficient statistics whose distributions are given by equations (7.1) and (7.2). Historically, this problem is known as the Behrens-Fisher problem. Linnik (1968) has shown that this problem has no exact fixed-level tests (conventional) based on the complete sufficient statistics. Exact conventional solutions based on other statistics and approximate solutions based the complete sufficient statistics do exist, however. Scheffe (1943) gave a class of exact solutions to the Behrens-Fisher problem and a number of other articles provide approximate solutions. (See, for example, Welch (1947), Lee and Gurland (1975), and Scheffe (1970))

It is known that Scheffe type solutions are inefficient in the sense that they do not use all the information in the data about the true value of the parameter. For instance, the expected length of confidence intervals given by the Scheffe solution is much larger than those given by approximate solutions. With the aid of generalized definitions given in Chapters 5 and 6, extending the results in Tsui and Weerahandi (1989), let us now derive the exact solutions to the Behrens-Fisher problem without this drawback. We shall do so by searching for exact procedures based on the complete sufficient statistics $\bar{X}$, $\bar{Y}$, S_x^2 and S_y^2.

Confidence intervals

First consider the problem of constructing interval estimates whose boundaries are functions of $\bar{x}$, $\bar{y}$, s_x^2, and s_y^2. In view of the fact that the difference in sample means is location invariant and that its distribution

$$\bar{X} - \bar{Y} \sim N\left(\theta\,,\ \frac{\sigma_x^2}{m} + \frac{\sigma_y^2}{n}\right)$$

depends on the parameter of interest θ and the nuisance parameter $\sigma_x^2/m + \sigma_y^2/n$, consider the potential pivotal quantity

$$R = (\bar{X} - \bar{Y} - \theta)\left[\frac{\sigma_x^2\, s_x^2/(m\, S_x^2) + \sigma_y^2\, s_y^2/(n\, S_y^2)}{\sigma_x^2/m + \sigma_y^2/n}\right]^{1/2} . \qquad (7.12)$$

The uniqueness (up to equivalent confidence intervals) of this pivotal can be established by invoking the notions of invariance and similarity. That is, we will show later that this generalized pivotal quantity can generate all invariant interval estimates similar in σ_y^2 and σ_x^2. Let us defer the derivation till the next subsection. Observe that, although such random quantities as $R' = s_x(\bar{X}-\mu_x)/S_x - s_y(\bar{Y}-\mu_y)/S_y$ are also generalized pivotals, they are not invariant under location changes of the data.

Let

$$Z = \frac{(\bar{X}-\bar{Y}-\theta)}{\sqrt{\sigma_x^2/m + \sigma_y^2/n}} , \; Y_x = \frac{mS_x^2}{\sigma_x^2} , \; Y_y = \frac{nS_y^2}{\sigma_y^2} ,$$

where $Z \sim N(0,1)$, $Y_x \sim \chi^2_{m-1}$ and $Y_y \sim \chi^2_{n-1}$ are all independent random variables. Moreover, the random variables

$$Y_{x+y} \,\Delta\, Y_x + Y_y \sim \chi^2_{m+n-2} ,$$

$$B \,\Delta\, \frac{Y_x}{Y_x+Y_y} \sim Beta(\frac{m-1}{2}, \frac{n-1}{2}) ,$$

and Z are also independently distributed (Exercise 7.4). The potential pivotal quantity R can be be expressed in terms of these random variables as

$$R = Z \left[\frac{s_x^2}{Y_x} + \frac{s_y^2}{Y_y} \right]^{1/2} \qquad (7.13)$$

$$= \frac{Z}{Y_{x+y}^{1/2}} \left[\frac{s_x^2}{B} + \frac{s_y^2}{1-B} \right]^{1/2} ,$$

That R is a generatized pivotal quantity is now clear from its representation given by (7.13). The observed value of R is $r_{obs} = \bar{x} - \bar{y} - \theta$.

Interval estimates of θ based on R can be obtained from appropriate probability statements about R. Let us first find a 100 γ% lower confidence bound for θ. We need to find a constant c_γ such that $Pr\,(\, R \le c_\gamma \,) = \gamma$. Let $T = Z(m+n-2)^{½} \,/\, Y_{x+y}^{½}$, a random variable having a Student's t distribution with $m+n-2$ degrees of freedom. In terms of this random variable the cumulative distribution of R can be expressed as

$$Pr\,(\, R \le r \,) = Pr \left(T \le r \left[\frac{m+n-2}{\left(s_x^2/B + s_y^2/(1-B) \right)} \right]^{½} \right)$$

$$= E\, G_{m+n-2} \left(r \left[\frac{m+n-2}{(\, s_x^2/B + s_y^2/(1-B))} \right]^{½} \right) ,$$

where G_{m+n-2} is the cumulative distribution function of T and the expectation is taken with respect to the beta random variable B. Hence, $c_\gamma = c_\gamma\,(s_x^2,\, s_y^2\,)$ needs to be found to satisfy the equation

$$E\, G_{m+n-2}\left(c_\gamma \left[(m+n-2)^{1/2} \left(\frac{s_x^2}{B} + \frac{s_y^2}{(1-B)} \right)^{-1/2} \right] \right) = \gamma. \quad (7.14)$$

Equation (7.14) can also be expressed in terms of an F-variate which is often encountered in the literature on the Behrens-Fisher problem, particularly that on the Behrens-Fisher solution (see, for instance, Barnard (1984)) to the problem:

$$E\, G_{m+n-2}\left(c_\gamma \left[\frac{(m+n-2)}{\left(\frac{s_x^2}{m-1} + \frac{s_y^2}{(n-1)F}\right)((m-1)+(n-1)F)} \right]^{1/2} \right) = \gamma, (7.15)$$

where $F = (m-1)Y_y/((n-1)Y_x)$ has an F- distribution with $n-1$ and $m-1$ degrees of freedom. The representation (7.14) is convenient and computationally more efficient, however.

Now it is clear that the desired $100\gamma\%$ generalized confidence interval of θ is $\left[(\bar{x}-\bar{y}) - c_\gamma(s_x^2, s_y^2)\ ,\ \infty \right]$. Other confidence intervals can be found in a similar manner. Of particular interest is the confidence interval which is symmetric about the point estimate $(\bar{x} - \bar{y})$ of θ:

$$\left[(\bar{x}-\bar{y}) - c_{(1+\gamma)/2}\,(s_x^2,\, s_y^2) \le \theta \le (\bar{x}-\bar{y}) + c_{(1+\gamma)/2}\,(s_x^2, s_y^2) \right], (7.16)$$

where γ is the confidence coefficient of the interval. This interval is also the shortest $100\gamma\%$ confidence interval for θ.

The value of c_γ satisfying Equation (7.14) can be found using a numerical algorithm to solve the underlying nonlinear equation. This in turn will require the computation of the left-hand side of Equation (7.14) for a large number of values of c_γ. For a given value of c_γ the left-hand side of the equation is computed by numerical integration. This procedure is readily integrated into the statistical software package XPro. XPro computes confidence intervals with and without the assumption of equal variances.

Observe that c_γ/s_y given by (7.14) depends on the data only through the ratio of sample variances s_x^2/s_y^2. Therefore, for given values of m and n it is convenient to compute and tabulate half lengths of the shortest confidence intervals for a scale invariant parameter such as θ/s_x rather than for $\theta = \mu_x - \mu_y$. For a selected set of values of sample sizes, Figure 7.1 shows how the half lengths of 95% confidence intervals for θ/s_y (i.e. $c_{.975}/s_y$) can vary with s_x^2/s_y^2. The lower set of curves of Figure 7.1 correspond to larger sample sizes. The curve corresponding to each pair of sample sizes is increasing in s_x^2/s_y^2; the increase tends to be smaller for larger sample sizes.

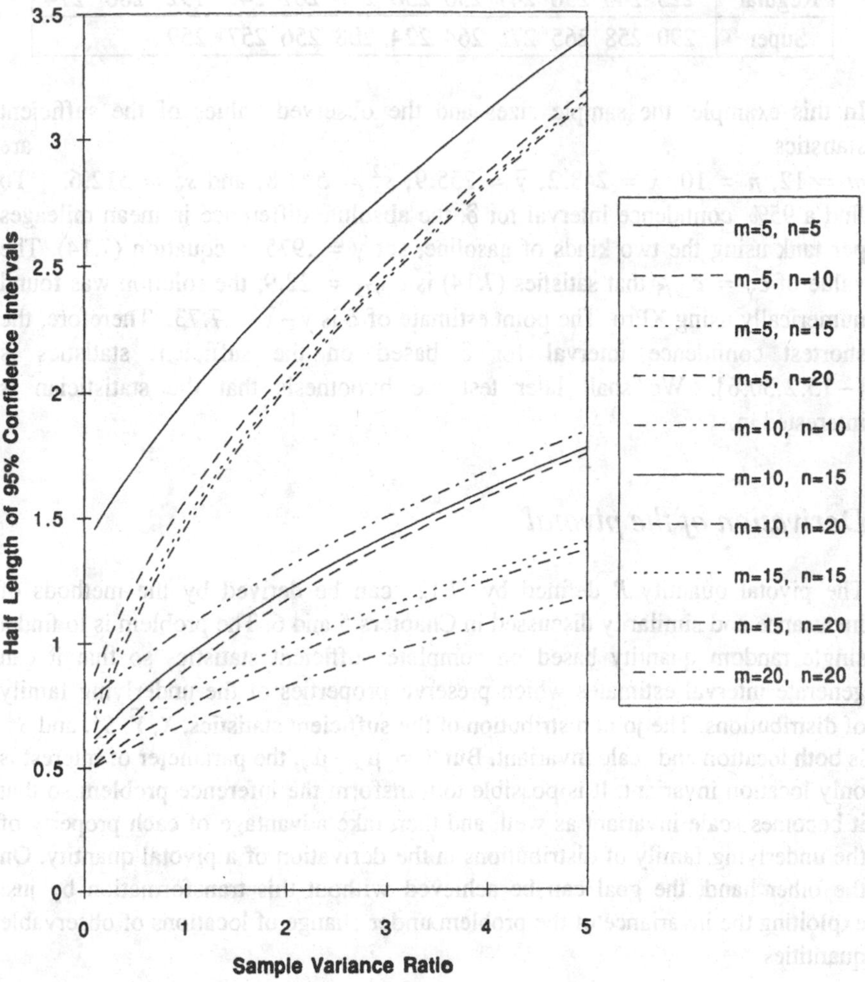

Figure 7.1. Half lengths of 95% confidence intervals for $(\mu_x - \mu_y)/s_x$ as a function of s_x^2/s_y^2.

Example 7.2. Mean increase in mileage

A statistician commutes to work by a car that requires only 'regular' gasoline according to the manufacturer's recommendation. Nonetheless, he wants to find out whether the higher price of a 'super' gasoline compared with the price of a 'regular' can be justified by the mean increase in mileage that super gasoline may yield in his daily commute to work. Table 7.2 shows a set of mileages he got from a full tank of gas of each kind.

Table 7.2. Mileage per tank using regular and super gasoline

Regular	225	246	256	247	230	256	270	261	241	192	280	274
Super	290	258	265	278	264	224	208	256	257	259		

In this example, the sample sizes and the observed values of the sufficient statistics are $m = 12$, $n = 10$, $\bar{x} = 248.2$, $\bar{y} = 255.9$, $s_x^2 = 543.8$, and $s_y^2 = 512.6$. To find a 95% confidence interval for δ, the absolute difference in mean mileages per tank using the two kinds of gasoline, set $\gamma = .975$ in equation (7.14). The value of $c_\gamma = c_{.975}$ that satisfies (7.14) is $c_{.975} = 22.9$; the solution was found numerically using XPro. The point estimate of δ is $\bar{y} - \bar{x} = 7.73$. Therefore, the shortest confidence interval for δ based on the sufficient statistics is $[-15.2, 30.6]$. We shall later test the hypothesis that the statistician is interested in.

Derivation of the pivotal

The pivotal quantity R defined by (7.12) can be derived by the methods of invariance and similarity discussed in Chapters 5 and 6. The problem is to find a single random quantity based on complete sufficient statistics so that it can generate interval estimates which preserve properties of the underlying family of distributions. The joint distribution of the sufficient statistics, $\bar{X}$, $\bar{Y}$, S_x^2, and S_y^2, is both location and scale invariant. But $\theta = \mu_x - \mu_y$, the parameter of interest is only location invariant. It is possible to transform the inference problem so that it becomes scale invariant as well, and then take advantage of each property of the underlying family of distributions in the derivation of a pivotal quantity. On the other hand, the goal can be achieved without this transformation by just exploiting the invariance of the problem under change of locations of observable quantities,

$$\bar{X} \to \bar{X} + c,\ \bar{Y} \to \bar{Y} + c,\ S_x \to S_x,\ S_y \to S_y\ ,$$

where c is any real number. The induced transformations on the parameters are $\mu_x \to \mu_x + c$, $\mu_y \to \mu_y + c$, $\sigma_x^2 \to \sigma_x^2$, and $\sigma_y^2 \to \sigma_y^2$.

Since $\bar{X} - \bar{Y}$, S_x^2, and S_y^2 are maximal invariants under location changes of data, our search for invariant confidence procedures can be based on these three quantities. We can further reduce the problem by applying Theorem 6.3. In order to facilitate this, consider the equivalent problem of constructing interval estimates based on the three random quantities, R, Y_x, and Y_y defined in Section 7.3. Recall that the distribution of each of these random variables is free of unknown parameters. Moreover, the observed value of Y_y depends on the nuisance parameter σ_y and it is independent of the data when $\sigma_y = s_y$. On the other hand, the observed value of $\mathbf{R} = (R, Y_x)$ does not depend on this nuisance parameter. Therefore, according to Theorem 6.3, all interval estimates similar in σ_y can be obtained from **R**. By applying Theorem 6.3 to further reduce the problem we can conclude that all confidence intervals similar in both σ_x and σ_y can be generated using R alone.

Testing of hypotheses

Next consider the problem of significance testing of hypotheses concerning the parameter $\theta = \mu_x - \mu_y$. First consider the hypotheses

$$H_0\colon \theta \le \theta_0 \quad \textit{versus} \quad H_1\colon \theta > \theta_0 \,.$$

The p-value for testing these hypotheses can be deduced directly from the generalized pivotal quantity defined by (7.12). The properties of a generalized pivotal quantity as defined in Section 6.2 are basically the same as the first two properties of a generalized test variable specified by Definition 5.1 and usually one can be deduced from the other. In this application, the properties of the latter can be achieved if $S = R - \theta$ is considered as the potential test variable. The observed value of S is $s_{obs} = \bar{x} - \bar{y}$. From (7.13) the distribution of T can be obtained from the representation

$$S = \frac{Z}{Y_{x+y}^{1/2}} \left[\frac{s_x^2}{B} + \frac{s_y^2}{1-B} \right]^{1/2} + \theta.$$

Since each of the three random variables B, Z, and Y_{x+y} (and hence the distribution of R) are free of unknown parameters, it is now clear that $Pr(S \le s;\, \theta) = F_R(s - \theta)$ is a decreasing function of θ, where F_R is the cdf of R. This means that T satisfies Property 3 of Definition 5.1 as well. Therefore, S is a generalized test variable.

Since S is stochastically increasing in θ, the generalized p-value appropriate for testing the left-sided null hypothesis is

$$\begin{aligned} p &= Pr(S \ge s_{obs} \mid \theta = \theta_0) \\ &= 1 - E\, G_{(m+n-2)}\left((\bar{x} - \bar{y} - \theta_0) \left[\frac{m+n-2}{(s_x^2/B + s_y^2/(1-B))} \right]^{1/2} \right) \\ &= E\, G_{(m+n-2)}\left(-(\bar{x} - \bar{y} - \theta_0) \left[\frac{m+n-2}{(s_x^2/B + s_y^2/(1-B))} \right]^{1/2} \right), \end{aligned} \qquad (7.17)$$

where G_{m+n-2} is the cumulative distribution function of the Student's t distribution with $m+n-2$ degrees of freedom and the expectation is taken with respect to the beta random variable $B \sim Beta((m-1)/2, (n-1)/2)$. The computation of exact p-values in this context can also be performed using the XPro statistical software package.

Also notice that if $\theta \neq \theta_0$, then

$$Z = \frac{(\bar{X}-\bar{Y}-\theta_0)}{\sqrt{\sigma_x^2/m + \sigma_y^2/n}} \sim N(\delta,1)\ ,$$

where

$$\delta = \frac{(\theta-\theta_0)}{\sqrt{\sigma_x^2/m + \sigma_y^2/n}}\ ,$$

Now it is evident that the power function of tests based on the p-value given by (7.17) is

$$\pi_0(\delta) = E\, G_{m+n-2,\delta}\left\{-(\bar{x}-\bar{y}-\theta_0)\left[\frac{m+n-2}{(s_x^2/B + s_y^2/(1-B))}\right]^{1/2}\right\}\ ,$$

where $G_{m+n-2,\delta}$ is the cumulative distribution function of the noncentral t distribution with $m+n-2$ degrees of freedom and the noncentrality parameter δ. It should be noted that, unlike the power function (7.8), this power function can not be employed for comparing alternative fixed level tests with the fixed-level test implied by (7.17).

The p-values for testing other types of hypotheses can be obtained in a similar manner. In particular, the p-value for testing point null hypotheses of the form H_0: $\theta = \theta_0$ is

$$p = 2E\, G_{(m+n-2)}\left\{-|\bar{x}-\bar{y}-\theta_0|\left[\frac{m+n-2}{(s_x^2/B + s_y^2/(1-B))}\right]^{1/2}\right\}.$$

For example, H_0 can be rejected at the .05 level if $p < .05$. As we established in Chapter 6, this is an exact fixed-level test of size .05.

Example 7.3. Comparing the efficiency of factory workers (continued)

In Example 7.1 we assumed the equality of the two variances of the underlying distributions so that we can construct a confidence interval for the difference in mean job completion times using formula (7.11). Now we are in a position to drop that assumption and compare the performances of the two workers using the foregoing results. Recall that the maximum likelihood estimates of the two means and the two variances are $\bar{x}_A = 3.584$, $\bar{x}_B = 3.694$, $s_A^2 = .3278$, and $s_B^2 = .6345$. Therefore, the p-value for testing the hypothesis of no difference between the mean job completion times is

$$p = 2EG_{18}\left(-.11\left[\frac{18}{.3278/B + .6345/(1-B)}\right]^{1/2}\right) = .7532,$$

where the expectation is taken with respect to the random variable $B \sim Beta(4.5,4.5)$; the p-value correct up to 4 decimal points was computed using XPro. Therefore, we can come to the same conclusion as before without relying on the assumption of equality of variances.

Example 7.4. Comparing the average costs of two kinds of gasoline

Consider again the problem of comparing super gas against the regular gas described in Example 7.2. The statistician's problem was to compare average cost of the two kinds of gas. Suppose it costs 12 dollars for a full tank of regular gas, whereas the cost is 14 dollars for a full tank of super. Let $\mu_x = \mu_{regular}$ and $\mu_y = \mu_{super}$ be the mean mileage obtained from a tank full of gasoline of regular and super, respectively. Then, the problem is to test the null hypothesis that the mean cost per mile using super gas is no more than the mean cost per mile using regular gas; that is to test the null hypothesis

$$H_0\colon a\mu_x \le b\mu_y \quad \text{against the alternative} \quad H_1\colon a\mu_x > b\mu_y\ ,$$

where in the present application $a\mu_x = 14\mu_{regular}$ and $b\mu_y = 12\mu_{super}$. It can be shown that (Exercise 7.5) the generalized p-value for testing these hypotheses is

$$p = E\, G_{m+n-2}\left(-(a\bar{x} - b\bar{y})\left[\frac{m+n-2}{(a^2 s_x^2/B + b^2 s_y^2/(1-B))}\right]^{1/2}\right),$$

where G_{m+n-2} is the cumulative distribution function of the Student's t distribution with $m+n-2$ degrees of freedom and the expectation is taken with respect to the beta random variable $B \sim Beta((m-1)/2,(n-1)/2)$. In the present application we have $m = 12$, $n = 10$, $\bar{x} = 248.2$, $\bar{y} = 255.9$, $s_x^2 = 543.8$, and $s_y^2 = 512.6$ and so the p-value for testing the desired hypotheses is

$$p = EG_{20}\left(-404.0\times\left[\frac{20}{106585/B + 73814/(1-B)}\right]^{1/2}\right) = .0053.$$

This shows strong evidence against the null hypothesis that super gas is economical. In fixed-level testing, H_0 can be rejected at the .01 level. In this application the p-value can be computed using XPro after a scale transformation of each data set; that is, by multiplying the two data sets by 14 and 12, respectively.

7.4. Inferences About the Ratio of Two Variances

In this section and in the next we will consider problems of constructing interval estimates and performing significance tests concerning the variances of two normal populations. On one hand, practitioners often need to test the equality of two variances so that the simple procedures in Section 7.2 can be employed in comparing two population means. On the other hand, in some real world applications, especially in quality improvement of production processes comparison of variances itself is the problem of primary interest (see, Phadke (1989), Taguchi (1986) and Taguchi, Elsayed, and Hsiang (1989)). For example, when a product with the same mean value of an important characteristic can be manufactured by two methods, the one with the smaller variability is preferred.

Consider the parameter $\rho = \sigma_x^2 / \sigma_y^2$, the ratio of the two variances. Inferences about ρ are to be based on the set of minimal sufficient statistics $(\bar{X}, \bar{Y}, S_x^2, S_y^2)$. The maximum likelihood estimator of ρ is

$$\hat{\rho} = \frac{\sum (X_i - \bar{X})^2 / m}{\sum (Y_j - \bar{Y})^2 / n} = \frac{S_x^2}{S_y^2}.$$

It follows from (7.2) that the distribution $\hat{\rho}$ is given by

$$R = \frac{(n-1)m}{(m-1)n} \frac{\hat{\rho}}{\rho} = \left[\frac{\tilde{S}_x}{\tilde{S}_y} \right]^2 \frac{1}{\rho} \sim F_{m-1,n-1}\,, \tag{7.18}$$

the F distribution with $m-1$ and $n-1$ degrees of freedom, where $\tilde{s}_x^2$ is the MVUE of σ_x^2 and $\tilde{s}_y^2$ is the MVUE of σ_y^2. Therefore, R is a pivotal quantity for ρ and its interval estimates can be based on R. Actually, this pivotal is the one given by the methods of invariance and similarity (Exercise 7.7).

Let F_γ be the γth quantile of the F distribution with $m-1$ and $n-1$ degrees of freedom. Then, it immediately follows from (7.18) that

$$\rho \le \left[\frac{\tilde{s}_x}{\tilde{s}_y} \right]^2 \frac{1}{F_\gamma} \quad \text{and} \quad \rho \ge \left[\frac{\tilde{s}_x}{\tilde{s}_y} \right]^2 \frac{1}{F_{1-\gamma}} \tag{7.19}$$

are, respectively, the $100\gamma\%$ upper and lower confidence bounds for ρ, the ratio of variances. The $100\gamma\%$ confidence interval for ρ given by equal tails of the F distribution is

$$\left[\frac{\tilde{s}_x}{\tilde{s}_y}\right]^2 \frac{1}{F_{(1+\gamma)/2}} \leq \rho \leq \left[\frac{\tilde{s}_x}{\tilde{s}_y}\right]^2 \frac{1}{F_{(1-\gamma)/2}} . \tag{7.20}$$

Now turning to the problem of significance testing concerning the ratio of the two variances, consider the hypotheses

$$H_0: \rho \leq \rho_0 \quad \text{and} \quad H_1: \rho > \rho_0 .$$

Obviously, $\hat{\rho}$ is a suitable test variable for testing these hypotheses. In fact this test statistic can be derived by applying the methods discussed in Chapters 2 and 5. Since $\hat{\rho}$ tends to be large for larger values of ρ, we can reject H_0 for large observed values of $\hat{\rho}$. In other words, the right tail of the F distribution corresponds to the extreme region on which the p-value for testing H_0 can be based. Hence, the p-value for testing the preceding hypotheses is

$$p = 1 - H_{m-1,n-1}\left(\left[\frac{\tilde{s}_x}{\tilde{s}_y}\right]^2 \frac{1}{\rho_0}\right), \tag{7.21}$$

where $H_{m-1,n-1}$ is the cdf of F distribution with $m-1$ and $n-1$ degrees of freedom. In fixed-level testing at the .05 level, H_0 is rejected if $p < .05$. The p-value for testing point null hypotheses about ρ can be obtained by applying Theorem 2.1 (Exercise 7.8). A convenient way to test this hypothesis at level α is to reject H_0 if the $100(1-\alpha)\,\%$ confidence interval does not contain the value of ρ specified by H_0. This is not the best fixed-level test, however.

Example 7.5. Comparing the variances of job completion times (Example 7.1 continued)

In Example 7.1 we assumed that the variances of the job completion times of the two workers are equal. We are now in a position to test whether that assumption is reasonable. Let σ_A^2 and σ_B^2 be the variances of the two underlying distributions and let $\rho = \sigma_A^2/\sigma_B^2$. Let us first find the equal tail confidence interval for ρ based on R defined by (7.18). The minimum variance unbiased estimates of the two variances are $\tilde{s}_A^2 = .3642$ and $\tilde{s}_B^2 = .7051$, respectively. In this application R has an F distribution with 9 and 9 degrees of freedom. Therefore, the 95% confidence interval for ρ given by (7.20) is

$$.517 \times \frac{1}{F_{.975}} \leq \rho \leq .517 \times \frac{1}{F_{.025}} ;$$

that is the 95% confidence interval for ρ is $[.128, 2.084]$. Observe that the value $\rho = 1$ suggested by the hypothesis of equal variances is well within this confidence interval. In other words, at the .05 level, the difference between the two variances is not statistically significant. The assumption made in Example

7.1 is reasonable.

7.5. Inferences About the Difference in Two Variances

In some applications, the difference in variances is of interest rather than the ratio of variances. Some hypotheses that we encounter in the area of statistical quality control and quality improvement can be expressed only as a difference in two variances. For example, in the nominal-the-best type tolerance designs, quality control engineers often need to deal with the difference in the variances of certain characteristics of products, parts, or components produced by two processes.

So, consider hypotheses of the form

$$H_0: \sigma_x^2 - \sigma_y^2 \le \delta_0 \quad \text{and} \quad H_1: \sigma_x^2 - \sigma_y^2 > \delta_0. \tag{7.22}$$

Assume, without loss of generality, that $\delta_0 \ge 0$. These hypotheses can be expressed as a ratio of variances only when $\delta_0 = 0$. The hypotheses are to be tested on the basis of the complete sufficient statistics $\bar{X}$, $\bar{Y}$, S_x^2, and S_y^2. Healy (1961) provided an exact solution (see Chapter 9) to this testing problem using an artificial randomization device in addition to the experimental data. Welch (1956) discussed the problem of obtaining approximate solutions to problems involving linear combinations of several variances.

To find a more powerful exact generalized test, notice that the testing problem is location invariant. Therefore, our search for invariant testing procedures can be confined to those based on the maximal invariants $\bar{X}-\bar{Y}$, S_x^2, and S_y^2. In order to reduce the problem further by means of similarity, set

$$S(\mathbf{X};\, \mathbf{x},\zeta) = \frac{\bar{X} - \bar{Y} - \theta}{(\sigma_x^2/m + \sigma_y^2/n)^{1/2}} \sim N(0,1)$$

in Theorem 5.2. This quantity satisfies the requirements that (i) the observed value of S depends on the nuisance parameter θ; (ii) the distribution of S is free of unknown parameters; and (iii) the observed value of S does not depend on the data when $\theta = (\bar{x}-\bar{y})$, of Theorem 5.2. Therefore, all testing procedures similar in the nuisance parameter θ can be obtained using S_x^2 and S_y^2 alone. Define

$$U = \frac{mS_x^2}{2} \sim G(\frac{m-1}{2},\sigma_x^2) \quad \text{and} \quad V = \frac{nS_y^2}{2} \sim G(\frac{n-1}{2},\sigma_y^2),$$

observable random variables with observed values $ms_x^2/2$ and $ns_y^2/2$. With this representation it is evident that the problem can be further reduced as in the problem of comparing exponential distributions that we tackled in Section 5.5. Obviously, the approach taken in Section 5.5 yields the generalized test variable

$$T = \frac{ms_x^2}{W_1} - \frac{ns_y^2}{W_2} - \delta, \tag{7.23}$$

where $\delta = \sigma_x^2 - \sigma_y^2$,

$$W_1 = \frac{mS_x^2}{\sigma_x^2} \sim \chi^2_{m-1} \quad \text{and} \quad W_2 = \frac{nS_y^2}{\sigma_y^2} \sim \chi^2_{n-1}.$$

The observed value of T is 0. Since T is stochastically decreasing in δ, the p-value appropriate for testing hypotheses in (7.22) is

$$p = Pr(\ T \le 0 \mid \delta = \delta_0\ )$$

$$= 1 - E\left[F_{W_1}\left[\frac{ms_x^2}{\delta_0 + ns_y^2/W_2}\right]\right], \tag{7.24}$$

where F_{W_1} is the cdf of W_1 and the expectation is taken with respect to the random variable W_2. It is of interest to note that when $\delta_0 = 0$, (7.24) reduces to

$$p = Pr\left(\frac{ms_x^2}{ns_y^2} \le \frac{W_1}{W_2}\right)$$

$$= 1 - H_{m-1,n-1}\left(\left[\frac{\tilde{s}_x}{\tilde{s}_y}\right]^2\right),$$

where $H_{m-1,n-1}$ is the cdf of F distribution with $m-1$ and $n-1$ the two variances equal to 1, degrees of freedom. This is the same as the p-value given by (7.21) when $\rho_0 = 1$.

Confidence intervals for δ can be deduced from (7.24). For instance, the right-sided $100\gamma\%$ confidence interval for the parameter δ is of the form $[\ k_\gamma(u,v)\ ,\infty)$, where

$$E\left[F_{W_1}\left(\frac{ms_x^2 W_2}{ns_y^2 + k_\gamma W_2}\right)\right] = \gamma \quad \text{if } k_\gamma \ge 0 \text{ and} \tag{7.25}$$

$$E\left[F_{W_2}\left(\frac{ns_y^2 W_1}{ms_x^2 - k_\gamma W_1}\right)\right] = 1-\gamma \quad \text{if } k_\gamma < 0\ .$$

Other confidence intervals can be found in a similar manner. Both the p-values and the confidence intervals can be computed using XPro.

Example 7.6. Comparison of deviations from the target

A quality improvement engineer wants to compare the mean deviations of the diameters of certain ball bearings manufactured by processes A and B. The engineer believes that the mean squared deviation from the target diameter for process A is at least 25% less than that for process B and that the absolute difference of two variances is more than 0.05. To test these and other necessary hypotheses he has obtained the following data:

Table 7.3. Deviations of ball bearing diameters (in μm) from the target

Process A:	.04 .01 -.08 -.17 -.31 .16 -.27 .18 .21 -.14 -.11 -.11
Process B:	.18 .35 .07 -.16 .26 -.18 .18 .39 -.13 -.18 -.26 -.60

The sample means and the sample variances of the two data sets are $\bar{x}_A = -.0492$, $\bar{x}_B = -.00667$, $s_A^2 = .0269$, $s_B^2 = .0789$, respectively.

Let us first examine whether these deviations have zero mean. The p-values for testing the hypotheses that each set of deviations have zero mean are .342 and .939, respectively for process A and process B. Therefore, there is no reason to doubt the hypothesis that each process produces ball bearings with mean at the target value. Now consider the two hypotheses that the engineer is interested in, namely,

$$H_{01}: \sigma_A^2 \leq .75\sigma_B^2 \quad \text{and} \quad H_{02}: \sigma_B^2 - \sigma_A^2 \geq .05.$$

The p-values for testing these hypotheses can be computed using the formulae (7.21) and (7.24). The p-values, computed using the XPro software package, are

$$p_1 = 1 - H_{11,11}(.4558) = .896 \quad \text{and}$$

$$p_2 = E\left(F_{\chi^2_{11}}\left(\frac{12\times.0789}{.05 + 12\times.0269/W}\right)\right) = .480\ ,$$

respectively, where the expectation is taken with respect to $W \sim \chi^2_{11}$. These are typical values that each p-value can assume if the two hypotheses H_{01} and H_{02} are true. Therefore, the data in Table 7.3 strongly support the engineer's claims. In fixed-level testing at the .05 level, for instance, the hypotheses are accepted. It should be noted that when the deviations are normally distributed with zero mean (or any other known mean) the p-values can be improved by incorporating this information. This is left as an exercise for the reader. (See Exercise 7.10 for a complete description of the problem).

7.6. Bayesian Inference

The purpose of this section is to develop Bayesian counterparts for comparing the parameters of two normal populations. Unlike most literature on these Bayesian procedures, we shall derive them with representations similar to those in preceding sections of this chapter. To do this, assume independent natural conjugate prior distributions for (μ_x, σ_x) and (μ_y, σ_y), given by

$$\mu_x \mid \sigma_x^2 \sim N(\nu_x, \frac{\sigma_x^2}{\lambda_x}), \quad \mu_y \mid \sigma_y^2 \sim N(\nu_y, \frac{\sigma_y^2}{\lambda_y}),$$

$$\tau_x = \frac{1}{\sigma_x^2} \sim G(\alpha_x, \beta_x^{-1}), \text{ and } \tau_y = \frac{1}{\sigma_y^2} \sim G(\alpha_y, \beta_y^{-1}),$$

where τ denotes the precision of the normal distribution and $G(\alpha, \beta^{-1})$ is the gamma distribution with hyper parameters α and β. Then, it follows from Theorem 3.1 that the posterior distributions are of the form

$$\mu_x \mid \sigma_x^2 \sim N(\tilde{\nu}_x, \frac{\sigma_x^2}{\tilde{m}}) \quad \text{where } \tilde{\nu}_x = \frac{\lambda_x \nu_x + m\bar{x}}{\lambda_x + m}, \quad \tilde{m} = m + \lambda_x$$

$$\mu_y \mid \sigma_y^2 \sim N(\tilde{\nu}_y, \frac{\sigma_y^2}{\tilde{n}}) \quad \text{where } \tilde{\nu}_y = \frac{\lambda_y \nu_y + n\bar{y}}{\lambda_y + n}, \quad \tilde{n} = n + \lambda_y$$

$$\tau_x = \frac{1}{\sigma_x^2} \sim G(\tilde{\alpha}_x, \tilde{\beta}_x^{-1}) \text{ and } \tau_y = \frac{1}{\sigma_y^2} \sim G(\tilde{\alpha}_y, \tilde{\beta}_y^{-1}),$$

where

$$\tilde{\alpha}_x = \alpha_x + \frac{m}{2}, \quad \tilde{\alpha}_y = \alpha_y + \frac{n}{2}$$

$$\tilde{\beta}_x = \beta_x + \frac{m s_x^2}{2} + \frac{m \lambda_x (\bar{x} - \nu_x)^2}{2(\lambda_x + m)},$$

and

$$\tilde{\beta}_y = \beta_y + \frac{n s_y^2}{2} + \frac{n \lambda_y (\bar{y} - \nu_y)^2}{2(\lambda_y + n)}$$

Difference between means

First consider the problem of making inferences about the parameter $\theta = \mu_x - \mu_y$. In order to obtain the posterior distribution of θ, notice that

$$\theta \mid \sigma_x^2, \sigma_y^2 \sim N\left(\tilde{\nu}_x - \tilde{\nu}_y, \frac{\sigma_x^2}{\tilde{m}} + \frac{\sigma_y^2}{\tilde{n}} \right) \tag{7.26}$$

Let

$$Z = \frac{(\theta - (\tilde{\nu}_x - \tilde{\nu}_y))}{\sqrt{\sigma_x^2/\tilde{m} + \sigma_y^2/\tilde{n}}}, \; Y_x = \frac{\tilde{\beta}_x}{\sigma_x^2}, \text{ and } Y_y = \frac{\tilde{\beta}_y}{\sigma_y^2}.$$

The joint distribution of these random variables is given by

$$Z \mid Y_x, Y_y \sim N(0,1),\ Y_x \sim G(\tilde{\alpha}_x, 1),\ \text{and } Y_y \sim G(\tilde{\alpha}_y, 1), \quad (7.27)$$

where Y_x and Y_y are independent random variables. Since the conditional distribution of Z given Y_x and Y_y does not depend Y_x and Y_y, all three random variables in (7.27) are independent. The random variables

$$S \triangleq Y_x + Y_y \sim G(\tilde{\alpha}_x + \tilde{\alpha}_y, 1) \text{ and}$$

$$B \triangleq \frac{Y_x}{Y_x + Y_y} \sim Beta(\tilde{\alpha}_x, \tilde{\alpha}_y),$$

are also independently distributed, and therefore, the posterior cumulative distribution of θ can be expressed as

$$\begin{aligned} F_\Theta(\theta) &= Pr(\Theta \le \theta) \\ &= Pr\left\{ Z \le \frac{\theta - (\tilde{\nu}_x - \tilde{\nu}_y)}{\sqrt{\sigma_x^2/\tilde{m} + \sigma_y^2/\tilde{n}}} \right\} \\ &= Pr\left\{ \frac{Z}{\sqrt{S/(\tilde{\alpha}_x + \tilde{\alpha}_y)}} \le \frac{(\theta - (\tilde{\nu}_x - \tilde{\nu}_y))(\tilde{\alpha}_x + \tilde{\alpha}_y)^{1/2}}{\left[\tilde{\beta}_x/(\tilde{m}B) + \tilde{\beta}_y/(\tilde{n}(1-B))\right]^{1/2}} \right\} \\ &= E\, G_{2(\tilde{\alpha}_x + \tilde{\alpha}_y)}\left\{ \frac{(\theta - (\tilde{\nu}_x - \tilde{\nu}_y))(\tilde{\alpha}_x + \tilde{\alpha}_y)^{1/2}}{\left[\tilde{\beta}_x/(\tilde{m}B) + \tilde{\beta}_y/(\tilde{n}(1-B))\right]^{1/2}} \right\} \end{aligned} \quad (7.28)$$

where $G_{2(\tilde{\alpha}_x + \tilde{\alpha}_y)}$ is the cumulative distribution function of the Student's t distribution with $2(\tilde{\alpha}_x + \tilde{\alpha}_y)$ (not necessarily an integer) degrees of freedom and the expectation is taken with respect to the beta random variable B.

Bayesian inferences concerning $\theta = \mu_x - \mu_y$ can now be made using the posterior distribution of θ given by (7.28). For example, the $100\gamma\%$ left-sided credible interval for θ is of the form $(-\infty, k_\gamma]$, where k_γ is the solution of the equation

$$E G_{2(\tilde{\alpha}_x + \tilde{\alpha}_y)}\left\{ [k_\gamma - (\tilde{\nu}_x - \tilde{\nu}_y)] \left[\frac{\tilde{\alpha}_x + \tilde{\alpha}_y}{\left[\tilde{\beta}_x/(\tilde{m}B) + \tilde{\beta}_y/(\tilde{n}(1-B))\right]} \right]^{1/2} \right\} = \gamma. \quad (7.29)$$

In practice, the expectation appearing in (7.29) can be evaluated by generating a large number of random digits of B and taking the average value of the integrand in (7.29). The integrand is well behaved so that the numerical integration is equally simple.

It can be deduced from foregoing results or shown directly that under the diffuse prior

$$d\pi(\mu_x,\mu_y,\sigma_x,\sigma_y) = d\,\mu_x\, d\,\mu_y\, \sigma_x^{-1} d\,\sigma_x\, \sigma_y^{-1} d\,\sigma_y \tag{7.30}$$

the posterior cdf reduces to

$$F_{\Theta}(\theta) = 1 - E\left[G_{m+n-2}\left(\frac{(\theta - (\bar{x} - \bar{y}))\sqrt{m+n-2}}{\sqrt{s_x^2/B + s_y^2/(1-B)}} \right) \right]$$

where the expectation is taken with respect to B which is distributed as $Beta\left(\frac{m-1}{2}, \frac{n-1}{2}\right)$ and G_{m+n-2} is the cdf of the Student's t-distribution with $m+n-2$ degrees of freedom. Now, it is evident that the p-value given in (7.17) is numerically the same as the posterior probability that the null hypothesis H_0: $\theta \le \theta_0$ is true. In testing point null hypotheses, however, the two approaches do not yield equivalent results, unless an approach such as that of Meng (1994) is taken in the Bayesian treatment. Similar numerical equivalences hold between interval estimates obtained by the two procedures as well. Since θ is considered as a random variable in the Bayesian treatment, the interpretation of results is different, however. Johnson and Weerahandi (1988) provide a Bayesian solution of the above form to the multivariate Behrens-Fisher problem.

Bayesian inferences about variances

Next consider the ratio of the two variances $\rho = \sigma_x^2/\sigma_y^2$. It follows from the marginal distributions of σ_x^{-2} and σ_y^{-2} that the posterior distribution of ρ is given by

$$\frac{\tilde{\beta}_x}{\tilde{\beta}_y} \frac{1}{\rho} \sim F_{2\tilde{\alpha}_x, 2\tilde{\alpha}_y}.$$

Bayesian inferences concerning ρ can be based on this result. For example, the null hypothesis H_0: $\rho \le \rho_0$ can be tested on the basis of its posterior probability

$$Pr(\rho \le \rho_0) = 1 - H_{2\tilde{\alpha}_x, 2\tilde{\alpha}_y}\left(\frac{\tilde{\beta}_x}{\tilde{\beta}_y} \frac{1}{\rho_0} \right),$$

where $H_{2\tilde{\alpha}_x, 2\tilde{\alpha}_y}$ is the cdf of F distribution with $2\tilde{\alpha}_x$ and $2\tilde{\alpha}_y$ degrees of freedom. Notice that this result is similar to the form of the p-value in (7.21). As a matter of fact, under the diffuse prior given by (7.30) the posterior probability of the null hypothesis is the same as the p-value.

Bayesian inferences about the difference in the two variances can be based on the posterior distribution of $\delta = \sigma_x^2 - \sigma_y^2$. The form of the posterior probability of one sided hypotheses is similar to the p-value given by (7.24).

7.7. Inferences About the Reliability Parameter

In this section and in the next an important problem that arises in the area of hardware reliability testing is considered. In this application, X represents the strength of a unit or a system and Y represents environmental stress to which the unit is subjected in its operation. We assume that both X and Y are normally distributed perhaps after a transformation. In many applications, the assumption of normality is easily achieved by means of the logarithmic transformation of actual random variables representing the stress and strength in their original form.

As discussed in detail by Johnson (1988), the problem in the context of stress-strength models is to make inferences concerning the reliability parameter, $R = P[X > Y]$ based on independent samples $X_1, \cdots, X_m$ and $Y_1, \cdots, Y_n$ obtained from the normal populations $N(\mu_x, \sigma_x^2)$ and $N(\mu_y, \sigma_y^2)$, respectively. The population variances are not necessarily equal.

Besides this application in reliability testing, even when comparing two treatments, it is sometimes considered more informative to specify hypotheses concerning the unit free quantity, R, rather than in the form $H: \mu_x - \mu_y \leq \delta$, as in Section 7.3. With the latter formulation, the practitioner may have to perform further analysis in order to understand the performance of the two treatments.

Since

$$R = P[X > Y] = \Phi\left(\frac{\mu_x - \mu_y}{\sqrt{\sigma_x^2 + \sigma_y^2}}\right) \tag{7.31}$$

where Φ is the standard normal cdf, the problem of testing

$$H_0 : R \leq R_0 \quad \text{versus} \quad H_1 : R > R_0$$

is equivalent to testing

$$H_0 : \theta \leq \theta_0 \quad \text{versus} \quad H_1 : \theta > \theta_0 \tag{7.32}$$

where

$$\theta = \frac{\mu_x - \mu_y}{\sqrt{\sigma_x^2 + \sigma_y^2}}$$

and $\theta_0 = \Phi^{-1}(R_0)$. Furthermore, if we first construct confidence intervals for θ, the confidence intervals for the reliability parameter can be trivially deduced.

By sufficiency, we can confine our attention to statistical procedures based on the maximum likelihood estimators $\overline{X}$, $\overline{Y}$, S_x^2, and S_y^2. The maximum likelihood estimator of the reliability parameter is

$$\hat{R} = \Phi\left(\frac{\bar{X} - \bar{Y}}{\sqrt{\sigma_x^2 + \sigma_y^2}} \right).$$

Downton (1973) derived the MVUE of R. Reiser and Guttman (1986) give an approximate confidence interval for this parameter.

The exact test is to reject H_0 if the p-value

$$p = 1 - E\left[G^{\delta}_{m+n-2}\left(\frac{(\bar{x} - \bar{y})\sqrt{m+n-2}}{\left(\frac{s_x^2}{B} + \frac{s_y^2}{1-B} \right)^{1/2}} \right) \right] \tag{7.33}$$

is too small, say less than .05. Here G is the cdf of the noncentral t distribution with $m + n - 2$ degrees of freedom and noncentrality parameter

$$\delta(B) = \theta_0 \left(\frac{m(1-B)s_x^2 + nBs_y^2}{(1-B)s_x^2 + Bs_y^2} \right)^{1/2},$$

and the expectation is taken with respect to B which is distributed as $Beta\left(\frac{m-1}{2}, \frac{n-1}{2}\right)$. Notice that ρ in (7.33) is free of nuisance parameters and thus produces a workable solution to the testing problem. In fixed-level testing H_0 is rejected at level α if $p \leq \alpha$. Since $Pr(X > Y) = Pr(log(X) > log(Y))$, the p-value in (7.33) can be employed when X and Y are lognormally distributed as well.

Derivation of the test

According to the definition of θ and the distribution of $(\bar{X}, \bar{Y}, S_x^2, S_y^2)$, we have that

$$V = \frac{m S_x^2}{\sigma_x^2} + \frac{n S_y^2}{\sigma_y^2} \sim \chi^2_{m+n-2}$$

$$B = \frac{m S_x^2}{\sigma_x^2 V} \sim Beta\left(\frac{m-1}{2}, \frac{n-1}{2}\right)$$

$$W = \bar{X} - \bar{Y} \sim N\left(\theta\sqrt{\sigma_x^2 + \sigma_y^2}, \frac{\sigma_x^2}{m} + \frac{\sigma_y^2}{n}\right)$$

and V, B, and W are independent by the result for sums and ratios of independent chi-squared variables.

Let $\bar{x}$, $\bar{y}$, s_x^2, and s_y^2 denote observed values of $\bar{X}$, $\bar{Y}$, S_x^2 and S_y^2, respectively. In order to calculate a significance value for testing H_0, as in the case of the Behrens-Fisher problem, it is necessary in this situation to consider random quantities that depend upon the parameters and the values of the observed sufficient statistics.

To proceed as in Section 7.3 and yet to remind ourselves of some basic principles in significance testing, consider the region of extreme outcomes

$$C = \{ (W , S_x^2 , S_y^2) \mid \frac{w - \theta_0 \sqrt{s_x^2 \sigma_x^2 / S_x^2 + s_y^2 \sigma_y^2 / S_y^2}}{\sqrt{s_x^2 \sigma_x^2 / (m S_x^2) + s_y^2 \sigma_y^2 / (n S_y^2)}} \le \frac{W - \theta_0 \sqrt{\sigma_x^2 + \sigma_y^2}}{\sqrt{\sigma_x^2 / m + \sigma_y^2 / n}} \}$$

$$= \{ (W , S_x^2 , S_y^2) \mid \frac{w - \theta_0 \sqrt{m s_x^2 / (B V) + n s_y^2 / ((1 - B) V)}}{\sqrt{s_x^2 / (B V) + s_y^2 / ((1 - B) V)}} \le Z \}$$

$$= \{ (W , S_x^2 , S_y^2) \mid 1 \le T \} \tag{7.34}$$

where

$$T = \frac{Z \sqrt{s_x^2 / (B V) + s_y^2 / ((1 - B) V)}}{w - \theta_0 \sqrt{m s_x^2 / (B V) + n s_y^2 / ((1 - B) V)}}$$

is a generalized test variable with the observed value $t = 1$ and

$$Z = \frac{W - \theta_0 \sqrt{\sigma_x^2 + \sigma_y^2}}{\sqrt{\sigma_x^2 / m + \sigma_y^2 / n}}$$

is distributed as $N (0 , 1)$ when $\theta = \theta_0$. As in Section 7.3, by imposing invariance and unbiasedness considerations, it is possible to establish that this choice of extreme region is unique up to equivalent testing procedures. Notice that, when the random variables (W , S_x^2 , S_y^2) are evaluated at the observed value (w , s_x^2 , s_y^2), $w = \bar{x} - \bar{y}$, equality occurs so that the observed value lies on the boundary. To establish that the p-value is $\rho = P [C]$ and that C is a reasonable choice for a region of extreme outcomes, note that

$$\rho = P \left[\frac{w \sqrt{m + n - 2}}{\sqrt{s_x^2 / B + s_y^2 / (1 - B)}} \le \frac{Z + \delta (B)}{\sqrt{V / (m + n - 2)}} \right]$$

$$= 1 - E \left[G_{m+n-2}^{\delta} \left(\frac{w \sqrt{m + n - 2}}{\sqrt{s_x^2 / B + s_y^2 / (1 - B)}} \right) \right] \tag{7.35}$$

where the expectation is taken with respect to B. Therefore, $\rho = Pr [C \mid \theta = \theta_0]$ is free of nuisance parameters. Notice that T and Z, appearing in (7.34) are stochastically increasing in θ. Consequently,

$$\rho = P [T \ge t \mid \theta_0]$$

serves to measure how strongly the data support H_0.

To further study the properties of tests based on ρ note that when $\theta > \theta_0$,

$$Z = \frac{W - \theta_0 (\sigma_x^2 + \sigma_y^2)^{1/2}}{(\sigma_x^2/m + \sigma_y^2/n)^{1/2}}$$
$$= \frac{W - \theta (\sigma_x^2 + \sigma_y^2)^{1/2}}{(\sigma_x^2/m + \sigma_y^2/n)^{1/2}} + (\theta - \theta_0)\left[\frac{\sigma_x^2 + \sigma_y^2}{\sigma_x^2/m + \sigma_y^2/n}\right]^{1/2}$$

is distributed as

$$N\left((\theta - \theta_0)\left[\frac{\lambda + 1}{\lambda/m + 1/n}\right]^{1/2}, 1\right)$$

where $\lambda = \sigma_x^2 / \sigma_y^2$. Consequently, the power function of the test is

$$\pi_0 = P\,[\, C \mid \theta = \theta_0]$$

$$= P\left[\frac{w\sqrt{m+n-2}}{\sqrt{s_x^2/B + s_y^2/(1-B)}} \le \frac{Z + \delta_\theta(B)}{\sqrt{V/(m+n-2)}}\right]$$

$$= 1 - E\left[G_{m+n-2}^{\delta_\theta}\left(\frac{w\sqrt{m+n-2}}{\sqrt{s_x^2/B + s_y^2/(1-B)}}\right)\right]$$

where

$$\delta_\theta(B) = \delta(B) + (\theta - \theta_0)\left[\frac{\lambda + 1}{\lambda/m + 1/n}\right]^{1/2}$$

is the noncentrality parameter of the t distribution since $Z + \delta(B)$ is distributed as $N(\delta_\theta(B), 1)$. This result is useful to study the behavior of the test for deviations of θ from θ_0. Notice in particular that, as expected, $P[\,C\,]$ is an increasing function of θ so that tests based on ρ are p-unbiased.

Example 7.7. Reliability of a rocket-motor at the highest operating temperature

Revisiting the rocket-motor experiment considered by Guttman, Johnson, Bhattacharyya, and Reiser (1988), and Weerahandi and Johnson (1992), denote the operating pressure by X and the chamber burst strength by Y. The authors established that these random variables are normally distributed at each operating temperature that they studied. Weerahandi and Johnson (1992) considered the problem of hypothesis testing about the reliability parameter $R = Pr(X > Y)$ at the highest operating temperature that they reported, namely 59 degrees centigrade. The data used in that study gave the observed sufficient statistics

$$\bar{x} = 16.485,\ \bar{y} = 7.789,\ s_x^2 = 0.3409, \text{ and } s_y^2 = 0.05414.$$

The sample sizes from X and Y distributions were $m = 17$ and $n = 24$, respectively.

Of equal importance is the problem of constructing confidence intervals for the reliability parameter; lower confidence bounds are of special importance. To do this, consider the same observed statistics. Confidence bounds can be deduced from the formula for the p-value given by (7.33). The exact 99% lower confidence bound for the parameter $\theta = \Pi^{-1}(R)$ is 8.54. This is computed by solving (7.33) numerically for θ_0 such that $p = .01$. This bound θ implies that the reliability parameter is equal to 1, that is, the probability of rocket motor failure due to operating pressure is 0, correct up to well over ten decimal places. Two-sided confidence intervals can be found based on probability statements of the form $Pr(a \le p \le b) = \gamma$. For example, [8.05 ,18.52] is the equal-tail 99% confidence interval for θ.

7.8. The Case of Known Stress Distribution

In this section we consider the problem of constructing confidence intervals and the problem of testing hypotheses about R when the stress distribution is known. This case naturally arises in some applications. For example, in the case of stress-strength problems concerning telephone poles and support towers for power-transmission cables, usually the wind loadings are so well known that the parameters of the stress distribution can be derived almost exactly.

Notice in this case that the random variables can be transformed by a common change of location and scale so that Y becomes a standard normal random variable. Therefore, in this section we assume, without loss of generality, that $X \sim N(\mu, \sigma^2)$ and $Y \sim N(0, 1)$. Then the parameter of interest is

$$\theta = \frac{\mu}{\sqrt{1 + \sigma^2}} \tag{7.36}$$

Let $X_1, \ldots, X_m$ be a random sample from the distribution of X. The problem is to construct procedures for making inferences about θ based on the MLEs $\bar{X}$ and $S^2 = S_x^2$, where the statistics are independent and their distributions are given by

$$Z = \frac{\bar{X} - \mu}{\sigma/\sqrt{m}} \sim N(0, 1) \quad \text{and} \quad U = mS^2/\sigma^2 \sim \chi^2_{m-1} \tag{7.37}$$

To test null hypotheses of the form $H_0\colon \theta \le \theta_0$, consider the potential generalized test variable

$$T = \frac{\bar{x}\sqrt{U} - \theta\sqrt{U + ms^2}}{Z} \tag{7.38}$$

with the observed value s. It is clear from (7.37) and (7.38) that (i) T is stochastically decreasing in θ; and (ii) the cdf of T does not depend on the nuisance parameter σ^2. Therefore, T is indeed a generalized test variable and the right tail of the distribution of T corresponds to the extreme region appropriate for testing H_0. The uniqueness of this generalized test variable immediately follows from Theorem 5.2. Hence, the p-value for testing H_0 is

$$p = Pr(T \le s \mid \theta = \theta_0)$$
$$= Pr\left[\frac{\bar{x}\sqrt{U} - \theta_0 \sqrt{U + ms^2}}{s} \le Z \right]$$
$$= E\left[\Phi\left(\frac{-\bar{x}\sqrt{U} + \theta_0 \sqrt{U + ms^2}}{s} \right) \right] \tag{7.39}$$

where Φ is the cdf of the standard normal distribution and the expectation is taken with respect to the chi-squared random variable U.

Confidence intervals for θ can be trivially deduced from (7.39). In reliability testing lower confidence bounds of θ are of special interest. The $100\gamma\%$ lower confidence bound for θ is θ_γ given by

$$E\left[\Phi\left(\frac{-\bar{x}\sqrt{U} + \theta_\gamma \sqrt{U + ms^2}}{s} \right) \right] = 1 - \gamma. \tag{7.40}$$

Other confidence intervals for θ can be derived in a similar manner.

A comparison with an approximate solution

Church and Harris (1970) gives an approximate solution to the problem of making inferences about θ defined in (7.36). The inferences are based on the statistic

$$V = \frac{\bar{X}}{\sqrt{1 + S^2}}.$$

It was shown by Church and Harris (1970) that the distribution of V is asymptotically normal with mean θ and variance

$$\bar{\sigma}^2 = \frac{\sigma^2}{1 + \sigma^2}\left[\frac{1}{m} + \frac{\theta^2}{2(m-1)} \frac{\sigma^2}{1 + \sigma^2} \right].$$

The parameter $\bar{\sigma}^2$ is estimated by

$$\hat{\sigma}^2 = \frac{s^2}{1 + s^2}\left[\frac{1}{m} + \frac{v^2}{2(m-1)} \frac{s^2}{1 + s^2} \right],$$

an increasing function of θ.

Consider the problem of testing hypotheses about θ. It can be noted that V is not necessarily stochastically increasing in θ even for large m. Therefore, V can not be used to define an extreme region. It can be used only to define an approximate fixed-level test. Let us now compare the approximate fixed-level test based on the statistic V with that based on the p-value given by (7.39). This can be accomplished by simulation methods. To do this, 10,000 simulation samples of size 8 were generated. The fixed-level test based on the results of Church and Harris (1970) is referred to as the C-H test, and the one based on the p-value is referred to as the exact test. At level α the C-H test rejects the null hypothesis if $v > \theta_0 + z_\alpha \hat{\sigma}$, where z_α is the $1-\alpha$th quantile of the standard normal distribution. Weerahandi and Johnson (1992) showed that the C-H test heavily underestimates the size of the test. For example, when $m = 8$, $\mu = 4$, and $\sigma = 1$, the actual size (probability of rejecting H_0 when $\theta = \theta_0$) of a test of intended size .05 was as large as .12. Therefore for the purpose of conducting a meaningful comparison of the two tests, the critical point v_c of the C-H test which gives the intended size is found by simulation; that is v_c is found such that $Pr(V > v_c) = .05$. Similarly, in order to facilitate the comparison using simulated samples, the critical value p_c of the test based on the p-value given by (7.39) is found as $Pr(p < p_c) = .05$.

To compare the power performance of the two tests, consider the particular test $H_0\colon \theta \leq .5$. Consider the problem of testing H_0 against $H_1\colon \theta > .5$ based on a sample of size 8. For each simulation sample v and the p-value are computed and H_0 is rejected if $v > v_c$ and $p > p_c$, respectively, according to the two tests. The power function of each fixed-level test is computed using the definition

$$\Pi(\theta) = Pr(\text{ rejecting } H_0 \mid \theta\ ).$$

In fixed-level testing of $H_0\colon \theta \leq \theta_0$, when two tests have the same size, the test giving larger values for $\Pi(\theta)$ for all $\theta > \theta_0$ is preferred, because $\Pi(\theta)$ represents the probability of rejecting H_0 when indeed it is false.

When $\mu = 1.5$, and when θ varies as a function of σ, Figure 7.2 displays the power function of each test. The test based on the p-value significantly outperforms the C-H test. It was shown by Weerahandi and Johnson (1992) that the power improvement is more dramatic for smaller values of μ and less significant for larger values of μ.

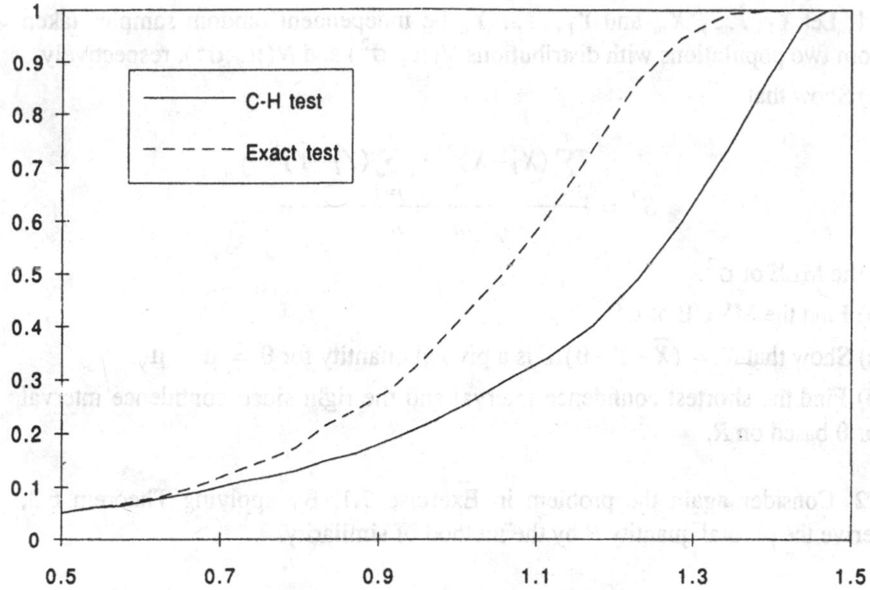

Figure 7.2. Power performance of the exact test and the C-H test.

Exercises

7.1. Let $X_1, \ldots, X_m$ and $Y_1, \ldots, Y_n$ be independent random samples taken from two populations with distributions $N(\mu_x, \sigma^2)$ and $N(\mu_y, \sigma^2)$, respectively.

(a) Show that

$$S^2 = \frac{\sum_{i=1}^{m}(X_i - \bar{X})^2 + \sum_{j=1}^{n}(Y_j - \bar{Y})^2}{m + n}$$

is the MLE of σ^2.

(b) Find the MVUE of σ^2.

(c) Show that $R = (\bar{X} - \bar{Y} - \theta)/S$ is a pivotal quantity for $\theta = \mu_x - \mu_y$.

(d) Find the shortest confidence interval and the right-sided confidence interval for θ based on R.

7.2. Consider again the problem in Exercise 7.1. By applying Theorem 6.3, derive the pivotal quantity R by the method of similarity.

7.3. Let $X_1, \ldots, X_m$ and $Y_1, \ldots, Y_n$ be independent random samples taken from two populations with distributions $N(\mu_x, \sigma_x^2)$ and $N(\mu_y, \sigma_y^2)$, respectively. Consider the problem of testing the hypotheses

$$H_0: \mu_x \leq \mu_y \quad \text{versus} \quad H_1: \mu_x > \mu_y \ .$$

(a) Show that the testing problem is both location and scale invariant.

(b) Show that all affine invariant testing procedures can be based on the two statistics $(\bar{X} - \bar{Y})/S_x$ and S_y/S_x.

(c) By applying Theorem 6.3, find a generalized test variable and hence find the p-value for testing the hypotheses.

7.4. Let X and Y be two independent gamma random variables with a common scale parameter β and shape parameters α_1 and α_2, respectively. Find the joint distribution of $S = X + Y$ and $R = X/(X + Y)$ and deduce that

(a) S and R are independently distributed;

(b) S has a gamma distribution with shape parameter $\alpha_1 + \alpha_2$;

(c) R has a beta distribution with parameters α_1 and α_2.

7.5. Let $X_1, \ldots, X_m$ and $Y_1, \ldots, Y_n$ be independent random samples taken from two populations with distributions $N(\mu_x, \sigma_x^2)$ and $N(\mu_y, \sigma_y^2)$, respectively. Consider the problem of testing the hypotheses

$$H_0: a\mu_x \leq b\mu_y \quad \text{versus} \quad H_1: a\mu_x > b\mu_y \ ,$$

where a and b are positive constants.

(a) Show that the generalized p-value based on complete sufficient statistics for testing these hypotheses is

$$p = 1 - E\, G_{m+n-2}\left((a\bar{x} - b\bar{y}) \left[\frac{m+n-2}{(a^2 s_x^2/B + b^2 s_y^2/(1-B))} \right]^{1/2} \right),$$

where G_{m+n-2} is the cumulative distribution function of the Student's t distribution with $m+n-2$ degrees of freedom, and the expectation is taken with respect to the beta random variable $B \sim Beta((m-1)/2,(n-1)/2)$.

(b) Deduce from the p-value, or find otherwise, a $100\gamma\%$ lower confidence bound for $\theta = a\mu_x - b\mu_y$.

(c) Deduce from the p-value, or find otherwise, the form of the shortest confidence interval for θ.

7.6. Consider again the two normal distributions in Exercise 7.5. Let $X_1, \ldots, X_m$ and $Y_1, \ldots, Y_n$ be two sets of independent observations taken from the two distributions. Find the minimum variance unbiased estimator of the ratio of the two variances, $\rho = \sigma_x^2/\sigma_y^2$ (Hint: the mean of an F distribution with M and N degrees of freedom is $N/(N-2)$, provided that $N > 2$. What is the variance of this estimator?

7.7. Let $X_1, \ldots, X_m$ and $Y_1, \ldots, Y_n$ be independent random samples from two populations with distributions $N(\mu_x, \sigma_x^2)$ and $N(\mu_y, \sigma_y^2)$, respectively. Consider the problem of constructing confidence intervals for $\rho = \sigma_x^2/\sigma_y^2$, the ratio of the two variances. Let $\bar{X}, \bar{Y}$, S_x^2, and S_y^2 be the sample means and the sample variances as defined in Section 7.1.

(a) Show that the problem is both location and scale invariant.

(b) By applying the location invariance, reduce the problem to confidence procedures based on $\bar{X} - \bar{Y}$, S_x^2, and S_y^2.

(c) By applying the scale invariance further reduce the problem to confidence procedures based on $(\bar{X} - \bar{Y})/S_y^2$ and S_x^2/S_y^2.

(d) By applying the principle of similarity further reduce the problem to confidence procedures based on S_x^2/S_y^2 only.

7.8. Consider again the samples in Exercise 7.7 and show that S_x^2/S_y^2 is a test statistic as well for testing H_0: $\rho = \rho_0$. By applying Theorem 2.1 or otherwise, find the form of the p-value for testing H_0 against H_1: $\rho \neq \rho_0$.

7.9. Consider again the samples in Exercise 7.7 and let $\theta = a\sigma_x^2 + b\sigma_y^2$ be the parameter of interest, where a and b are real numbers. Establish procedures for making inferences about θ.

7.10. Let $X_1, \ldots, X_m$ and $Y_1, \ldots, Y_n$ be independent random samples from two populations with distributions $N(0,\sigma_x^2)$ and $N(0,\sigma_y^2)$, respectively. Let $\delta = \sigma_x^2 - \sigma_y^2$ be the parameter of interest.

(a) Find the set of complete sufficient statistics for the parameters of the distributions.

(b) Establish procedures based on these statistics for testing hypotheses about δ.

(c) Establish procedures for constructing interval estimates for δ.

7.11. Consider again the samples from normal populations as in Exercise 7.7. Let the reliability parameter $R = Pr(X > Y)$ be the parameter of interest. In a Bayesian treatment, assume independent conjugate prior distributions as given in Section 7.6. Find the form of the Bayesian credible regions for R. Also show that under the diffuse prior $\pi(\mu_x,\mu_y,\sigma_x,\sigma_y) = \sigma_x^{-1}\sigma_y^{-1}$, the posterior probability that the null hypothesis H_0: $R \le R_0$ is true is the same as the p-value found in Section 7.7.

7.12. Two sections of a Calculus course are taught by two instructors. The first instructor has 15 students in his or her class and the second instructor has 19. The score on which the grade is based is calculated for each student in each class using the same formula applied to marks given for homework, midterm tests, and the final exam. The average of scores given by first instructor is 62.5 and that for the second instructor is 69.8. The standard deviations of the scores are 12.2 and 14.1 respectively. Assume that the scores are approximately normally distributed and that the two sets of scores are independent.

(a) Test whether it is reasonable to assume that the variances of the scores that the two instructors give are equal.

(b) Test the hypothesis that, compared to the first instructor, the second tends to give higher grades.

(c) Find the shortest confidence interval for the difference in mean scores that the two instructors give.

7.13. A famer wants to compare two brands of fertilizers applied to his corn field. To perform a test, the farmer selects 20 plots of similar soil and environmental conditions, plants the same number of trees with the same design, and eight plots are treated with fertilizer A and the remaining plots are treated with fertilizer B. The yield from each plot is shown below in certain units.

Fertilizer	Yield
A	63, 59, 62, 68, 63, 60, 57, 61
B	56, 61, 58, 62, 61, 63, 57

Construct a 95% confidence interval for the difference in mean yields. Is there significant evidence to conclude that Fertilizer A is better than Fertilizer B? Also find a 95% confidence interval for the probability that Fertilizer A yields more than Fertilizer B.

7.14. The data in the following table represent the gains in weight of 17 hogs of same age that were fed with one of two experimental diets over a certain fixed period of time:

Diet	Gain in weight (in pounds)
X	99, 62, 69, 88, 95, 72, 84, 89, 91
Y	81, 63, 77, 82, 78, 85, 68, 74

Let μ_x and μ_y be the mean gain in weight due to Diet X and Diet Y, respectively. Test the hypothesis that $\mu_x \geq \mu_y$ assuming that the the weight gain due to each diet is normally distributed and that the variances of the underlying distributions are equal. Test whether the assumption of equal variances reasonable. Carry out the test without the assumption of equal variances. Also construct a 99% confidence interval for the difference in the two variances.

7.15. Consider again the two diets in Exercise 7.14. A farmer who currently uses Diet Y for his hogs considers switching to Diet X. Let X and Y be the random variables representing the gain in weight using Diet X and Diet Y, respectively. The farmer wishes to switch only if the probability $Pr(X > Y)$ is no smaller than .6. Do the data in Exercise 7.14 provide sufficient evidence for the farmer to switch to Diet X? Suppose a pound of Diet X costs 0.12 and a pound of Diet Y costs 0.11. Is it economical for the famer to switch to Diet X?

7.16. A brick manufacturer makes bricks which are 20 cm long. The bricks are burnt in batches and there are two designs available for setting up the batches for burning. Each design yields bricks of average length of 20 cm but it is not known whether the variance of the deviations are equal. The lengths of 20 bricks made by each design is measured. The data from one design gives a standard deviation of .7 centimeters and the other design gives a standard deviation of .6 centimeters. Is there a significant difference between the standard deviations of bricks made by the two designs? Construct 95% confidence intervals for the ratio of the two variances and for the difference in the variances.

7.17. In order to compare the durability of two brands of tires, 36 families who own a certain car model are randomly selected from a city and are offered free tires for participating in an experiment. Front wheels of 20 cars are outfitted with Brand A tires and front wheels of 16 cars are outfitted with Brand B tires. After three months of driving, the amount of wear in all 72 tires is measured.

Brand A tires gave a sample average of 17.8 (in certain units) and a standard deviation of 4.1, whereas the same statistics for Brand B tires were 18.6 and 4.6, respectively. Specifying the assumptions you make, construct 95% confidence intervals for each of the following and compare the two brands of tires:

(a) the difference between mean wear of Brand A and Brand B tires;

(b) the ratio of standard deviations of wear of Brand A and Brand B tires;

(c) the difference between variances of wear of Brand A and Brand B tires;

(d) the probability that Brand A tires will last longer than Brand B tires.

Chapter 8

Analysis of Variance

8.1. Introduction

We devoted Chapter 7 to tackling problems in comparing two normal populations. The purpose of this chapter is to establish procedures for comparing more than two normal populations. In practical applications, often we need to compare several population means. For example, an investigator might wish to compare several brands of a product, or diet plans, or methods of instruction, or advertising methods, and so on. When samples are taken from underlying populations, the sample means will almost always be different regardless of whether the population means are equal or not. The question that we need to answer is whether or not some of these differences are relatively large due to differences in the population means rather than due to sampling variation.

In testing the equality of several means it is tempting to carry out pairwise comparisons using the results in Chapter 7. Not only is this approach inefficient in the sense that we have to carry out a large number of tests, but also this will seriously increase the chance of rejecting the hypothesis of equal means when it is true. For example, consider the null hypothesis H_0: $\mu_1 = \mu_2 = \mu_3 = \mu_4$ that arises in comparing four populations, where μ_i, $i=1, 2, 3, 4$ are the means of the populations. If we are to test this hypothesis using the results from Chapter 7, we will have to run a sequence of tests on each of the six null hypotheses, $\mu_1 = \mu_2$, $\mu_1 = \mu_3$, $\mu_1 = \mu_4$, $\mu_2 = \mu_3$, $\mu_2 = \mu_4$, and $\mu_3 = \mu_4$. Suppose H_0 is true. If we perform each test at fixed size .05, the size of the combined test will be considerably larger than .05. As the number of means being compared goes up, the size of the combined test will increase to prohibitive levels. The size of the combined test cannot be computed exactly since the six tests are not independent; if they were independent the size would be as large as $1-.95^6 = .26$. The point is that even when there are no differences in the population means the chance of observing significantly different sample means can be large. Therefore, even when the ultimate goal is to identify the population with the largest mean, the hypothesis of equal means should be first tested by a procedure which utilizes all the information in the data. Once the evidence against the hypothesis of equal means is shown to be fairly strong, one can proceed to multiple comparisons and eventually deal with absolute differences between population means.

It will become clear that the combined test will involve various sample variances rather than sample means. Therefore, the procedure is commonly known as *Analysis of Variance*. It is customarily abbreviated as ANOVA. Fisher introduced this technique and applied it heavily in agricultural research. Now it has found applications in almost all areas of scientific research, including biomedical research, economics, marketing, psychology, sociology, and industrial engineering.

In this chapter and in the following chapters, less emphasis will be given to the derivation of statistical procedures and more emphasis will be given to their applications. These procedures are extensions of the results that we established in previous chapters and therefore appropriate test variables and pivotal quantities can be found intuitively. While we will argue that the procedures have necessary and desirable properties, we will not attempt to establish their uniqueness. In many situations, the derivations based on notions such as sufficiency, invariance, and unbiasedness are similar to those in the previous chapters.

8.2. One-way Layout

Consider the problem of comparing k populations. We sometimes refer to the populations being compared as *treatments* and the variable being measured as *response variable*. Suppose a random sample of size n_i is available from Treatment i, $i=1,\ldots,k$. Suppose the following observations are available from the k populations:

Treatment 1: $X_{11}, X_{12}, \cdots, X_{1j}, \cdots, X_{1n_1}$
Treatment 2: $X_{21}, X_{22}, \cdots, X_{2j}, \cdots, X_{2n_2}$
.
.
Treatment i: $X_{i1}, X_{i2}, \cdots, X_{ij}, \cdots, X_{in_i}$
.
.
Treatment k: $X_{k1}, X_{k2}, \cdots, X_{kj}, \cdots, X_{kn_k}$

Assume that the random variables are all mutually independent and that they are all normally distributed. The observed data are denoted as $x_{ij},\ i=1,...,k,\ j=1,...,n_i$.

The sample means and the sample variances of the k treatments are denoted as

$$\bar{X}_{1.},\ \bar{X}_{2.},\ \cdots,\ \bar{X}_{k.}$$

and

$$S_1^2,\ S_2^2,\ \cdots,\ S_k^2,$$

respectively, where

$$\bar{X}_{i.} = \frac{\sum_{j=1}^{n_i} X_{ij}}{n_i} \quad \text{and} \quad S_i^2 = \frac{\sum_{j=1}^{n_i} (X_{ij} - \bar{X}_{i.})^2}{n_i},\ i=1,...,k. \tag{8.1}$$

In one-way ANOVA, the treatment sample means are also denoted as $\bar{X}_i,\ i=1,...,k$, as this will cause no confusion. The observed values of these random variables are $\bar{x}_i,\ s_i^2,\ i=1,...,k$, respectively. Let $\bar{X}$ be the *grand mean* of all of the observations. Note that the grand mean can be obtained from treatment sample means as $\bar{X} = \sum n_i \bar{X}_i / \sum n_i$. To develop statistical procedures in one-way ANOVA, it will be useful to introduce the following notation for various sums of squares that we will have to deal with:

$$S_E = \sum_{i=1}^{k}\sum_{j=1}^{n_i} (X_{ij} - \bar{X}_{i.})^2 = \sum_{i=1}^{k} n_i S_i^2, \tag{8.2}$$

$$S_B = \sum_{i=1}^{k} n_i (\bar{X}_{i.} - \bar{X})^2, \tag{8.3}$$

$$S_T = \sum_{i=1}^{k}\sum_{j=1}^{n_i} (X_{ij} - \bar{X})^2. \tag{8.4}$$

In the literature on Analysis of Variance, the sum of squares S_E is called the *within sum of squares* or the *error sum of squares*, whereas, S_B and S_T are called the *between sum of squares* and the *total sum of squares*, respectively. The between sum of squares is also known as the *treatment sum of squares*. The observed values of these sums of squares are denoted as s_E, s_B, and s_T, respectively.

These sums of squares are related by the equation

$$S_T = S_E + S_B. \tag{8.5}$$

This relationship is easily derived from the identities

$$\sum_{i=1}^{k}\sum_{j=1}^{n_i} (X_{ij} - \bar{X})^2 = \sum_{i=1}^{k}\sum_{j=1}^{n_i} [(X_{ij} - \bar{X}_{i.}) + (\bar{X}_{i.} - \bar{X})]^2,$$

$$\sum_{i=1}^{k}\sum_{j=1}^{n_i}[(X_{ij} - \bar{X}_{i.})(\bar{X}_{i.} - \bar{X})] = \sum_{i=1}^{k}(\bar{X}_{i.} - \bar{X}) \sum_{j=1}^{n_i}(X_{ij} - \bar{X}_{i.}) = 0.$$

The total sum of squares and the between sum of squares can be also computed using the formulas

$$S_T = \sum_{i=1}^{k}\sum_{j=1}^{n_i} X_{ij}^2 - N\bar{X}^2 \quad \text{and} \quad S_B = \sum_{i=1}^{k} n_i \bar{X}_{i.}^2 - N\bar{X}^2,$$

where $N = \sum_{i=1}^{k} n_i$ is the total number of observations. Then, the error sum of squares can be computed using the formula $S_E = S_T - S_B$ given by (8.5).

Example 8.1. Comparing the mean yield of corn hybrids

A researcher at an experimental agriculture site is interested in comparing four hybrids of corn. The four corn hybrids were planted in a random order in 22 plots of equal size and fairly homogeneous soil conditions. Table 8.1 presents the data (yield per plot in bushels) obtained from the experiment.

Table 8.1. Yield of corn from four hybrids

Hybrid A	7.4, 6.6, 6.7, 6.1, 6.5, 7.2
Hybrid B	7.1, 7.3, 6.8, 6.9, 7.0
Hybrid C	6.8, 6.3, 6.4, 6.7, 6.5, 6.8
Hybrid D	6.4, 6.9, 7.6, 6.8, 7.3

In this example corn yield is the response variable. The problem is to compare the four hybrids of corn in terms of mean values of this variable. Let us compute sample means, sample standard deviations, and various sums of squares, which we will utilize later to carry out the comparison. Table 8.2 displays some of these statistics for each hybrid.

Table 8.2. Sample sizes, means and standard deviations

Population	n_i	$\bar{x}_{i.}$	s_i
Hybrid A	6	6.750	.435
Hybrid B	5	7.020	.172
Hybrid C	6	6.583	.195
Hybrid D	5	7.000	.415

In this example the number of treatments being compared is $k = 4$ and the total sample size is $N = 22$. Using the formulae (8.2) and (8.3), the error sum of squares and the between sum of squares can be computed from the information in Table 8.2 as $s_E = 2.372$ and $s_B = 0.728$, respectively. The total sum of squares is $s_T = s_E + s_B = 3.1$. These quantities can also be computed using the alternative formulae given prior to the example.

8.3. Testing the Equality of Means

Consider the problem of comparing the equality of the means of k populations. Let μ_i be the population mean of treatment i. In this section we also assume that the variances of the underlying distributions are equal. Since we have also assumed that observations are normally distributed we have the following linear model

$$X_{ij} = \mu_i + \varepsilon_{ij}, \text{ with } \varepsilon_{ij} \sim N(0,\sigma^2),\ i=1,\ldots,k,\ j=1,\ldots,n_i, \quad (8.6)$$

where the *error terms* $\varepsilon_{ij},\ i=1,\ldots,k,\ j=1,\ldots,n_i$ are independently distributed and σ^2 is the common variance. It can be noted immediately that

$$\bar{X}_{i.} \sim N\left(\mu_i, \frac{\sigma^2}{n_i}\right) \quad \text{and} \quad \frac{n_i S_i^2}{\sigma^2} \sim \chi^2_{n_i - 1}\ ,\quad i=1,\ldots,k \quad (8.7)$$

and that all $2k$ random variables are independent.

Having observed $x_{ij},\ i=1,\ldots,k,\ j=1,\ldots,n_i$ and the summary statistics $\bar{x}_i,\ s_i^2,\ i=1,\ldots,k$, we wish to test the null hypothesis

$$H_0\colon \mu_1 = \mu_2 = \cdots = \mu_k$$

against the alternative

$$H_1\colon \text{not all the population means are equal}\ .$$

It follows from (8.2) that

$$E(S_E) = \sum_{i=1}^{k} n_i E(S_i^2) = (N-k)\sigma^2$$

so that the *mean sum of squares*

$$MSE = \frac{S_E}{N-k} \quad (8.8)$$

is an unbiased estimator of the error variance σ^2. Furthermore,

$$E(S_B) = \sum_{i=1}^{k} n_i E\bar{X}_{i.}^2 - NE\bar{X}^2$$

$$= \sum_{i=1}^{k} n_i \left[\frac{\sigma^2}{n_i} + \mu_i^2 \right] - N\left[\frac{\sigma^2}{N} + \left[\frac{\sum_{i=1}^{k} n_i \mu_i}{N} \right]^2 \right]$$

$$= (k-1)\sigma^2 + \sum_{i=1}^{k} n_i(\mu_i - \bar{\mu})^2, \tag{8.9}$$

where $\bar{\mu} = \sum_{i=1}^{k} n_i \mu_i / N$. Note that if H_0 is true, then the mean sum of squares

$$MSB = \frac{S_B}{k-1} \tag{8.10}$$

is also an unbiased estimator for σ^2. Otherwise, the expected value of MSB is greater than σ^2. Consequently, the random variable

$$F = \frac{MSB}{MSE} = \frac{S_B/(k-1)}{S_E/(N-k)} \tag{8.11}$$

tends to take values closer to 1 under H_0 and values greater than 1 otherwise. This quantity is referred to as the F-Statistic for one-way ANOVA. In fact, under H_0 this random variable has an F distribution with $k-1$ and $N-k$ degrees of freedom, because (8.7) implies that

$$\frac{S_B}{\sigma^2} \sim \chi^2_{k-1} \quad \text{and} \quad \frac{S_E}{\sigma^2} \sim \chi^2_{N-k}$$

and that they are independently distributed. Moreover, it can be shown (Exercise 8.1) that

$$F \sim F_{k-1,N-k}(\delta);$$

that is F has an F distribution with $k-1$ and $N-k$ degrees of freedom, and noncentrality parameter δ defined as

$$\delta = \frac{1}{(k-1)\sigma^2} \sum_{i=1}^{k} n_i(\mu_i - \bar{\mu})^2.$$

Since F tends to take larger values under H_1 than under H_0, the left tail of its distribution yields an unbiased test. In fact, F is stochastically increasing in each $|\mu_i - \bar{\mu}|$, $i = 1,\ldots,k$. Therefore, the p-value appropriate for testing H_0 is

$$p = Pr\left[F \geq \frac{s_B/(k-1)}{s_E/(N-k)} \right]$$

$$= 1 - H_{k-1,N-k}\left[\frac{s_B/(k-1)}{s_E/(N-k)} \right], \tag{8.12}$$

where $H_{k-1,}N-k$ is the cdf of the F distribution with $k-1$ and $N-k$ degrees of freedom. At fixed-level α, H_0 is rejected if $p \leq \alpha$. Equivalently, we can reject H_0 at level α if the observed value of the F-statistic is greater than $F_{k-1,N-k}(\alpha)$, the $(1-\alpha)$th quantile of the F distribution with $k-1$ and $N-k$ degrees of freedom. This test will be referred to as the conventional F-test.

Various sums of squares, their degrees of freedom, the mean sums of squares, and the F-statistic that we need to compute to perform an F-test can be neatly set out in an *analysis of variance table*. Table 8.3 shows the general form of a one-way analysis of variance table.

Table 8.3. One-way ANOVA table

Source of Variation	D.F.	Sum of Squares	Mean Sum of Squares	F-Statistic
Between	$k-1$	s_B	*MSB*	MSB/MSE
Error	$N-k$	s_E	*MSE*	
Total	$N-1$	s_T		

Example 8.2. Comparing the mean yield of corn hybrids (continued)

Let μ_i, $i = 1,2,3,4$ be the mean yield of the four respective corn hybrids considered in Example 8.1. Consider the null hypothesis H_0: $\mu_1 = \mu_2 = \mu_3 = \mu_4$ and the alternative hypothesis that at least two means are different. With the foregoing results we are now in a position to test whether or not the data in the Table 8.1 supports the null hypothesis. Using the sums of squares computed in Example 8.1 we can compute the mean sums of squares and the F-statistic. The results are displayed in Table 8.4.

Table 8.4. Anova table for comparing the four hybrids

Source of Variation	D.F.	Sum of Squares	Mean Sum of Squares	F-Statistic
Between	3	0.728	.2427	1.841
Error	18	2.372	.1318	
Total	21	3.1		

Under the assumption of equal error variances we can use the F-test to compare the four hybrids; in Section 8.4 we shall revisit this problem and try to attack it without this assumption. The 95th percentile of the F distribution with 3 and 18 degrees of freedom is 3.16. Therefore, the null hypothesis of equal means cannot be rejected at the .05 level. The p-value computed from (8.12) is $p = 1 - H_{3,18}(1.841) = .176$ which further confirms that, if the population variances are equal, then the data do not provide sufficient evidence to doubt the null hypothesis H_0.

8.4. ANOVA with Unequal Error Variances

Now consider the problem of comparing the equality of the k means without the assumption of equal error variances. In most practical applications this is typically the case. In any case, if the data does not support the assumption of equal variances the results in Section 8.3 are no longer valid. Section 8.6 develops an exact procedure for testing the equality of variances. Use of the F-test developed in Section 8.3 in a situation with significantly different population variances can lead to serious repercussions. For example, if the variances are quite different and the sample sizes are small one can mistakenly conclude that the differences in means are statistically significant even when the observed differences in sample means are entirely due to sampling variation. In comparing a number of proposed treatments with a currently used treatment, this in turn may lead to the recommendation of a worse treatment than the one currently being used. On the other hand, an equally serious situation occurs when the F-test does not reject the hypothesis of equal means when a set of hard earned data actually provides strong evidence in favor of a better treatment than the one currently being used or a placebo.

Welch (1951) gave an approximate test for the one-way ANOVA problem under heteroscedasticity. Krutchkoff (1988) provided an extensive study to show the poor power performance of the classical F-test and provided a simulation based solution to overcome its drawbacks. Some practitioners use Fisher-Pitman test (cf. Boik (1987)) as an alternative to the classical F-test. Bradbury (1988) obtains an easily computable approximation of this test which is related to the F-test. Based on the findings of a simulation study, however, Boik (1987) recommends the use of approximate tests such as that of Welch (1951) over the permutation test when the assumption of normal distributions is reasonable. Krutchkoff (1988) argues that the Kruskal-Wallis test is not an alternative for the F-test when the variance heterogeneity is serious.

Rice and Gaines (1989) extended the p-value for the Behrens-Fisher problem given by Barnard (1984) to the one-way ANOVA case. Weerahandi (1994) obtained a different representation of the same test which is computationally more efficient than the representation given by Rice and Gaines (1989). The former also provides a simple proof of the fact that the p-value given by Rice and Gaines (1989) is indeed the exact probability of a well defined and unbiased extreme region. This test will be referred to as the generalized F-test.

To present the generalized F-test consider the one-way ANOVA model in (8.6) without the assumption of equal variances. Hence, we now have the following linear model:

$$X_{ij} = \mu_i + \varepsilon_{ij}, \text{ with } \varepsilon_{ij} \sim N(0,\sigma_i^2), \quad i=1,\ldots,k,\ j=1,\ldots,n_i,$$

where σ_i^2 is the variance of population i. The model can also be expressed as

$$X_{ij} = \mu + \alpha_i + \varepsilon_{ij}, \quad i=1,\ldots,k,\ j=1,\ldots,n_i, \tag{8.13}$$

where $\mu = \sum \mu_i / k$ and $\alpha_i = \mu_i - \mu$ so that $\sum_{i=1}^{k} \alpha_i = 0$. The α_i, $i=1,\ldots,k$ are called a set of *contrasts* of means. In terms of this notation the null hypothesis can be rewritten as

$$H_0\colon \alpha_1 = \alpha_2 = \cdots = \alpha_k = 0. \tag{8.14}$$

The p-value in (8.12) can be used to test this null hypothesis if the variances are not significantly different. But often we encounter situations where that assumption is not reasonable. In any case, the null hypothesis of equal means can be tested by a test similar to the one given by (8.12) without relying on the assumption of equal variances. To do this, define the standardized between sum of squares

$$\tilde{S}_B = \tilde{S}_B(\sigma_1^2,\ldots,\sigma_k^2) = \sum_{i=1}^{k} \frac{n_i \bar{X}_i^2}{\sigma_i^2} - \frac{1}{\sum_{i=1}^{k} n_i/\sigma_i^2} \left[\sum_{i=1}^{k} n_i \bar{X}_i / \sigma_i^2 \right]^2 . \tag{8.15}$$

Let $\tilde{s}_B$ be the observed value of $\tilde{S}_B$ obtained by replacing $\bar{X}_i$, $i=1,\ldots,k$, terms appearing in (8.15), by $\bar{x}_i$, $i=1,\ldots,k$. Then, the p-value for testing the null hypothesis in (8.15) is given by

$$p = 1 - E\Big(H_{k-1,N-k}\Big(\frac{N-k}{k-1}\tilde{s}_B\Big[\frac{n_1 s_1^2}{B_1 B_2 \cdots B_{k-1}}, \frac{n_2 s_2^2}{(1-B_1)B_2 \cdots B_{k-1}},$$

$$\frac{n_3 s_3^2}{(1-B_2)B_3 \cdots B_{k-1}}, \ldots, \frac{n_k s_k^2}{(1-B_{k-1})}\Big]\Big)\Big), \tag{8.16}$$

where $H_{k-1,N-k}$ is the cdf of F distribution with $k-1$ and $N-k$ degrees of freedom, and the expectation is taken with respect to the independent beta random variables

$$B_j \sim Beta\left(\sum_{i=1}^{j} \frac{(n_i-1)}{2}, \frac{n_{j+1} - 1}{2} \right), \ j = 1,2,\ldots,k-1. \tag{8.17}$$

The p-value serves to measure the evidence in favor of H_0. In generalized fixed-level testing at level α, H_0 is rejected if $p < \alpha$.

In significance testing, the p-value given by (8.16) provides an exact unbiased test based on sufficient statistics. Exact fixed-level tests (conventional) based on sufficient statistics do not exist. In this case, the generalized fixed-level test provides an excellent approximate test. Therefore, the generalized F-test is extremely useful regardless of whether one prefers conventional fixed-level testing or significance testing.

A derivation of the test is given following the two examples. We saw in Chapter 7 that for the problem of comparing two normal populations, the test is unique (up to equivalent p-values) among all affine invariant and unbiased procedures based on minimal sufficient statistics. In view of results in Khuri (1990) and Zhou and Mathew (1994), however, for the general test given by (8.16) the uniqueness of the test is not expected unless further conditions are imposed.

The p-value can be computed by numerical integration. The integrand here is so well behaved so that in a typical practical application there is no difficulty evaluating the integral correct up to a reasonable degree of accuracy. If the number of treatments being compared is very large, the expectation in (8.16) can also be well approximated by a Monte Carlo method. In this method equal (large) numbers of random numbers from each beta random variable are generated, the cdf $H_{k-1,N-k}$ is evaluated at each set, and then the expectation is estimated by their sample mean. The accuracy of the approximation can also be assessed. For example, if the sample (simulated) standard deviation of H values is σ_h, then with probability .999, the estimated p-value is accurate up to about $3\times\sigma_h/L^{1/2}$, where L is the number of simulated sample sets. Both the exact (up to 4 decimal points) numerical integration and the Monte Carlo procedure are integrated into the one-way ANOVA tool of the XPro software package.

Note that for large sample sizes the degenerated beta random variables give rise to the same p-value as the one implied by (8.12) with known error variances. That the above formula for the p-value is actually symmetric with respect to the population indices will become clear in the derivation of the result. In other words, we will get the same p-value irrespective of how the treatments are indexed.

Example 8.3. Comparing the mean yield of corn hybrids (continued)

Consider again the data in Table 8.1 and the summary statistics in Table 8.2. In Example 8.2 by applying the F-test we concluded that there is no sufficient evidence to reject the null hypothesis that the mean yields of the four corn hybrids are equal. Let us now drop the assumption of equal error variances and retest the hypothesis. In this application the p-value given by (8.16) reduces to

$$p = 1 - E(H_{3,18}(6\tilde{s}_B[\frac{6\times.189}{B_1B_2B_3}, \frac{5\times.030}{(1-B_1)B_2B_3}, \frac{6\times.038}{(1-B_2)B_3}, \frac{5\times.172}{(1-B_3)}]))$$
$$= .049$$

correct up to three decimal points, where the expectation is taken with respect to the beta random variables $B_1 \sim Beta(2.5,2)$, $B_2 \sim Beta(4.5,2.5)$ and $B_3 \sim Beta(7,2)$. The exact p-value is computed using XPro. The p-value estimated by generating 1,000 sets of random numbers from Beta distributions, using the Monte Carlo option of XPro is .048, which is in close agreement with the exact p-value computed by numerical integration. Hence, the null hypothesis can be rejected at the .05 level. This means that, although the simple F-test in Section 8.3 failed to detect it, the data in Table 8.1 do provide sufficient evidence to conclude that the observed differences of the mean yields from the four hybrids are actually significant and cannot be attributed to just the sampling variation. Therefore, the experimenter can proceed with multiple comparisons and deal with hypotheses concerning absolute differences in mean yields.

Recall that the p-value given by the conventional F-test was .176 which is much larger than the p-value obtained above under milder assumptions. This means that when the assumption of equal variances is not reasonable, the test given by (8.16) is much more powerful than the test given by (8.12). The failure of the conventional F-test to detect mean differences in this kind of application can be considered very serious. This is especially true in biomedical experiments where one does not usually get large samples and cannot afford to resort to less efficient statistical procedures for the sake of simplicity. The need for large samples for the conventional F-test to work well may also result in a long delay and a large cost before the experiment can be concluded with significant results. Therefore, the simple F-test is recommended only when the assumption of equal variances is reasonable. In Section 8.6 we will study how to test the assumption of equal variances.

Example 8.4. Type I error performance of the F-test

What we observed in Example 8.3 is basically the performance of the F-test with respect to Type II error; that is, accepting the null hypothesis when the alternative hypothesis is true. Let us now study the performance of the F-test with respect to Type I error; that is rejecting the null hypothesis when it is actually true. Since the test has the exact specified size when the variances are equal we study its performance when the variances are quite different. We shall do this with a set of simulated data from normal distributions. Consider the problem of comparing three means. The mean of each distribution is taken to be ten so that the null hypothesis H_0: $\mu_1 = \mu_2 = \mu_3 = 10$ is true. When the sample sizes are equal the F-test is known to be fairly robust against the assumption of equal variances. The most serious situation occurs when the sample sizes are negatively correlated with the variances. We consider this situation only; a reader interested in details about other situations is referred to Anderson and McLean (1974).

Table 8.5 shows the results of a simulated experiment in which data are generated from normal distributions with mean ten and four values of the variance, σ_i^2, $i=1,...,5$. The table provides the summary statistics, namely the sample size n, the sample mean $\bar{x}$, and the sample standard deviation s for each value of σ.

Table 8.5. Standard deviations and summary statistics of simulated data

Population	σ_i	n_i	$\bar{x}_{i.}$	s_i
Treatment A	1	16	10.03	1.08
Treatment B	2	12	9.57	1.90
Treatment C	3	8	8.70	2.46
Treatment D	4	6	7.92	3.34
Treatment E	4	4	12.96	3.40

Three treatments out of the five in Table 8.5 are to be compared at a time. The p-values are computed with and without the assumption of equal variances. These p-values computed using the formulae (8.12) and (8.16) are denoted by p_e and p_u, respectively. The findings of the study are displayed in Table 8.6. It can be noted that the conventional F-test tends to reject the null hypothesis if the variances of the populations are substantially different or when they are moderately different and the sample sizes are small. Even in comparing treatments B, C, and E, this test suggests that we have strong evidence to reject the null hypothesis although the data on which the evidence is based are generated when the hypothesis is actually true. This cannot be attributed to sampling variation since the disposition of the equal variances assumption rectifies the problem. The sample sizes and the standard deviations in this study are quite typical of many situations so that even this behavior of the F-test is unacceptable.

Table 8.6. P-values with and without equal variances assumption

Treatments compared	p_e	p_u
(B, C, and D)	.448	.525
(A, B, and C)	.258	.376
(A, B, and D)	.112	.367
(A, B, and E)	.016	.330
(B, C, and E)	.038	.251

Interestingly, p_u, the p-value valid under unequal variances, takes large values under each of the five scenarios considered and does not at all suggest the rejection of the null hypothesis. Recall that in Example 8.3, in which the variation of sample variances was at about the same level, the two p-values p_e and p_u had the opposite order of magnitude. Compared to the conventional F-test, the procedure given by (8.16) seem to provide a much more efficient procedure in detecting the significance of mean differences.

Derivation of the test

Consider the standardized between sum of squares in (8.15) and the standardized error sum of squares defined as

$$\tilde{S}_E = \sum_{i=1}^{k} n_i S_i^2 / \sigma_i^2 \,.$$

It is easily seen (Exercise 8.2) that $\tilde{S}_E$ is the error sum of squares (for the transformed problem with specified variances) having a chi-squared distribution with $N-k$ degrees of freedom, and $\tilde{S}_B$ is the between sum of squares having an independent chi-squared distribution with $k-1$ degrees of freedom when H_0 is true. Moreover,

$$\tilde{S}_T = \sum_{i=1}^{k}\sum_{j=1}^{n_i} \frac{1}{\sigma_i^2}(X_{ij} - \bar{\bar{X}})^2$$

is the total sum of squares such that $\tilde{S}_t = \tilde{S}_E + \tilde{S}_B$, where $\bar{\bar{X}} = (\sum_{i=1}^{k} \bar{X}_i n_i/\sigma_i^2)/(\sum_{i=1}^{k} n_i/\sigma_i^2)$. Hence as in one-way ANOVA with equal variances we get under H_0,

$$\frac{\tilde{S}_B/(k-1)}{\tilde{S}_E/(N-k)} \sim F_{k-1,N-k}.$$

Let

$$B_j = \frac{\sum_{i=1}^{j} n_i S_i^2/\sigma_i^2}{\sum_{i=1}^{j+1} n_i S_i^2/\sigma_i^2}\,, \quad j=1,\ldots,k-1. \qquad (8.18)$$

It can be shown (Exercise 8.4) that B_j is a beta random variable with parameters $\sum_{i=1}^{j}(n_i - 1)/2$ and $(n_{j+1} - 1)/2$ and that $\tilde{S}_E$, B_j, $j = 1,2,\ldots,k-1$ are all independent random variables. Note also that the chi-squared random variables $n_i S_i^2/\sigma_i^2$ can be expressed as

$$\frac{n_1 S_1^2}{\sigma_1^2} = \tilde{S}_E B_1 B_2 \cdots B_{k-1},$$

$$\frac{n_i S_i^2}{\sigma_i^2} = \tilde{S}_E(1-B_{i-1})B_i \cdots B_{k-1} \quad \text{for } i=2,...,k, \tag{8.19}$$

$$\text{and } \frac{n_k S_k^2}{\sigma_k^2} = \tilde{S}_E(1 - B_{k-1}),$$

In view of these results, define a potential generalized test variable as

$$T = \frac{\tilde{S}_B(\sigma_1^2,...,\sigma_k^2)}{\tilde{s}_B(s_1^2\sigma_1^2/S_1^2,...,s_k^2\sigma_k^2/S_k^2)}. \tag{8.20}$$

The observed value of T is $t_{obs} = 1$. Moreover, it will become clear that under the null hypothesis its distribution does not depend on any nuisance parameters, and just like the F-statistic used in Section 8.3, T tends to take larger values for deviations from H_0. Hence, T is indeed a generalized test variable. The p-value given by this test variable can be found as

$$\begin{aligned} p &= Pr(T \geq t_{obs}) \\ &= Pr\left[\tilde{S}_B(\sigma_1^2,...,\sigma_k^2) \geq \tilde{s}_B(\frac{s_1^2\sigma_1^2}{S_1^2},...,\frac{s_k^2\sigma_k^2}{S_k^2})\right] \\ &= Pr\left[\tilde{S}_B \geq \tilde{s}_B(\frac{n_1 s_1^2}{B_1 B_2 \cdots B_{k-1}\tilde{S}_E},...,\frac{n_k s_k^2}{(1-B_{k-1})\tilde{S}_E})\right]. \end{aligned}$$

But it follows from the definition of $\tilde{S}_B$ that it has the property that, for any given positive constant c and a vector $\mathbf{y}$ with positive elements,

$$\tilde{S}_B(cx_1,...,cx_k) = \tilde{S}_B(x_1,...,x_k)/c.$$

Therefore, the p-value can be expressed as

$$\begin{aligned} p &= Pr\left[\tilde{S}_B \geq \tilde{S}_E \tilde{s}_B(\frac{n_1 s_1^2}{B_1 B_2 \cdots B_{k-1}},...,\frac{n_k s_k^2}{(1-B_{k-1})})\right] \\ &= Pr\left[\frac{\tilde{S}_B/k-1}{\tilde{S}_E/N-k} \geq \frac{N-k}{k-1}\tilde{s}_B(\frac{n_1 s_1^2}{B_1 B_2 \cdots B_{k-1}},...,\frac{n_k s_k^2}{(1-B_{k-1})})\right] \\ &= 1 - E(H_{k-1,N-k}\left(\frac{N-k}{k-1}\tilde{s}_B\left[\frac{n_1 s_1^2}{B_1 B_2 \cdots B_{k-1}},...,\frac{n_k s_k^2}{(1-B_{k-1})}\right]\right)) \end{aligned}$$

as claimed.

8.5. Multiple Comparisons

Now suppose a hypothesis of equal treatment means has been rejected at a certain nominal level. In many applications, the problem is not completely solved. We still need to identify the means which are significantly different from others. One can of course do this with pairwise comparisons. Although the concerns about pairwise comparisons expressed in the introduction of this chapter still hold, the issue is no longer very serious since the null hypothesis has already been rejected. In any event, the appropriate procedure to carry out multiple comparisons depends on how we want to control Type I error. A reader interested in various pre-planned and post-hoc procedures valid under different control conditions is referred to Woolson (1987). Here we consider only (i) the problem of controlling the error rate for each pre-planned experiment separately; and (ii) Scheffe's procedure in which Type I error is on a *per-experiment* basis.

First consider the problem of making pairwise comparisons under the conditions of the former procedure. This is appropriate only for a couple of planned comparisons. For example, having concluded that the differences in treatment means are statistically significant, an experimenter may wish to compare only two or three treatments with the largest sample means, or to compare the treatment having the largest mean with the placebo treatment. If the variances are equal, as assumed in Section 8.3, the common variance can be estimated as in (8.8). Let $\hat{\sigma}^2 = S_E/(N-k)$ be the unbiased estimator which utilizes all the information about the common variance σ^2. The distribution of this random variable is given by

$$\frac{(N-k)\hat{\sigma}^2}{\sigma^2} \sim \chi^2_{N-k}.$$

With this result we can test hypotheses about contrasts of the two means by employing the results developed in Section 7.2. For example, one-sided hypotheses of two means, say $H_0 : \mu_1 \leq \mu_2$ can be tested on the basis of the p-value

$$p = G_{N-k}\left(-\frac{\hat{\mu}_2 - \hat{\mu}_1}{\hat{\sigma}(1/m + 1/n)^{1/2}} \right), \tag{8.21}$$

where $\hat{\mu}_1$ and $\hat{\mu}_2$ are the sample means of the two treatments being compared, m and n are the sample sizes, and G_{N-k} is the cdf of the Student's t distribution with $N-k$ degrees of freedom. Absolute differences of the two means can be tested as in equation (7.9).

If the variances are not assumed equal, the results in Section 7.3 can be readily employed to make inferences about the difference in two means. In this case, the procedure remains the same, as no additional information about the variances are available.

To illustrate the procedure consider the summary statistics of an experiment given by the following table:

Treatment	n_i	$\bar{x}_i$	s_i^2
A	7	40.7	8.9
B	6	37.2	4.6
C	8	46.6	18.9

The sum of squares and the F-statistics can be computed using formulae given in Sections 8.2 and 8.3. The results are summarized in the following ANOVA table:

Source of Variation	D.F.	Sum of Squares of Squares	Mean Sum	F-Statistic
Between	2	319.3	159.6	11.92
Error	18	241.1	13.39	
Total	20	560.4		

Clearly there is strong evidence to reject the hypothesis of equal means; the observed levels of significance are .0005 and .0017, respectively with and without the assumption of equal variance. Now let us perform pairwise comparisons, each at .05 level. By applying (8.21) it is seen that the mean of Treatment C is greater than those of B and A, but the difference in means between *A* and *B* is not quite statistically significant at the 5% level; the p-values of the hypotheses $\mu_B \le \mu_C$ and $\mu_B \le \mu_A$ are unchanged even if the assumption of equal variances is dropped; the p-values of the two hypotheses are .051 and .011, respectively. Therefore, we can conclude at the .05 level that

$$\mu_A < \mu_C \text{ and that } \mu_B < \mu_C.$$

In other words the mean of treatment C is significantly larger than the means of other two treatments.

Scheffe's Procedure

This is an experiment-wise multiple comparison method which is closely related to the F-test described in Section 8.3. This method allows us to perform simultaneous tests and confidence intervals for linear combinations of means. Among all multiple comparison methods it has a unique advantage, namely, it detects one or more significant linear contrasts if and only if the F-test is significant.

To describe the procedure, consider the null hypothesis

$$H_0: \sum_{i=1}^{k} c_i \mu_i = 0 \text{ for all } c_i \text{ such that } \sum_{i=1}^{k} c_i = 0, \tag{8.22}$$

which states that all contrasts of means are zero. However, in applications, we may be interested only in some contrasts. As a result, the true size of a test designed for (8.22) and applied to a subset of contrasts will be less than the specified nominal size.

Testing of (8.22) and simultaneous confidence intervals for the linear contrasts can be deduced from results of Section 8.3 and the following lemma, which is an implication of the Schwarz inequality.

Lemma 8.1. Given a set of real numbers (x_i, μ_i), $i = 1,...,k$ and a nonnegative constant θ_0, $\sum_{i=1}^{k}(x_i - \mu_i)^2 \le \theta_0$ if and only if $(\sum_{i=1}^{k} a_i x_i - \sum_{i=1}^{k} a_i \mu_i)^2 \le \theta_0 \sum_{i=1}^{k} a_i^2$ for all real numbers a_i, $i = 1,...,k$.

Since, $Var(\sum_{i=1}^{k} c_i \bar{X}_i) = \sigma^2 \sum_{i=1}^{k} c_i^2 / n_i^2$ it follows from Lemma 8.1 (see Scheffe (1953) for details) and results in Section 8.3 that

$$Pr\{(\sum_{i=1}^{k} c_i \bar{X}_i - \sum_{i=1}^{k} c_i \mu_i)^2 \le \theta_0 (k-1) \hat{\sigma}^2 \sum_{i=1}^{k} \frac{c_i^2}{n_i}$$

$$\text{for all } c_i \text{ such that } \sum_{i=1}^{k} c_i = 0\} = H_{k-1,N-k}(\theta_0) .$$

It is now evident that the null hypothesis in (8.22) can be tested on the basis of the p-value

$$p = 1 - H_{k-1,N-k} \left[\frac{N-k}{k-1} \frac{(\sum_{i=1}^{k} c_i \bar{x}_i)^2}{s_E \sum_{i=1}^{k} c_i^2 / n_i} \right] . \tag{8.23}$$

A set of $100(1-\alpha)\%$ simultaneous confidence intervals for the linear contrasts is given by

$$\sum_{i=1}^{k} c_i \bar{x}_i - g \le \sum_{i=1}^{k} c_i \mu_i \le \sum_{i=1}^{k} c_i \bar{x}_i + g , \tag{8.24}$$

where

$$g^2 = (k-1)\hat{\sigma}^2 (\sum_{i=1}^{k} c_i^2 / n_i) F_{k-1,N-k}(\alpha),$$

and $F_{k-1,N-k}$ is the $(1-\alpha)$th quantile of the F distribution with $k-1$ and $N-k$

degrees of freedom. Note that when applied to a particular contrast such as $\mu_1-\mu_2$ by setting $c_1 = 1$, $c_2 = -1$ and $c_i = 0$ for $i = 3,...,k$, the length of the confidence interval given by (8.24) can be much larger than the $100(1-\alpha)\%$ confidence interval given by (8.21).

Fixed-level tests of (8.22) with specified values of c_i's can be based on (8.23) or on (8.24); with the latter, the null hypothesis is rejected if the interval does not contain zero.

To illustrate the procedure, consider again the preceding data set. Consider the problem of making simultaneous inferences about the three differences in means. Let us construct 90% confidence intervals by applying formula (8.24) so that the one-sided tests implied by them will have size .05. Setting $c_i=1$, $c_j=-1$ and the third c parameter to be equal to 0, we have

$$g^2 = 2\times 13.39\left(\frac{1}{n_i}+\frac{1}{n_j}\right)F_{2,18}(.1)\ .$$

Hence, the simultaneous confidence intervals given by (8.24) are as follows:

Parameter	$\bar{x}_i - \bar{x}_j$	Confidence Interval
$\mu_A - \mu_B$	3.5	(-1.16, 8.16)
$\mu_C - \mu_B$	9.4	(4.88, 13.92)
$\mu_C - \mu_A$	5.9	(1.56, 10.24)

Observe that the first interval contain zero while the other two do not. Thus we come to the same conclusion as before; that is, while the difference between the means of treatments A and B is not statistically significant, the mean of treatment C is significantly larger.

As in Section 8.3, Scheffe's procedure for multiple comparisons can be easily extended to the case of unequal population variances. This is accomplished by considering the standardized observations $\tilde{X}_{ij} = X_{ij}/\sigma_i$, $i=1,...,k$, $j=1,...,n_i$, defining a test variable parallel to (8.20) in terms of $\sum_{i=1}^{k} c_i\bar{\tilde{X}}_i$, and utilizing the above results. Then, it can be shown (Exercise 8.5) that the p-value appropriate for testing (8.22) is given by

$$p = 1 - E\left(H_{k-1,N-k}\left(\frac{N-k}{k-1}\,\frac{(\sum_{i=1}^{k} c_i\bar{x}_i)^2}{\sum_{i=1}^{k} c_i^2 s_i^2/Y_i}\right)\right), \qquad (8.25)$$

where the expectation is taken with respect to the random variables

$$Y_i = (1-B_{i-1})B_i\cdots B_{k-1};\ i=2,...,k-1,$$

$$Y_1 = B_1 B_2 \cdots B_{k-1}, \text{ and } Y_k = (1 - B_{k-1}),$$

where B_i, $i = 1, ..., k-1$ are the beta random variables defined by (8.17). A test based on this p-value is referred to as a generalized Scheffe test. For a desired set of contrasts, the statistical software package XPro provides p-values to carry out the test with and without the assumption of equal variances.

Simultaneous generalized confidence intervals for contrasts of means can also be deduced from (8.25). The result is an extension of formula (7.16). A pre-planned comparison involving any desired contrast can also be performed in a similar manner. The test is based on the fact that, under the null hypothesis, $\sum_{i=1}^{k} c_i^2 \bar{x}_i$ is normally distributed. It is now evident that the hypothesis can be tested based on the p-value

$$p = 1 - E(H_{1,N-k}\left(N-k \frac{(\sum_{i=1}^{k} c_i \bar{x}_i)^2}{\sum_{i=1}^{k} c_i^2 s_i^2 / Y_i} \right)).$$

XPro provides p-values for the pre-planned version as well. This is also given both under the homoscedasticity and the heteroscedasticity. Continuing with our illustrative example, suppose the desired contrast placed on the three treatments is given by $c_1 = 8$, $c_2 = -1$ and $c_3 = -7$. The p-values (computed using XPro) for testing this contrast obtained under various scenarios is shown in the following table:

Comparison	Equal variances	Unequal variances
Pre-planned	.0170	.0417
Post-hoc	.0537	.1133

Although the sample variances in this application are quite typical of many applications, both in the pre-planned comparison and in the post-hoc comparison, the assumption of equal variances seems to be quite serious. Therefore, one should not make this assumption unless it is actually a reasonable assumption that can be justified by providing additional evidence.

8.6. Testing the Equality of Variances

In preceding sections we obtained procedures for testing the equality of treatment means with and without the assumption of equal variances. While the former has the advantage of being simple and easy to compute, the latter is more general than the former. In either case we also assumed that the underlying random variables are normally distributed. According to the studies carried out by Cochran and Cox (1957), the F-test is relatively robust to minor departures from the underlying assumptions. In many applications the assumption of normality can be achieved for practical purposes by a suitable transformation. For example, random variables taking only positive values and having a unimodal distribution can be often transformed into approximate normal random variables by applying the logarithmic transformation. For more details about the use of transformations, the reader is referred to Bartlett (1947). Except for a few special cases, however, unless the variances are known it is not usually possible to simultaneously obtain the approximate normality and equal variances by this method. For an interesting discussion of this issue and other related issues, the reader is referred to Krutchkoff (1988).

The F-test is fairly robust with respect to small departures from the the assumption of a common variance as long as an equal number of observations are available from each treatment. According to studies, the worst situation occurs when the sample sizes are negatively correlated with the variances. Unfortunately, this is the case when some old treatments are being compared with some new treatments, because compared with old treatments, new treatments tend to have larger means and variances while the available sample sizes tend to be smaller. When the assumption is not reasonable, as we saw in Example 8.4, the actual size (probability of rejecting the hypothesis of equal means when the hypothesis is true) of the F-test tends to be larger than the intended nominal level. Consequently, one may conclude that the means are significantly different when the variances are actually the parameters which are significantly different. On the other hand, as we saw in Example 8.3, the F-test may fail to detect differences in means which are actually statistically significant. Therefore, it is important to test the assumption of equal variances if the conventional F-test is to be entertained. This is a task worth undertaking because if the variances are not very different, the F-test might be more powerful than the one which does not make the assumption.

As we discussed in Chapter 7, comparison of variances itself is the main objective of some studies. In any event, consider the problem of testing the equality of variances. There are a number of approximate procedures to test this hypothesis. Anderson and McLean (1974) describe these tests in some detail. Perhaps the most widely used test is the one due to Bartlett (see, for instance, Bartlett and Kendall (1946)). The drawback of these tests is that they require large samples. On the other hand, when large samples are available it is argued that there is little point in testing the equality of means or variances (as large samples can easily detect even small differences among parameters) and one can proceed directly to multiple comparisons and should deal with absolute differences between means. Cohen and Strawderman (1971) studied a class of exact tests which are unbiased when the sample sizes are equal. Of particular interest is the exact test given by the generalized likelihood ratio method. The test is presented here in terms of the beta random variables that we encountered in developing the generalized F-test.

We wish to test the null hypothesis

$$H_0: \sigma_1^2 = \sigma_2^2 = \cdots = \sigma_k^2 \qquad (8.26)$$

against the alternative

$$H_1: \text{at least one variance is different.}$$

The hypotheses are to be tested based on the linear model (8.13) and the sufficient statistics given by (8.1), namely the sample means and the sample variances. When $k = 2$ an exact test is given by results in Section 7.4. For any value of k in general, the null hypothesis can be tested on the basis of the p-value

$$p = Pr\left[\prod_{i=1}^{k} Y_i^{n_i} < \prod_{i=1}^{k}\left(\frac{n_i s_i^2}{\sum n_j s_j^2}\right)^{n_i}\right], \qquad (8.27)$$

where Y_i random variables are defined as

$$Y_i = (1-B_{i-1})B_i \cdots B_{k-1};\ i=2,\dots,k-1,$$
$$Y_1 = B_1 B_2 \cdots B_{k-1},\ Y_k = (1-B_{k-1}),$$

and B_i, $i=1,\dots,k-1$ are the beta random variables defined by (8.17). The probability is computed with respect to the independent beta random variables. XPro computes the p-value both by the exact formula given by (8.27) and by Barlett's asymptotic approximation. We shall show later that the test given by the above p-value is nothing but the test given by the generalized likelihood ratio method described in Section 2.4. This test is exact, but like Bartlett's test, it is not quite unbiased unless the sample sizes are equal. However, it has the desirable property that, as was the case with tests for comparing means, the p-value reduces to one when the sample variances are equal and it tends to zero as any one sample variance tends to infinity or to zero.

At fixed-level α, H_0 is rejected if $p < \alpha$. The p-value can be computed by numerical integration with respect to the beta random variables. A good approximation can also be obtained by generating a large number of beta random numbers and then finding the frequency, say $\hat{p}$, with which the inequality in (8.27) is satisfied. The accuracy of the approximation can be assessed as in Section 8.4. In this case, the Monte Carlo variance of the estimate of p can be easily computed using the formula

$$Var(\hat{p}) = \frac{p(1-p)}{L},$$

where L is the number of Monte Carlo samples used to estimate p.

Example 8.5. Testing the equality of variances in comparing corn hybrids
Consider the hypothesis that the variances of the distributions of yields from the

four corn hybrids are equal. The test can be performed using the sample sizes and the sample variances in Table 8.2. The p-value based on the formula (8.27) for testing the equality of all four variances, computed using XPro is .206. Therefore, the data in Table 8.1 do not provide enough evidence to reject the null hypothesis at a reasonable level. It is remarkable that, even though the variances are not significantly different in this application, the test which does not assume the equality of the variances is more powerful than the F-test, as we saw in Example 8.3. The F-test applied to one-sided pairwise comparisons provides some indication of different variances. The ratio between the variances of Hybrid A and Hybrid B is significant at the .05 level. This is the most significant pairwise comparison of variances. Although this suggests some weak evidence against the equality of variances, as pointed out in Section 1, drawing conclusions based on pairwise comparisons can increase the Type I error to prohibitive levels.

Derivation of the generalized likelihood ratio test

Consider the hypothesis in (8.26). Defining $\zeta = \sum \sigma_i^2$ and $\rho_i = \sigma_i^2/\zeta$, we can rewrite the null hypothesis as

$$H_0;\ \rho_i = \frac{1}{k},\quad i=1,\ldots,k.$$

Then, we can consider $\rho_i,\ i=1,\ldots,k$ as the parameters of interest and $\mu_i,\ i=1,\ldots,k,\ \zeta$ as nuisance parameters.

In terms of the observations from the one-way layout, the likelihood function can be expressed as

$$L = \prod_{i=1}^{k}\prod_{j=1}^{n_i} \frac{1}{(2\pi\sigma_i^2)^{½}} \exp\left[-\frac{1}{2}\frac{(x_{ij}-\mu_i)^2}{\sigma_i^2}\right] \tag{8.28}$$

$$= \prod_{i=1}^{k}(2\pi\zeta\rho_i)^{-n_i/2} \exp\left[-\frac{1}{2\zeta}\frac{n_i}{\rho_i}(s_i^2 + (\bar{x}_i-\mu_i)^2)\right].$$

If the likelihood function is maximized with respect to all unknown parameters the maximum occurs at the MLE's of means and variances given by (8.1). If it is maximized with respect to only the nuisance parameters, the maximum occurs at

$$\hat{\zeta} = \frac{\sum_{i=1}^{k} n_i s_i^2/\rho_i}{\sum_{i=1}^{k} n_i},\quad \hat{\mu}_i = \bar{x}_i,\ i=1,\ldots,k.$$

Therefore, the generalized likelihood ratio (GLR) defined by (2.12) reduces to

$$G(\mathbf{x};\rho) = \frac{\left[\prod_{i=1}^{k}(2\pi\hat{\zeta}\rho_i)^{-n_i/2}\right] e^{-\sum n_i/2}}{\left[\prod_{i=1}^{k}(2\pi s_i^2)^{-n_i/2}\right] e^{-\sum n_i/2}}$$

$$= \prod_{i=1}^{k}\left[\frac{s_i^2}{\rho_i}\frac{\sum n_j}{\sum n_j s_j^2/\rho_j}\right]^{n_i/2} = \prod_{i=1}^{k}\left[\frac{s_i^2}{\sigma_i^2}\frac{\sum n_j}{\sum n_j s_j^2/\sigma_j^2}\right]^{n_i/2} \qquad (8.29)$$

When computed under H_0, small values of GLR discredit the null hypothesis. Therefore, the p-value of the GLR test is

$$p = Pr(G(\mathbf{X};\rho) < G(\mathbf{x};\rho \mid H_0))$$

$$= Pr\left[\prod_{i=1}^{k}\left[\frac{n_i S_i^2/\sigma_i^2}{\sum n_j S_j^2/\sigma_j^2}\right]^{n_i} < \prod_{i=1}^{k}\left[\frac{n_i s_i^2}{\sum n_j s_j^2}\right]^{n_i}\right]. \qquad (8.30)$$

Finally, the representation (8.27) of the p-value immediately follows from (8.19) and (8.30).

8.7. Two-way ANOVA without Replications

To set the stage for the two-way analysis of variance that we will undertake in Section 8.8 and to outline some ideas in the design of experiments, let us briefly consider the two-way layout with no replications. In this design, the available experimental units are divided into nearly homogeneous groups called blocks. Suppose there are n homogeneous (within) blocks. In the case of balanced complete blocks, experiments designed to compare k treatments based on equal numbers of observations, each homogeneous block is divided into k units and the k treatments are randomly assigned to the k units. If the particular blocks are of no particular interest and they are chosen from a larger population, then block effects should be treated as *random effects* and ANOVA should be carried out under the setting of mixed models that we will study in Chapter 9. In this section we consider only the case where the block effects can be treated as *fixed effects* and we perform the crossed two-factor design to reduce the mean error sum of squares.

The arrangement of N experimental units in n blocks in this manner is called a *randomized block design*. If the treatments are randomly assigned in N experimental units, as was the case in the applications considered in previous sections, then the design is said to be a *completely randomized design*. In agricultural experiments, the blocks can be plots of land having similar soil conditions within each block, soil conditions can vary across blocks. Blocks can also be groups of animals of the same type (e.g. pigs of the same weight and age grouped together), patients of various age groups, ethnic groups, counties of a state, and so on. The idea in the randomized block design is to reduce the variability. When the experimental units or subjects are highly heterogeneous, blocks are introduced in order to minimize the heterogeneity within blocks. Although comparison of treatments is of primary interest, in some applications the comparison of block effects is also of interest. This is especially true when the data are classified according to two characteristics and the experimenter is interested in studying the effect of each characteristic. In the literature, the characteristics of classification are known as *factors*.

So, in general, suppose we are interested in the fixed effects of two factors, A and B; in the case of the randomized block design, A stands for treatment effects and B stands for block effects. Let $A_1, A_2, \ldots, A_k$ be the levels of factor A, and $B_1, B_2, \ldots, B_n$ be the levels of factor B. Distinct values of a factor are called the *levels* of the factor. In the current setting only a single observation is allowed for each combination of levels of A and B; this restriction will be relaxed in the following section. The available data can bc set out as in Table 8.7.

Table 8.7. Layout of a randomized block design

	Blocks				
Treatments	B_1	$\cdots$	B_j	$\cdots$	B_n
A_1	x_{11}	$\cdots$	x_{1j}	$\cdots$	x_{1n}
A_2	x_{21}	$\cdots$	x_{2j}	$\cdots$	x_{2n}
$\cdot$	$\cdot$	$\cdot$	$\cdot$	$\cdot$	$\cdot$
A_i	x_{i1}	$\cdots$	x_{ij}	$\cdots$	x_{in}
$\cdot$	$\cdot$	$\cdot$	$\cdot$	$\cdot$	$\cdot$
A_k	x_{k1}	$\cdots$	x_{kj}	$\cdots$	x_{kn}

Assuming a linear model of the form (8.13), we have

$$X_{ij} = \mu + \alpha_i + \beta_j + \varepsilon_{ij}, \; i=1,\ldots,k, \; j=1,\ldots,n, \tag{8.31}$$

where α_i is the ith effect of factor A and β_j is the jth effect of factor B standardized such that $\sum_{i=1}^{k} \alpha_i = 0$ and $\sum_{j=1}^{n} \beta_j = 0$, respectively. In this context we are interested in testing the null hypotheses

$$H_{0A}\colon \alpha_1 = \alpha_2 = \cdots = \alpha_k = 0 \tag{8.32}$$

and

$$H_{0B}: \beta_1 = \beta_2 = \cdots = \beta_n = 0 \tag{8.33}$$

against the natural alternative hypotheses.

Let $\bar{X}_{i.}$ be the sample mean of factor A_i and let $\bar{X}_{.j}$ be the sample mean of factor B_j. Let $\bar{X}$ be the grand mean and let $N = nk$. The minimum variance unbiased estimators of the parameters α_i and β_j are $\hat{\alpha}_i = \bar{X}_{i.} - \bar{X}$ and $\hat{\beta}_j = \bar{X}_{.j} - \bar{X}$, respectively. As before testing procedures can be based on a decomposition of the total sum of squares $S_T = \sum\sum(X_{ij} - \bar{X})^2$. Consider the decomposition

$$S_T = S_A + S_B + S_E,$$

implied by the indentity

$$(X_{ij} - \bar{X}) = (\bar{X}_{i.} - \bar{X}) + (\bar{X}_{.j} - \bar{X}) + (X_{ij} - \bar{X}_{i.} - \bar{X}_{.j} + \bar{X}),$$

where

$$S_A = \sum_{i=1}^{k}\sum_{j=1}^{n}(\bar{X}_{i.} - \bar{X})^2 = n\sum_{i=1}^{k}\bar{X}_{i.}^2 - N\bar{X}^2,$$

$$S_B = \sum_{i=1}^{k}\sum_{j=1}^{n}(\bar{X}_{.j} - \bar{X})^2 = k\sum_{j=1}^{n}\bar{X}_{.j}^2 - N\bar{X}^2,$$

$$S_E = \sum_{i=1}^{k}\sum_{j=1}^{n}(X_{ij} - \bar{X}_{i.} - \bar{X}_{.j} + \bar{X})^2$$

are the sums of squares due to the sources A, B, and error, respectively.

Suppose that the error terms are independently and normally distributed. Let

$$MSA = \frac{S_A}{(k-1)}, \quad MSB = \frac{S_B}{(n-1)}, \quad \text{and} \quad MSE = \frac{S_E}{(n-1)(k-1)}$$

be the mean sums of squares obtained by dividing each sum of squares by its degrees of freedom. Assume as in (8.6) that the error variances are all equal. If we have more than one observation from each cell we can allow the variances to be different; we will study this in Section 8.9.

By the orthogonality of the above decomposition, the sums of squares S_A, S_B, and S_E are all independently distributed and

$$\frac{S_E}{\sigma^2} \sim \chi^2_{(n-1)(k-1)},$$

where σ^2 is the common error variance. Moreover,

$$\frac{S_A}{\sigma^2} \sim \chi^2_{k-1} \quad \text{and} \quad \frac{S_B}{\sigma^2} \sim \chi^2_{n-1}$$

under H_{0A} and H_{0B}, respectively.

Hence, under H_{0A},

$$F_A = \frac{MSA}{MSE} = \frac{S_A/(k-1)}{S_E/(n-1)(k-1)} \sim F_{k-1,(n-1)(k-1)}; \tag{8.34}$$

that is, the random variable F_A has an F distribution with $k-1$ and $(n-1)(k-1)$ degrees of freedom. Further, F_A tends to take large values for deviations from H_{0A}. Therefore, H_{0A} can be tested on the basis of the p-value

$$p_a = 1 - H_{a,e}\left(\frac{s_A/a}{s_E/e}\right), \tag{8.35}$$

where $a = k-1$, $e = (n-1)(k-1)$, $H_{a,b}$ is the cdf of the F distribution with a and b degrees of freedom. At fixed-level α, H_{0A} is rejected if $p \le \alpha$. Equivalently, we can reject H_{0A} at level α if the observed value of the F_A statistic is greater than $F_{a,b}(\alpha)$, the $(1-\alpha)$th quantile of the F distribution with a and b degrees of freedom. Similarly, when H_{0B} is true, we have

$$F_B = \frac{MSB}{MSE} = \frac{S_B/(n-1)}{S_E/(n-1)(k-1)} \sim F_{n-1,(n-1)(k-1)} \tag{8.36}$$

and large observed values of the random variable discredits H_{0B}. This hypothesis is tested based on the p-value

$$p_b = 1 - H_{b,e}\left(\frac{s_B/b}{s_E/e}\right), \tag{8.37}$$

where $b = n-1$, $e = (n-1)(k-1)$. As illustrated by Table 8.8, the calculations involved in the computation of the F-statistics can be summarized by an ANOVA table.

Table 8.8. Two-way ANOVA for testing effects of factors A and B

Source of Variation	D.F.	Sum of Squares	Mean Sum of Squares	F-Statistic
A	$k-1$	s_A	MSA	MSA/MSE
B	$n-1$	s_B	MSB	MSB/MSE
Error	$(n-1)(k-1)$	s_E	MSE	
Total	$nk-1$	s_T		

Example 8.6. Comparison of IQ scores

In order to test whether there is no sex difference in average intelligence of men and women, an IQ test is administered in the form of a complete two factor design with no replications. The test scores and the row and column means are as follows:

Education level	1	2	3	4	5	6	Mean
Men	45	52	51	59	56	63	54.33
Women	52	50	54	56	57	64	55.50
Mean	48.5	51	52.5	57.5	56.5	63.5	54.92

In this example $n=6$ and $k=2$. The sums of squares and the F-statistics for testing the hypothesis of interest and the effect of educational level are given in the following table.

Source of Variation	D.F.	Sum of Squares	Mean Sum of Squares	F-Statistic
Sex	1	4.08	4.08	.629
Education	5	290.4	58.1	8.95
Error	5	32.4	6.48	
Total	11	326.9		

The 95% percentiles of the corresponding F distributions are $F_{1,5}(.05) = 6.61$ and $F_{5,5}(.05) = 5.05$, respectively. Therefore, we can conclude that there is no difference between the mean IQ scores of men and women. Of course the IQ scores does depend on the education level.

8.8. ANOVA in a Balanced Two-Way Layout with Replications

Now consider experiments involving two factors of interest, say A and B, in which two or more observations are available from each combination of factor levels. For example, this is the case when a number of drugs are compared and for each drug a number of safe dosage levels are also being tested, or when several diets are given to hogs of various age groups. In the second example the two sets of factor levels are the diets and the age groups. We consider only the case where equal number of observations are available from each cell. A reader interested in the unbalanced case is referred to Scheffe (1959) and Searle (1987).

Let x_{ijk}, $i=1,...,I$, $j=1,...,J$, $k=1,...,K$ be the available data, where K is the number of observations available from each combination of factor levels. Define the vectors $\mathbf{x}_{ij} = (x_{ij1}, x_{ij2}, ..., x_{ijK})$, $i=1,...,I$, $j=1,...,J$. Then, the data can be set out as in Table 8.9.

Table 8.9. General two-way layout

	B_1	$\cdots$	B_j	$\cdots$	B_J
A_1	$\mathbf{x}_{11}$	$\cdots$	$\mathbf{x}_{1j}$	$\cdots$	$\mathbf{x}_{1J}$
A_2	$\mathbf{x}_{21}$	$\cdots$	$\mathbf{x}_{2j}$	$\cdots$	$\mathbf{x}_{2J}$
$\cdot$	$\cdot$	$\cdot$	$\cdot$	$\cdot$	$\cdot$
A_i	$\mathbf{x}_{i1}$	$\cdots$	$\mathbf{x}_{ij}$	$\cdots$	$\mathbf{x}_{iJ}$
$\cdot$	$\cdot$	$\cdot$	$\cdot$	$\cdot$	$\cdot$
A_I	$\mathbf{x}_{I1}$	$\cdots$	$\mathbf{x}_{Ij}$	$\cdots$	$\mathbf{x}_{IJ}$

Let μ_{ij} be mean of the random variables representing observations in the ijth cell in Table 8.9. Define

$$\mu = \frac{\sum_{i=1}^{I}\sum_{j=1}^{J}\mu_{ij}}{IJ}, \ \mu_{i.} = \frac{\sum_{j=1}^{J}\mu_{ij}}{J}, \text{ and } \mu_{.j} = \frac{\sum_{i=1}^{I}\mu_{ij}}{I}.$$

and assume the linear model:

$$X_{ijk} = \mu + \alpha_i + \beta_j + \gamma_{ij} + \varepsilon_{ijk},\ i=1,\ldots,I,\ j=1,\ldots,J,\ k=1,\ldots,K \quad (8.38)$$

where $\alpha_i = \mu_{i.} - \mu$ and $\beta_j = \mu_{.j} - \mu$ are the standardized *main effects* respectively due to A and B, and ε_{ijk} are the error terms. The term γ_{ij} is called the *interaction effect* of factor A at level i and factor B at level j and is defined as

$$\gamma_{ij} = \mu_{ij} - \mu - \alpha_i - \beta_j,\ i=1,\ldots,I,\ j=1,\ldots,J.$$

Note that $\sum_{i=1}^{I}\gamma_{ij} = 0$ and that $\sum_{j=1}^{J}\gamma_{ij} = 0$. Further, assume that, $\varepsilon_{ijk} \sim N(0,\sigma^2)$; the assumption of common error variance will be relaxed later.

Let $\bar{x}$ be the grand mean and let $\bar{x}_{ij} = \bar{x}_{ij.}$ be the mean of the sample corresponding to the factor levels (A_i, B_j), that is, in terms of random variables,

$$\bar{X} = \frac{\sum_{i=1}^{I}\sum_{j=1}^{J}\sum_{k=1}^{K}X_{ijk}}{N} \text{ and } \bar{X}_{ij} = \frac{\sum_{k=1}^{K}X_{ijk}}{K},$$

where $N = IJK$ is the total sample size. Let $\bar{x}_{i.}$ be the observed mean of all the data corresponding to factor A_i and let $\bar{x}_{.j}$ be that of factor B_j; that is

$$\bar{X}_{i.} = \frac{\sum_{j=1}^{J}\sum_{k=1}^{K}X_{ijk}}{JK} \text{ and } \bar{X}_{.j} = \frac{\sum_{i=1}^{I}\sum_{k=1}^{K}X_{ijk}}{IK}.$$

The MLEs of the parameters α_i, β_j, and γ_{ij} are $\hat{\alpha}_i = \bar{X}_{i.} - \bar{X}$, $\hat{\beta}_j = \bar{X}_{.j} - \bar{X}$, and $\hat{\gamma}_{ij} = \bar{X}_{ij} - \bar{X}_{i.} - \bar{X}_{.j} + \bar{X}$, respectively.

It is of interest to test whether there is no interaction effect; that is the hypothesis

$$H_{0AB}: \gamma_{ij} = 0 \text{ for all } i=1,...,I,\ j=1,...,J. \tag{8.39}$$

If this hypothesis is not rejected, it is also of interest to test the equality of the main effects; that is the null hypotheses H_{0A} and H_{0B} defined by (8.32) and (8.33), respectively. As before, the testing of hypotheses can be based on a decomposition of the total sum of squares into sums of squares that can be attributed to factor A, factor B, the interaction between factors A and B, and the random error. Consider the partition

$$S_T = S_A + S_B + S_I + S_E, \tag{8.40}$$

where

$$S_T = \sum_{i=1}^{I}\sum_{j=1}^{J}\sum_{k=1}^{K}(X_{ijk} - \bar{X})^2,$$

$$S_A = JK\sum_{i=1}^{I}(\bar{X}_{i.} - \bar{X})^2, \quad S_B = IK\sum_{j=1}^{J}(\bar{X}_{.j} - \bar{X})^2,$$

$$S_I = K\sum_{i=1}^{I}\sum_{j=1}^{J}(\bar{X}_{ij} - \bar{X}_{i.} - \bar{X}_{.j} + \bar{X})^2$$

and

$$S_E = \sum_{i=1}^{I}\sum_{j=1}^{J}\sum_{k=1}^{K}(X_{ijk} - \bar{X}_{ij})^2 = K\sum_{i=1}^{I}\sum_{j=1}^{J}S_{ij}^2;$$

S_{ij}^2 being the sample variance (MLE of σ_{ij}^2) of the the data in ijth cell.

By the orthogonality of the above decomposition, the sums of squares S_A, S_B, and S_E are all independently distributed and

$$\frac{S_E}{\sigma^2} \sim \chi^2_{N-IJ}.$$

Moreover,

$$\text{under } H_{0A},\quad \frac{S_A}{\sigma^2} \sim \chi^2_{I-1},\quad \text{under } H_{0B},\quad \frac{S_B}{\sigma^2} \sim \chi^2_{J-1},$$

and

$$\text{under } H_{0AB},\quad \frac{S_I}{\sigma^2} \sim \chi^2_{(I-1)(J-1)}.$$

Hence, under H_{0AB}, we have

$$F_I = \frac{MSI}{MSE} = \frac{S_I/(I-1)(J-1)}{S_E/(N-IJ)} \sim F_{(I-1)(J-1),N-IJ}; \tag{8.41}$$

that is F_I has an F distribution with $(I-1)(J-1)$ and $N-IJ$ degrees of freedom. Moreover, F_I tends to take large values for deviations from H_{0AB}. Therefore, H_{0AB} can be tested on the basis of the p-value

$$p_A = 1 - H_{i,e}\left(\frac{s_I/i}{s_E/e}\right), \tag{8.42}$$

where $i = (I-1)(J-1)$, $e = N-IJ$, in general the notation $H_{a,b}$ stands for the cdf of the F distribution with a and b degrees of freedom. At fixed-level α, H_{0AB} is rejected if the observed value of the F_I statistic is greater than $F_{i,e}(\alpha)$, the $(1-\alpha)$th quantile of the F distribution with i and e degrees of freedom.

Similarly, testing of the hypotheses H_{0B} and H_{0AB} can be based on the p-values

$$p_A = 1 - H_{a,e}\left(\frac{s_A/a}{s_E/e}\right) \text{ and } p_B = 1 - H_{b,e}\left(\frac{s_B/b}{s_E/e}\right), \tag{8.43}$$

respectively, where $a = I-1$ and $b = J-1$. The main effects and interactions can also be tested jointly. For instance, to test whether H_{A0} and H_{0AB} are both true, the sum of squares $s_A + s_I$ and its degrees of freedom $J(I-1)$ can be used in the numerator of the appropriate F-statistic.

The computation of the F-statistics and the p-values can be facilitated by an analysis of variance table as shown in Table 8.10. Each mean squared error term appearing in the fourth column of the table is obtained by dividing the corresponding sum of squares by its degrees of freedom.

Table 8.10. Two-way ANOVA for testing main effects and interactions

Source of Variation	D.F.	Sum of Squares	Mean Sum of Squares	F-Statistic
A	$I-1$ s_A	MSA	MSA/MSE	
B	$J-1$ s_B	MSB	MSB/MSE	
Interac.	$(I-1)(J-1)$	s_I	MSI	MSI/MSE
Error	$N-IJ$	s_E	MSE	
Total	$N-1$	s_T		

Example 8.7. Effect of catalyst and temperature

In order to study the effect of two catalysts and the temperature on the yield of a chemical process, an experiment is carried out using one catalyst at a time, each under three temperatures. The numbers in Table 8.11 represent yields obtained in four runs under each condition.

Table 8.11. Yield of chemical process

	Catalyst A	Catalyst B
Temp. 1	53, 56, 62, 58	59, 63, 65, 57
Temp. 2	58, 59, 57, 64	58, 62, 67, 66
Temp. 3	59, 54, 61, 60	64, 55, 61, 58

The data can be summarized by sample means and sample variances (MLEs of variances). Table 8.12 shows these quantities for each cell. Also shown in Table 8.12 are the the values of $\bar{x}_{i.}$ for $i = 1,2,3$ and the values of $\bar{x}_{.j}$ for $j = 1,2$.

Table 8.12. Sample means and sample variances

	Catalyst A		Catalyst B		
	$\bar{x}_{i1}$	s^2_{i1}	$\bar{x}_{i2}$	s^2_{i2}	$\bar{x}_{i.}$
Temp. 1	57.25	10.69	61.00	10.00	59.13
Temp. 2	59.50	7.25	63.25	12.69	61.38
Temp. 3	58.50	7.25	59.50	11.25	59.00
$\bar{x}_{.j}$	58.42		61.25		

Table 8.13 presents the sum of squares computed using the data in Table 8.12. Corresponding mean sum of squares and the F-statistics are also displayed in the ANOVA table.

Table 8.13. Two-way ANOVA for testing main effects and interactions

Source of Variation	D.F.	Sum of Squares of Squares	Mean Sum	F-Statistic
Tempera.	2	28.59	14.30	1.09
Catalyst	1	48.16	48.16	3.67
Interac.	2	10.08	5.040	0.38
Error	18	236.5	13.14	
Total	23	323.3		

The 95th quantile of F distribution with 2 and 18 degrees of freedom is 3.55. The observed F-statistic corresponding to the interaction effect is .384 (the p-value is .687) suggesting that there is no interaction between the temperature and the catalyst. Therefore, we can proceed to test for the main effects. Obviously, the main effect due to the temperature is also not statistically significant. The p-value of the observed F-statistic corresponding to the catalysts is .072. So, the data provides some evidence in favor of the hypothesis that the effect of the two catalysts are different; it is not significant at the 5% level, however.

8.9. Two-way ANOVA under Heteroscedasticity

Now consider the problem of testing the interaction effect and the main effects of a factorial design with two factors when the error variances are unequal. In other words, consider the linear model (8.38) under the milder assumption that $\varepsilon_{ijk} \sim N(0,\sigma_{ij}^2)$. Inferences can still be based on the the sufficient statistics $\bar{X}_{ij}$, S_{ij}^2, $i=1,...,I$, $j=1,...,J$. As in Section 8.4, appropriate tests can be deduced from the case of known variances. Appropriate sums of squares leading to F-statistics can be deduced, for instance, from results in Fujikoshi (1993) for unbalanced models.

To test H_{0AB}, consider the standardized interaction sum of squares and the error sum of squares

$$\tilde{S}_I(\sigma_{11}^2,...,\sigma_{IJ}^2) = K\sum_{i=1}^{I}\sum_{j=1}^{J}\sigma_{ij}^{-2}(\bar{X}_{ij} - \hat{\alpha}_i - \hat{\beta}_j - \hat{\mu})^2, \tag{8.44}$$

and

$$\tilde{S}_E = \sum_{i=1}^{I}\sum_{j=1}^{J}\sum_{k=1}^{K}\sigma_{ij}^{-2}(X_{ijk} - \bar{X}_{ij})^2 = K\sum_{i=1}^{I}\sum_{j=1}^{J}S_{ij}^2/\sigma_{ij}^2,$$

where

$$\hat{\mu} = \frac{\sum_{i=1}^{I}\sum_{j=1}^{J}\sigma_{ij}^{-2}\bar{X}_{ij}}{\sum_{i=1}^{I}\sum_{j=1}^{J}\sigma_{ij}^{-2}}$$

and $\hat{\alpha}_i$ and $\hat{\beta}_j$ are the solutions of the linear equations

$$\sum_{j=1}^{J} \sigma_{ij}^{-2}(\bar{X}_{ij} - \hat{\mu} - \hat{\alpha}_i - \hat{\beta}_j) = 0,\ i=1,...,I, \tag{8.45}$$

$$\sum_{i=1}^{I} \sigma_{ij}^{-2}(\bar{X}_{ij} - \hat{\mu} - \hat{\alpha}_i - \hat{\beta}_j) = 0,\ j=1,...,J \tag{8.46}$$

and the constraints

$$\sum_{i=1}^{I}\sum_{j=1}^{J} \sigma_{ij}^{-2}\hat{\alpha}_i = 0 \quad \text{and} \quad \sum_{i=1}^{I}\sum_{j=1}^{J} \sigma_{ij}^{-2}\hat{\beta}_j = 0\ . \tag{8.47}$$

Noting that the standardized error sum of squares $\tilde{S}_E$ is the summation of the independent chi-squared random variables

$$V_{ij} = \frac{KS_{ij}^2}{\sigma_{ij}^2} \sim \chi_{K-1}^2,\ i=1,...,I,\ j=1,...,J,$$

the p-value of H_{0AB} can be obtained as in Section 8.4. The hypothesis of no interaction effects can be tested on the basis of the p-value

$$p = 1 - E(G_i(\tilde{s}_I[\frac{Ks_{11}^2}{V_{11}}, \frac{Ks_{12}^2}{V_{12}}, ..., \frac{Ks_{IJ}^2}{V_{IJ}}])), \tag{8.48}$$

where G_i is the cdf of the chi-squared distribution with i $(I-1)(J-1)$ degrees of freedom and the expectation is taken with respect to the chi-squared random variables V_{ij}. The p-value can also be expressed as

$$p = 1 - E(H_{i,e}(\frac{e}{i}\tilde{s}_I[\frac{Ks_{11}^2}{B_{11}B_{12}\cdots B_{IJ-1}}, \frac{Ks_{12}^2}{(1-B_{11})B_{12}\cdots B_{IJ-1}},$$
$$\frac{Ks_{13}^2}{(1-B_{12})B_{13}\cdots B_{IJ-1}}, ..., \frac{Ks_{IJ}^2}{(1-B_{IJ-1})}])), \tag{8.49}$$

where $H_{i,e}$ is the cdf of F distribution with $i = (I-1)(J-1)$ and $e = N - IJ$ degrees of freedom and the expectation is taken with respect to the independent beta random variables defined parallel to (8.18). For details and for a formal derivation of foregoing results, the reader is referred to Ananda and Weerahandi (1997).

Now consider the problem of testing the hypotheses H_{0A} and H_{0B}. As in unbalanced models with unequal cell frequencies (see, for instance, Lindman (1992)), the two-way ANOVA model with unequal and known variances does not yield orthogonal terms leading to a sums of squares decomposition parallel to (8.40). Moreover, F-statistics can be defined in alternative ways using different constraints for α_i and β_j main effects. Consequently, there is no common agreement about how the main effects should be tested in the presence of interactions. Here we employ a widely used method of computing the sums of squares due to main effects when the model is balanced. For results concerning the unbalanced (due to unequal cell frequencies) model under heteroscedasticity, the reader is referred to Ananda and Weerahandi (1997). The statistical software package XPro computes the p-values for the unbalanced case, in general.

To obtain the standardized main effects sums of squares first define the sums of squares

$$S_A(\sigma_{11}^2,...,\sigma_{IJ}^2) = K\sum_{i=1}^{I}\sum_{j=1}^{J}\sigma_{ij}^{-2}(\bar{X}_{ij} - \bar{X}_{i.})^2 \qquad (8.50)$$

and

$$S_B(\sigma_{11}^2,...,\sigma_{IJ}^2) = K\sum_{i=1}^{I}\sum_{j=1}^{J}\sigma_{ij}^{-2}(\bar{X}_{ij} - \bar{X}_{.j})^2 \qquad (8.51)$$

as in the two-way ANOVA model without an interaction term. Also define

$$\tilde{S}_{T-E} = \tilde{S}_T - \tilde{S}_E = K\sum_{i=1}^{I}\sum_{j=1}^{J}\sigma_{ij}^{-2}(\bar{X}_{ij} - \hat{\mu})^2. \qquad (8.52)$$

Then, the standardized sums of squares due to main effects A and B are defined as

$$\tilde{S}_A(\sigma_{11}^2,...,\sigma_{IJ}^2) = \tilde{S}_{T-E} - \tilde{S}_I - S_B \qquad (8.53)$$

and

$$\tilde{S}_B(\sigma_{11}^2,...,\sigma_{IJ}^2) = \tilde{S}_{T-E} - \tilde{S}_I - S_A, \qquad (8.54)$$

respectively. Now, when the variances are unknown, the main effects can be tested as before. The p-value for testing H_{0A} is

$$p = 1 - E(H_{a,e}(\frac{e}{a}\tilde{s}_A[\frac{Ks_{11}^2}{B_{11}B_{12}\cdots B_{IJ-1}}, \frac{Ks_{12}^2}{(1-B_{11})B_{12}\cdots B_{IJ-1}}, \frac{Ks_{13}^2}{(1-B_{12})B_{13}\cdots B_{IJ-1}}, ..., \frac{Ks_{IJ}^2}{(1-B_{IJ-1})}])), \qquad (8.55)$$

where $\tilde{s}_a$ is the observed value of $\tilde{S}_A$ and $a = I - 1$ is the degrees of freedom of the underlying F distribution. Similarly, the p-value for testing H_{0B} is obtained by replacing $\tilde{s}_A$ by $\tilde{s}_B$ and replacing a by the corresponding degrees of freedom $b = J - 1$. The statistical software package XPro computes these p-values under various scenarios. Ananda (1996) has obtained counterparts of above results for nested designs with unequal cell frequencies and unequal variances.

Example 8.8. Effect of catalyst and temperature (continued)

Consider again the data presented in Table 8.11 and the summary statistics reported in Table 8.12. In Example 8.7 we tested the main effects of the catalyst and the temperature and their interaction under the assumption that data from each cell have equal variances, an assumption made for convenience rather than anything else. Now we are in a position to carry out the tests without that assumption. The p-value computed from formula (8.45) using XPro is .78. So we can make the same conclusion that there is no significant interaction between the catalyst and the temperature. The p-values appropriate for testing the main effects of the temperature and the catalyst, computed by applying (8.54) are .47 and .15, respectively. We had concluded in Example 8.7 that there is some evidence to suggest that the main effect of the catalyst is statistically significant. The present results without the assumption of equal variances do not support that conclusion.

Exercises

8.1. In the one-way ANOVA model given in Section 8.3 consider the statistic

$$F = \frac{S_B/(k-1)}{S_E/(N-k)},$$

where S_B is the between sum of squares and S_E is the error sum of squares. Find the distribution of $\frac{S_B}{\sigma^2}$ and $\frac{S_E}{\sigma^2}$ when the means of the underlying normal populations are not necessarily equal. Hence show that

$$F \sim F_{k-1,N-k}(\delta);$$

that is F has an F distribution with $k-1$ and $N-k$ degrees of freedom, and the noncentrality parameter

$$\delta = \frac{1}{(k-1)\sigma^2}\sum_{i=1}^{k} n_i(\mu_i - \bar{\mu})^2,$$

where σ^2 is the common variance, μ_i is the mean of population i, and $\bar{\mu} = \sum_{i=1}^{k} n_i\mu_i/N.$

8.2. Consider the linear model defined by (8.13). Show that

$$\tilde{S}_B = \sum_{i=1}^{k}\frac{n_i\bar{X}_i^2}{\sigma_i^2} - \frac{1}{\sum_{i=1}^{k} n_i/\sigma_i^2}\left[\sum_{i=1}^{k} n_i\bar{X}_i/\sigma_i^2\right]^2 \sim \chi^2_{k-1},$$

$$\tilde{S}_E = \sum_{i=1}^{k} n_i S_i^2/\sigma_i^2 \sim \chi^2_{N-k},$$

and that $\tilde{S}_B$ and $\tilde{S}_E$ are independent random variables. Show also that

$$\tilde{S}_B + \tilde{S}_E = \tilde{S}_t = \sum_{i=1}^{k}\sum_{j=1}^{n_i}\frac{1}{\sigma_i^2}(X_{ij} - \bar{\bar{X}})^2,$$

where $\bar{\bar{X}} = (\sum_{i=1}^{k}\bar{X}_i n_i/\sigma_i^2)/(\sum_{i=1}^{k} n_i/\sigma_i^2).$

8.3. Show that the right tail of the test variable (8.20) yields unbiased procedures for testing the hypothesis H_0 of equality of means; that is

$$Pr(T \ge t \mid H_0) \le Pr(T \ge t \mid H_1).$$

8.4. Let X_i, $i = 1,2,3$ be three independent random variables distributed as

$$X_i \sim Gamma(\alpha_i, \beta), i = 1,2,3.$$

(a) Deduce from Exercise 4 of Chapter 7 that the random variables

$$B_1 = \frac{X_1}{X_1 + X_2}, B_2 = \frac{X_1 + X_2}{X_1 + X_2 + X_3}, \text{ and } S = X_1 + X_2 + X_3$$

are pairwise independent and that

$$B_1 \sim Beta(\alpha_1, \alpha_2), B_2 \sim Beta(\alpha_1 + \alpha_2, \alpha_3),$$

and

$$S \sim Gamma(\alpha_1 + \alpha_2 + \alpha_3, \beta).$$

(b) Show that B_1, B_2 and S are in fact mutually independent random variables.

8.5. Consider the problem of multiple comparisons when the populations variances are not equal. Show that the null hypothesis in (8.22) can be tested on the basis of the p-value given by (8.25). Deduce the form of the simultaneous confidence intervals for all contrasts.

8.6. Consider the linear model defined by (8.38). Show that the minimum variance unbiased estimators of the parameters α_i, β_j, and γ_{ij} are $\hat{\alpha}_i = \bar{X}_{i.} - \bar{X}$, $\hat{\beta}_j = \bar{X}_{.j} - \bar{X}$, and $\hat{\gamma}_{ij} = \bar{X}_{ij} - \bar{X}_{i.} - \bar{X}_{.j} + \bar{X}$, respectively. Show that these are also the MLEs of the parameters. Find the minimum variance unbiased estimator of the variance of ε_{ij}.

8.7. Prove equation 8.40. (Hint: use the identity

$$X_{ijk} - \bar{X} = (X_{ijk} - \bar{X}_{ij}) + (\bar{X}_{i.} - \bar{X}) + (\bar{X}_{.j} - \bar{X}) + (\bar{X}_{ij} - \bar{X}_{i.} - \bar{X}_{.j} + \bar{X}),$$

square it, and then show that each cross product is equal to zero.)

8.8. The Board of Education of a certain state wishes to investigate whether there are significant differences in the performances of different schools across the state. For this purpose the board obtains the mean SAT scores that students of a sample of schools obtained. The following table shows the mean values of percent SAT scores that a sample of schools from urban, suburban, and rural schools had obtained.

School district	Mean percent score
Urban	63.5, 70.4, 61.8, 64.6, 62.8, 68.8
Suburban	63.4, 68.9, 77.5, 72.4, 65.9, 72.3, 69.8
Rural	58.3, 68.6, 61.3, 64.5, 61.5

(a) Assuming that the variances of the three sets of scores are equal test whether

there is a significant difference between the mean scores of the three districts.
(b) Carry out the above hypothesis without the assumption of equal variances.
(c) Test whether data provides sufficient evidence to indicate that the variances of the scores are not the same.
(d) Under appropriate assumptions, carry out multiple comparisons to compare the performances of three school districts.

8.9. A farm with somewhat nonhomogeneous soil conditions was divided into 5 blocks of faily homogeneous conditions. Then, each block was divided into 4 plots of equal size and four concentration levels of a fertilizer were allotted at random within each block. The following table show the wheat crop (in bushels) obtained from each plot.

	B1	B2	B3	B4	B5
Level 1	22.3	16.7	18.9	20.4	17.8
Level 2	24.5	17.5	20.4	29.6	23.2
Level 3	22.7	18.5	21.4	20.5	19.8
Level 4	21.5	17.3	14.6	17.8	18.7

At the .05 level of significance compare the four levels of fertilizer concentrations in their effect on the yield of wheat. Also compare the block effects. Give point estimates and interval estimated for the effects of concentration levels.

8.10. In a comparison of four brands of bologna sold in packs, a number of packs of a certain fixed weight from each brand are tested for their fat content. In certain units the following table presents the data obtained from the study.

Brand	Fat content
A	26, 32, 33, 38, 35, 38
B	33, 37, 36, 34, 35, 32
C	36, 31, 24, 32, 27
D	27, 28, 32, 31, 29

At the .05 level of significance test the hypothesis that the variances of the four distributions of fat contents are equal. Test the equality of the means of the distributions with and without this assumption and compare the results.

8.11. In a two-way factorial design, certain experimental animals with induced high blood pressure were allocated to one of three levels of dosages of three drugs. The reductions in blood presure due to the drugs are shown in the following table:

	Drug 1	Drug 2	Drug 3
Dosage 1	14, 18, 13	11, 22, 24	15, 9, 11
Dosage 2	15, 20, 19	29, 28, 14	18, 14, 16
Dosage 3	14, 11, 15	21, 12, 18	7, 13, 8

Perform an analysis of variance of this data under the assumption of equal error variances. Repeat the analysis without the assumption and compare the results.

Chapter 9
Mixed Models

9.1. Introduction

In Chapter 8, we were interested in the effects of certain treatments, certain brands of a product, certain diet plans, and so on. The levels (e.g. competing treatments) of each factor were considered as deliberate choices of an experimenter and the models we encountered in that chapter allowed the experimenter to compare the effects of competing treatments. The models that we studied in the previous chapter are called *fixed effects models*. In some other applications, the levels of each factor used in an experimental design are not of particular interest and they are selected at random from a large population of potential levels. Models that incorporate this feature are called *random effects models*. In yet other applications, the levels of some factors are of special interest and deliberately chosen and the levels of other factors are randomly selected. As the underlying models in this situation have some features in common with each of aforementioned model types, such models are referred to as *mixed models*. In many industrial experiments it is important to identify and control major sources of variation, and in such situations random effects models and mixed models arise naturally.

To describe this more clearly, consider an experiment in which a comparison of three brands of tires for compact cars is of primary interest. In order to compare the durability of tire brands, suppose that for each brand, ten compact cars are randomly selected and all four wheels of each car are outfitted with the brand. Suppose the drivers are also randomly assigned and the wear of each tire is measured after a certain fixed mileage of driving. The three brands of tires are to be compared based on the findings of the experiment. In this experiment, a certain amount of variation in the observed data on wear of tires will be due to drivers and cars which were randomly selected. The main objective is to compare the performance of tire brands when they are used in a large population of compact cars driven by a population of drivers. Therefore, effects due to drivers as well as to cars need to be treated and modeled as random effects and the effects of brands must be treated as fixed effects. The effects of individual drivers and cars need to be considered as observations from the underlying populations.

As in the case of fixed effects models, each random effect term can be normalized to have zero mean so that the effect of each factor with a random effect can be measured and characterized by its variance alone. The variances of random effect terms in mixed models are sometimes called *variance components*. In random effect models, inferences concerning variance components are of primary interest. In mixed models, usually it is important to make inferences about both the means of fixed effect terms and the variance components. For example, in many industrial processes usually there are a number of sources of variation causing random variations on various characteristics of products. Then, it is important to study the magnitudes of their variance components so that controlled conditions can be established and measures can be taken to reduce the variation of a characteristic from its target value.

9.2. One-way Layout

The simplest model in this context is the random effect model involving just one factor and no fixed effect terms. In other words, this model has only two sources of variation, namely the variation due to randomly selected factor levels and the overall sampling variation. Then, consider the case where there are n observations corresponding to each level of the factor. (We will briefly consider the unbalanced case in Sections 9.3 and 9.4; the readers interested in more details of the unbalanced case are referred to Searle(1987), Searle, Casella, and McCulloch (1992) and Burdick and Graybill (1992)). Let X_{ij} be the response corresponding to the jth data point from A_i, the ith level of the factor. Then, the model can be formulated as

$$X_{ij} = \mu + \alpha_i + \varepsilon_{ij}, \quad i=1,\dots,k, \quad j=1,\dots,n, \tag{9.1}$$

where μ is the grand mean, α_i is the mean deviation (from the grand mean) of observations from A_i, and ε_{ij} is the error term representing the deviation of the response of the jth observation from the mean of observations from A_i. As in the fixed-effect one-way model in Chapter 8, we need to make further assumptions so we can make inferences about underlying parameters. The role of the error term in (9.1) is similar to that in fixed effect models and so similar assumptions can be made. However, now we must treat α_i as a random effect term whose individual values are of no particular interest. What is most important is their population variance and so the assumptions must be made accordingly within the context of normal theory. Therefore, we make the following additional assumptions:

$$\varepsilon_{ij} \sim N(0,\sigma_\varepsilon^2),$$
$$\alpha_i \sim N(0,\sigma_\alpha^2),$$

α_i, $i=1,...,k$, and ε_{ij}, $i=1,...,k$, $j=1,...,n$ are jointly independent.

Foregoing assumptions imply that

$$E(X_{ij} \mid \alpha_i) = \mu + \alpha_i, \quad E(X_{ij}) = \mu, \tag{9.2}$$

and that

$$Var(X_{ij}) = \sigma_\alpha^2 + \sigma_\varepsilon^2. \tag{9.3}$$

Moreover, X_{ij} is normally distributed. It is relationships such as (9.3) which gave variances such as σ_α^2 and σ_ε^2 the name variance components. Following the notation in the previous chapter, let $\bar{x}_{i.}$, $i=1,...,k$ be the sample means corresponding to the k random effects and let $\bar{x} = \bar{x}_{..}$ be the mean of all nk observations. The corresponding random variables are denoted by $\bar{X}_{i.}$ and $\bar{X}$, respectively. The distribution of these random variables can be easily obtained from the identity

$$\bar{X}_{i.} = \frac{1}{n}\sum_{j=1}^{n}(\mu + \alpha_i + \varepsilon_{ij})$$
$$= \mu + \alpha_i + \bar{\varepsilon}_{i.}$$

Since X_{ij} is normally distributed and the variance of $\bar{\varepsilon}_{i.}$ is σ_ε^2/n, we have

$$\bar{X}_{i.} \sim N\left(\mu,\ \sigma_\alpha^2 + \sigma_\varepsilon^2/n\right),\ i=1,...,k. \tag{9.4}$$

Obviously, these sample means are independently distributed and consequently

$$\bar{X} \sim N\left(\mu,\ \frac{1}{k}(\sigma_\alpha^2 + \frac{\sigma_\varepsilon^2}{n})\right).$$

Furthermore, from standard results valid for sampling within each group, we get k independent and identically distributed chi-squared random variables:

$$\frac{\sum_{j=1}^{n}(X_{ij} - \bar{X}_{i.})^2}{\sigma_\varepsilon^2} \sim \chi_{n-1}^2,\ i=1,...,k. \tag{9.5}$$

Now let

$$S_E = \sum_{i=1}^{k}\sum_{j=1}^{n}(X_{ij} - \bar{X}_{i.})^2$$

and

$$S_B = n\sum_{i=1}^{k}(\bar{X}_{i.} - \bar{X})^2$$

be, respectively, the within group sum of squares and the between group sum of squares that played an important role in ANOVA, as we saw in Chapter 8. As before, the total sum of squares is defined by

$$S_T = S_E + S_B = \sum_{i=1}^{k}\sum_{j=1}^{n}(X_{ij} - \bar{X})^2.$$

It immediately follows from (9.4) and (9.5) that

$$U = \frac{S_E}{\sigma_\varepsilon^2} \sim \chi^2_{k(n-1)} \tag{9.6}$$

and that

$$V = \frac{S_B}{\sigma_\varepsilon^2 + n\sigma_\alpha^2} \sim \chi^2_{k-1}. \tag{9.7}$$

Moreover, these random variables are independently distributed. Inferences about the grand mean μ can be easily conducted using the t-random variable implied by the distribution of $\bar{X}$ and (9.7):

$$\frac{\sqrt{kn}\,(\bar{X} - \mu)}{\sqrt{S_B/(k-1)}} \sim t_{k-1}. \tag{9.8}$$

Point estimation

From the foregoing results it is clear that $\hat{\mu} = \bar{X}$ is the MLE of μ. The estimator is also unbiased. Equations (9.6) and (9.7) are particularly useful in obtaining unbiased estimates for the variance components. The two equations imply that

$$E\left[\frac{S_E}{k(n-1)}\right] = \sigma_\varepsilon^2 \text{ and that } E\left[\frac{S_B}{k-1}\right] = \sigma_\varepsilon^2 + n\sigma_\alpha^2,$$

respectively. This means that the *mean within group sum of squares*

$$MSE = \frac{\sum_{i=1}^{k}\sum_{j=1}^{n}(X_{ij} - \bar{X}_{i.})^2}{k(n-1)} \tag{9.9}$$

is an unbiased estimator of σ_ε^2 and that the *mean between group sum of squares*

$$MSB = \frac{n\sum_{i=1}^{k}(\bar{X}_{i.} - \bar{X})^2}{k - 1}$$

is an unbiased estimator of $\sigma_\varepsilon^2 + n\sigma_\alpha^2$. These calculation can be conveniently set out in the form of an analysis of variance table. Table 9.1 shows an appropriate ANOVA table which displays both the calculation of mean sums of squares and their expected values.

Table 9.1. ANOVA in the random effect model

Source	D.F.	SS	MS	E(MS)
Between	$k - 1$	s_B	MSB	$\sigma_\varepsilon^2 + n\sigma_\alpha^2$
Within	$k(n - 1)$	s_E	MSE	σ_ε^2
Total	$nk - 1$	s_T		

An unbiased estimator of the variance component σ_α^2 can also be deduced from (9.8) and (9.9):

$$\hat{\sigma}_\alpha^2 = \frac{MSB - MSE}{n}. \tag{9.10}$$

These estimates of σ_ε^2 and σ_α^2 are also the method of moments estimators. Note that there is no guarantee that the unbiased estimate of σ_α^2 obtained in this manner is necessarily positive. As a matter of fact, formula (9.10) frequently yields negative estimates. It is embarrassing to give a negative estimate for a parameter which is known to be nonnegative. Since the parameter is supposed to be nonnegative, one can simply set $\hat{\sigma}_\alpha^2 = 0$ in (9.10), whenever $MSB < MSE$. We can also avoid negative estimates by resorting to the method of maximum likelihood estimation or to the restricted maximum likelihood estimation (abbreviated RMLE) method, where the likelihood function is maximized subject to the constraint $\sigma_\varepsilon^2 \geq 0$. The MLEs and RMLEs of σ_ε^2 and σ_α^2 are closely related to the unbiased estimates given above. The readers interested in the formulae and their derivations are referred to Searle, Casella, and McCulloch (1992). The MLE and RMLE do not quite address the underlying problem which causes negative unbiased estimates. In Section 9.4, we shall revisit the problem of estimation in variance components in search of interval estimates which give a clearer picture of the situation.

Example 9.1. Weight distribution at birth

A scientist of a health department of a certain country is interested in studying the current distribution of the weight of newborns in the country . She selects a random sample of eight hospitals from among all hospitals in the country and obtains the required data from 20 children born at each hospital. She assumes a one-way random effect model of the form (9.1) to account for the nonhomogeneous economic conditions across the country and to study the underlying variance components. Table 9.2 shows the summary statistics, namely the sample means and sample variances (unbiased).

Table 9.2. Sample means and sample variances of weights in pounds

	Hospital							
	H1	H2	H3	H4	H5	H6	H7	H8
$\overline{x}_{i.}$	7.6	8.3	7.5	7.8	8.5	7.9	7.8	7.2
s_i^2	2.1	2.3	2.2	1.9	2.1	2.0	2.1	2.0

The grand mean of the data is $\overline{x} = 7.825$, which is an unbiased estimate of the mean of all newborns in the country. The error sum of squares and the between group sum of squares are

$$s_E = \sum_{i=1}^{8} 19 s_i^2 = 317.3$$

and

$$s_B = 20 \sum_{i=1}^{8} (\overline{x}_{i.} - \overline{x})^2 = 24.7$$

Hence, we can compute the ANOVA table as follows:

Source	D.F.	SS	MS	E(MS)
Between	7	24.7	3.53	$\sigma_\varepsilon^2 + 20\sigma_\alpha^2$
Within	152	317.3	2.09	σ_ε^2
Total	159	342.0		

Thus, the unbiased estimates of of the variance components σ_ε^2 and σ_α^2 given by (9.9) and (9.10) are

$$\hat{\sigma}_\varepsilon^2 = 2.09 \quad \text{and} \quad \hat{\sigma}_\alpha^2 = 0.072 \,,$$

respectively.

9.3. Testing Variance Components

Making inferences about the error variance σ_ε^2 is an easy matter. Confidence intervals and tests of hypotheses for this parameter follow directly from (9.6) (see Exercise 9.3). Inferences concerning the variance component σ_α^2 are more important in practical applications. The purpose of this section is to undertake various problems of testing hypotheses concerning this variance component. The problem of interval estimation will be addressed in the following section.

Testing the proportion of variance due to α

Recall that $\sigma_t^2 = \sigma_\varepsilon^2 + \sigma_\alpha^2$ is the total variance in the entire population from which samples are taken. The proportion of variance due to the random effect with variance component σ_α^2 is

$$\rho_\alpha = \frac{\sigma_\alpha^2}{\sigma_t^2} = \frac{\sigma_\alpha^2}{\sigma_\alpha^2 + \sigma_\varepsilon^2}.$$

This parameter is also known as the *intraclass correlation.* Consider the hypotheses:

$$H_0\colon \rho_\alpha \le \rho \quad \text{versus} \quad H_1\colon \rho_\alpha > \rho\,. \tag{9.11}$$

The hypotheses can also be written equivalently as

$$H_0\colon \sigma_\alpha^2 \le \delta\sigma_\varepsilon^2 \quad \text{versus} \quad H_1\colon \sigma_\alpha^2 > \delta\sigma_\varepsilon^2\,,$$

where $\delta = \rho/(1-\rho)$. So, $\sigma_\alpha^2/\sigma_\varepsilon^2$ can be treated as the parameter of interest.

As in ANOVA with fixed effects, equations (9.6) and (9.7) can be exploited to define an F-statistic. Observe that

$$F = \frac{1}{\eta}\,\frac{S_B/(k-1)}{S_E/(kn-k)} \sim F_{k-1,kn-k}\,, \tag{9.12}$$

where

$$\eta = 1 + n\frac{\sigma_\alpha^2}{\sigma_\varepsilon^2}.$$

In view of this result, consider the test statistic

$$T = \frac{S_B/(k-1)}{S_E/(kn-k)}.$$

That T is stochastically increasing in $\sigma_\alpha^2/\sigma_\varepsilon^2$, the parameter of interest, follows from the fact that the cdf of T,

$$Pr(T \le t) = Pr(F \le \eta^{-1} t)$$
$$= H_{k-1,k(n-1)}(\eta^{-1} t)$$

is a decreasing function of η where $H_{k-1,k(n-1)}$ is the cdf of the F distribution with $k - 1$ and $k(n - 1)$ degrees of freedom. Hence, extreme regions can be defined using the right tail of the F distribution and the hypotheses in (9.11) can be tested using the p-value

$$p = Pr(T \ge t_{obs} \mid \sigma_\alpha^2 = \delta\sigma_\varepsilon^2)$$
$$= 1 - H_{k-1,k(n-1)}\left[\frac{s_B/(k-1)}{(1+n\delta)s_E/(kn-k)}\right]. \tag{9.13}$$

The unbalanced case

Wald (1940) provided an exact solution to the problem of making inferences about the intraclass correlation ρ_α in the case of unequal class frequencies. To obtain the p-value implied by his results, now suppose n_i observations are available from the i-th class. The total sample size is $N = \sum_{i=1}^{k} n_i$. The underlying model is

$$X_{ij} = \mu + \alpha_i + \varepsilon_{ij}, \quad i=1,\ldots,k, \quad j=1,\ldots,n_i,$$

with the distributional assumptions made in Section 9.2. Let $\bar{X}_{i.}$ denote the sample mean of observations from i-th class. Define $\delta_\alpha = \sigma_\alpha^2/\sigma_\varepsilon^2$ and a set of weights

$$w_i = \frac{n_i}{1 + n_i\delta_\alpha}, \quad i=1,\ldots,k.$$

In terms of these weights Wald (1940) defined a weighted between group sum of squares as

$$S_{WB} = \sum_{i=1}^{k} w_i\left[\bar{X}_{i.} - \left(\sum_{i=1}^{k} w_i\bar{X}_{i.}\right) / \sum_{i=1}^{k} w_i\right]^2$$

and proved that

$$\tilde{V} = \frac{S_{WB}}{\sigma_\varepsilon^2} \sim \chi_{k-1}^2.$$

Moreover, S_{WB} and

$$\tilde{U} = \frac{S_E}{\sigma_\varepsilon^2} = \frac{\sum\sum (X_{ij} - \bar{X}_{i.})^2}{\sigma_\varepsilon^2} \sim \chi_{N-k}^2$$

are independently distributed, thus leading to an F-variable on which inferences on σ_α can be based. The parallel between this distributional result and (9.12) immediately implies the following p-value for testing (9.11):

$$p = 1 - H_{k-1,N-k}\left(\frac{s_{WB}(\delta)/(k-1)}{s_E/(N-k)}\right), \tag{9.14}$$

where $s_{WB}(\delta)$ is the value of the weighted between group sum of squares when the weighted are computed by equating δ_α to δ. This test is referred to as the *Wald test.*

Example 9.2. Weight distribution at birth (continued)

Consider again the data in Table 9.2. Now suppose the scientist is interested in testing the hypothesis that the percent variation of the data due to the random effect term representing nonhomogeneous conditions across the country is less than 5%; i.e. the null hypothesis of interest is $H_0 : \rho_\alpha \le .05$. The observed value of the test statistic is $t_{obs} = 1.6903$ and $\delta = .0476$. Therefore, the observed level of significance is

$$p = 1 - H_{7,152}\left(\frac{1.6903}{2.0526}\right) = .569.$$

It is now evident that there is no reason to doubt the null hypothesis.

Testing the variance component

In many applications the magnitude of the error variance is of secondary importance and what we really need to deal with are hypotheses about the absolute value of the variance component σ_α^2 regardless of the magnitude of σ_ε^2. Left-sided null hypotheses of the form

$$H_0\colon \sigma_\alpha^2 \le \sigma_0^2 \tag{9.15}$$

are of particular interest, where σ_0^2 is a specified constant. First, consider the case where $\sigma_0^2 = 0$ so that (9.15) would reduce to

$$H_0\colon \sigma_\alpha^2 = 0 .$$

Noting that this is actually a special case of (9.11), the p-value can be readily deduced from (9.13):

$$p = 1 - H_{k-1,k(n-1)}\left(\frac{s_B/(k-1)}{s_E/(kn-k)}\right).$$

Therefore, as expected, the F-test for testing the hypothesis of zero variance component is identical to that of testing the equality of means in a one-way fixed effects model. The results are quite different in the general case, however.

Exact tests based on test statistics (constructed from experimental data only) are not available for testing more general hypotheses such as that in (9.15); Healy (1961) gave an exact test which utilizes an artificial randomization device in addition to the experimental data. Nevertheless, more powerful exact tests can be easily constructed using generalized test variables. A test variable which can define extreme regions for the general null hypothesis defined by (9.15) is

$$T = V\left(\frac{1}{U} + \frac{n\sigma_\alpha^2}{s_E}\right)$$

where U and V are the chi-squared random variables defined by (9.6) and (9.7), respectively. It is evident from its composition that T is indeed stochastically increasing in σ_α^2, the parameter of interest. Moreover, the observed value of the test variable is $t_{obs} = s_B / s_E$ so that the extreme region given by T is

$$E = \left[(S_B, S_E) \mid S_B(\sigma_\varepsilon^2 \frac{s_E}{S_E} + n\sigma_\alpha^2) \geq s_B(\sigma_\varepsilon^2 + n\sigma_\alpha^2) \right]$$

The p-value of the test is the maximum probability of this region under the null hypothesis. It is computed as

$$p = Pr(T \geq t_{obs} \mid \sigma_\alpha^2 = \sigma_0^2) = Pr(V(s_E/U + n\sigma_0^2) \geq s_B)$$
$$= 1 - E\left[G_{k-1}\left(\frac{s_B}{n\sigma_0^2 + s_E/U} \right) \right], \qquad (9.16)$$

where G_{k-1} is the cdf of the chi-squared distribution with $k-1$ degrees of freedom and the expectation is taken with respect to the chi-squared random variable $U \sim \chi^2_{k(n-1)}$.

The uniqueness (up to equivalent p-values) of this test can be established by invoking either the principle of invariance or the principle of unbiasedness. For a formal derivation of the result, the reader is referred to Weerahandi (1991). It is of interest to observe that when $\sigma_0^2 = 0$, the p-value given by (9.16) reduces to that given by the simple F-test established above. Therefore, the test given by (9.16) is indeed a generalization of the F-test.

The expectation appearing in (9.16) can also be expressed as an integral over a subset of the interval [0, 1] with respect to a beta random variable. To do this, define the random variables

$$S = U + V \sim \chi^2_{nk-1} \quad \text{and} \quad B = \frac{U}{U+V} \sim Beta\left(\frac{k(n-1)}{2}, \frac{k-1}{2}\right),$$

which are also independently distributed. Then, the p-value can be expressed as

$$p = Pr\left\{\frac{s_E}{SB} + n\sigma_0^2 \geq \frac{s_B}{S(1-B)}\right\}$$

$$= 1 - \int_{\frac{s_E}{s_E+s_B}}^{1} G_{nk-1}\left[\frac{1}{n\sigma_0^2}\left(\frac{s_B}{1-b} - \frac{s_E}{b}\right)\right] f_B(b)\,db, \qquad (9.17)$$

where G_{nk-1} is the cdf of the chi-squared random variable with $nk-1$ degrees of freedom and $f_B(b)$ is the pdf of the random variable B. In either case, the p-value can be evaluated by numerical integration or it can be well approximated using a large number of random numbers generated from the underlying distribution with respect to which the expectation or the integration is to be evaluated. The statistical software package XPro computes the p-value by exact numerical integration.

Box and Tiao (1973) provided the Bayesian solution to the problem of making inferences about σ_α^2, under the noninformative prior given by Jeffreys' rule. This prior does not take into account the possibility that s_E and s_B may have been observed from a situation where $\sigma_\alpha^2 = 0$. Therefore, the p-value given by (9.17) is not quite equal to the posterior probability of the null hypothesis under that prior. It is possible however to find a noninformative prior (with an appropriately chosen probability mass placed on σ_0^2) so that the p-value is numerically equivalent to the posterior probability under that prior.

The unbalanced case

As in the case of testing intraclass correlation, the foregoing results easily extend to the unbalanced one-way random effect model. This is accomplished by using the weighted between group sum of squares $S_{WB}(\delta_\alpha)$ defined above. Following the derivation of (9.16) we get

$$p = Pr\left[\frac{1}{\sigma_\varepsilon^2} S_{WB}\left(\frac{\sigma_\alpha^2}{\sigma_\varepsilon^2}\right) \geq \frac{1}{\sigma_\varepsilon^2} s_{WB}\left(\frac{\sigma_\alpha^2}{\sigma_\varepsilon^2}\right) \mid \sigma_\alpha^2 = \sigma_0^2\right]$$

$$= Pr\left\{\tilde{V} \geq \frac{\tilde{U}}{s_E} s_{WB}\left(\sigma_0^2 \frac{\tilde{U}}{s_E}\right)\right\} = 1 - EG_{k-1}\left[\frac{\tilde{U}}{s_E} s_{WB}\left(\sigma_0^2 \frac{\tilde{U}}{s_E}\right)\right] \qquad (9.18)$$

The test of (9.15) is now based on this p-value. The XPro software package can be employed to compute the p-value in the unbalanced case as well.

Example 9.3. Weight distribution at birth (continued)

Considering again the data in Table 9.2, suppose the scientist is now interested in testing the hypothesis that the variance of the random effect term representing nonhomogeneous conditions across the country is less than .01. In other words, the null hypothesis of interest is $H_0 : \sigma_\alpha^2 \le .01$. The observed level of significance computed from (9.17) by using XPro is

$$p = 1 - \int_{.928}^{1} G_{159}\left[5\left(\frac{24.7}{1-b} - \frac{317.3}{b}\right)\right] f_B(b)\,db = .156,$$

where $f_B(b)$ is the pdf of $B \sim Beta(76.0,3.5)$. Therefore, the null hypothesis can be accepted as the data do not provide sufficient evidence to reject it.

9.4. Confidence Intervals

Interval estimation in random effect models is of special interest and importance, because, as discussed in Section 9.2, the point estimates of variance components given by standard methods often become negative or zero when the true values of the variance components are small. In such situations it is more informative and sensible to report confidence intervals (preferably, intervals which do not suffer from the same drawback) rather than point estimates.

A number of articles including those of Satterthwaite (1946), Welch (1956), Bulmer (1957), and Samaranayake and Bain (1988) provided approximate confidence intervals for variance components. Wald (1940) provided exact confidence intervals for the ratio of variances of the unbalanced one-way model and Thomas and Hultquist (1978), and Burdick and Graybill (1984), provided approximate intervals for various parameters.

Confidence intervals for the intraclass correlation

Confidence intervals for the ratio of the two variances as well as the intraclass correlation, $\rho_\alpha = \sigma_\alpha^2/(\sigma_\alpha^2 + \sigma_\varepsilon^2)$, can be obtained directly from the F-statistic defined by (9.12). For instance, one sided intervals for ρ_α are obtained using one-sided probability statements on the F-random variable defined by (9.12) and by appropriate transformation. It is easily seen that the exact $100(1-\beta)\%$ confidence interval given by the the choice of equal tails of that F distribution is

$$\frac{A-1}{A+n-1} \le \rho_\alpha \le \frac{B-1}{B+n-1}, \tag{9.19}$$

where

$$A = \frac{MSB}{F_{1-\beta/2}MSE} \quad \text{and} \quad B = \frac{MSB}{F_{\beta/2}MSE},$$

F_γ being the γ quantile of the F distribution with $k-1$ and $kn-k$ degrees of freedom. If a confidence limit given by (9.19) is negative it is adjusted to be zero.

For the unbalanced case due to unequal sample sizes n_i, Wald (1940) provided exact confidence intervals for $\delta_\alpha = \sigma_\alpha^2/\sigma_\varepsilon^2$, which in turn imply intervals for ρ_α. The confidence limits for δ_α obtained by this method are equivalent to solving Equation (9.14) when δ is considered as a confidence limit for δ_α and p is replaced by the appropriate F-tail probability. For example, the lower $100\gamma\%$ confidence bound for δ_α is the solution (for δ) of the nonlinear equation

$$\frac{s_{WB}(\delta)/(k-1)}{s_E/(N-k)} = F_\gamma .$$

The confidence limits can be conveniently computed using the XPro software package.

Confidence intervals for the variance components

Tukey (1951) and Williams (1962) constructed a confidence interval for σ_α^2 where the confidence coefficient is no larger than the specified level. The construction is based on simultaneous probability statements based on the chi-squared statistic given by (9.7) and the F-statistic given by (9.6) and (9.7). Using these statistics, Williams (1962) proved that the interval

$$\frac{1}{n\chi^2_{1-\beta/2}}\left[s_B - \frac{(k-1)s_E F_{1-\beta/2}}{k(n-1)}\right] \leq \sigma_\alpha^2 \leq \frac{1}{n\chi^2_{\beta/2}}\left[s_B - \frac{(k-1)s_E F_{\beta/2}}{k(n-1)}\right] \quad (9.20)$$

is a confidence interval with confidence coefficient greater than $1 - 2\beta$, where χ^2_γ is the γ quantile of the chi-squared distribution with $k-1$ degrees of freedom and F_γ is the γ quantile of the F distribution with $k-1$ and $k(n-1)$ degrees of freedom. If the lower (or upper) confidence bound is negative, it is replaced by zero. Samaranayake and Bain (1988) provided some improvement of the Tukey-Williams confidence intervals. Improving upon the confidence level of the Tukey-Williams confidence interval, Wang (1990) argued that for small (as is the case in practice) values of β, $1-\beta$ (instead of $1-2\beta$) is a lower bound for the confidence coefficient of the test given by (9.20). Unlike the generalized intervals, neither the original Tukey-Williams interval nor this adjustment is available for unbalanced models and for a large class of problems which cannot be reduced to one involving two chi-squared random variates.

Healy (1961) constructed an exact confidence interval for σ_α^2 using an auxiliary experiment. To implement this method an independent random sample of size $N = k(n-1)$ needs to be generated from the standard normal distribution. Let $z_1,\ldots,z_N$ be the auxiliary observations. Healy (1961) used this auxiliary random sample to define a pivotal quantity involving the chi-squared random variables U and V as well as σ_α^2, the parameter of interest. The exact two-sided $100(1-2\beta)\%$ confidence interval constructed from the pivotal quantity is (see Healy (1961) for a derivation of the result) given by

$$\left\{\left[\max\{0,\frac{1}{s_z}\left[\frac{k(n-1)s_B}{(k-1)F_{1-\beta}}-s_E\right]\}+\left[\frac{z_1\sqrt{s_E}}{s_z}\right]^2\right]^{1/2}-\frac{z_1\sqrt{s_E}}{s_z}\right\}^2 \le n\sigma_\alpha^2 \le$$

$$\left\{\left[\frac{1}{s_z}\left[\frac{k(n-1)s_B}{(k-1)F_\beta}-s_E\right]+\left[\frac{z_1\sqrt{s_E}}{s_z}\right]^2\right]^{1/2}-\frac{z_1\sqrt{s_E}}{s_z}\right\}^2, \qquad (9.21)$$

where $s_z = \sum z_i^2$. Neither this procedure nor the previous procedure quite addresses the problem of negative estimates.

As generalized confidence intervals are just extensions of classical confidence intervals they also suffer from the drawback of negative estimates unless a generalized pivotal quantity is especially designed to avoid this. Before we discuss how to obtain interval estimators which do not yield negative estimates, let us first establish the interval estimator implied by (9.17), which tends to be more efficient (see Example 9.4) than the intervals reported above. The generalized confidence interval implied by the p-value is the same as the one given by the potential pivotal quantity

$$R = (\sigma_\varepsilon^2 + n\sigma_\alpha^2)\frac{s_B}{S_B} - \sigma_\varepsilon^2\frac{s_E}{S_E} = \frac{s_B}{V} - \frac{s_E}{U},$$

where U and V are the chi-squared random variables defined by (9.6) and (9.7). Note that R is related to the generalized test variable that we employed in the testing problem. Obviously, the probability distribution of R is free of unknown parameters. Moreover, at the observed values of the data this variable reduces to $r_{obs} = n\sigma_\alpha^2$. It is now evident that R is indeed a generalized pivotal quantity that can yield interval estimates for σ_α^2.

We can now find generalized confidence intervals for σ_α^2 by writing down the appropriate probability statement concerning the pivotal R. To illustrate this, consider the problem of constructing left-sided confidence intervals based on R. If a $100\gamma\%$ confidence interval is desired, we need to find the value of $k = k(s_E, s_B) > 0$ such that

$$\gamma = Pr(R \le k) = Pr\left[\frac{s_B}{V} - \frac{s_E}{U} \le k\right].$$

$$= 1 - \int_{\frac{s_E}{s_E+s_B}}^{1} G_{nk-1}\left[\frac{1}{k}\left(\frac{s_B}{1-b} - \frac{s_E}{b}\right)\right] f_B(b)\,db, \qquad (9.22)$$

where G_{nk-1} is the cdf of the chi-squared random variable S and $B \sim Beta(k(n-1)/2, (k-1)/2)$. Since $G_{nk-1}(x) = 0$ for all $x \le 0$, (9.22) can also be written as

$$\gamma = 1 - EG_{nk-1}\left[\frac{1}{k}\left(\frac{s_B}{1-B} - \frac{s_E}{B}\right)\right],$$

where the expectation is taken with respect to the beta random variable B. If there exists no $k \geq 0$ satisfying (9.22) (that is, when k satisfying the original probability statement is negative), the confidence bound is taken to be 0. Hence, the $100\gamma\%$ confidence interval for σ_α^2 given by (9.22) is

$$\left[0, \frac{k(s_E, s_B)}{n}\right], \tag{9.23}$$

where k is the solution of equation (9.22). Other confidence intervals can be found by taking a parallel approach. In particular, the $100\gamma\%$ lower confidence bound σ_0^2 of the variance component is the solution of the equation

$$\gamma = \int_{\frac{s_E}{s_E+s_B}}^{1} G_{nk-1}\left[\frac{1}{n\sigma_0^2}\left(\frac{s_B}{1-b} - \frac{s_E}{b}\right)\right] f_B(b)\,db.$$

According to some simulation studies (c.f. Weerahandi and Amaratunga (1996)), in repeated sampling, equal-tail confidence intervals obtained by this method guarantees the intended confidence level.

The value of k satisfying Equation (9.22) can be found using a numerical algorithm to solve the underlying nonlinear equation. This in turn will require the computation of the left hand side of Equation (9.22) for a large number of values of k. For each value of k the right hand side of the equation is computed by numerical integration. This procedure is readily integrated into the statistical software package XPro. XPro also provides confidence intervals found by applying Tukey-Williams' method and Healy's method.

As with classical confidence intervals, generalized intervals with different properties can be constructed with the same data and the same coefficient of confidence; a confidence statement make sense only when the random quantity (on which the interval is based) is specified along with the intended properties. It is possible to avoid negative confidence limits or obtain strictly positive confidence bounds by appropriate choice of procedure. For example, if a nonnegative pivotal quantity is defined as $\tilde{R} = \max(R,0)$, then we get the same confidence intervals as above without having to deal with negative values of k. Foregoing confidence intervals are desirable if $\sigma_\alpha^2 = 0$ is a possible value in a given application. On the other hand, if it is known that $\sigma_\alpha^2 > 0$, it is desirable to obtain strictly positive confidence limits (except for specified numerical accuracy). This is the case considered by Box and Tiao (1973) in their Bayesian treatment of the problem. A generalized confidence interval with strictly positive confidence limits can be obtained by the method of conditional inference (see, for instance, Kiefer (1977) and Lehmann (1986)) based on R; the conditioning is made with respect to the event that $R > 0$. It is easily seen, by following the steps in the above results, that the 100γ% upper confidence bound for σ_α^2 found by this method is the value of $\bar{\sigma}^2$ satisfying the equation

$$\gamma = Pr(R \le \bar{\sigma}^2 \mid R > 0)$$

$$= 1 - \int_{\frac{s_E}{s_E+s_B}}^{1} G_{nk-1}\left[\frac{1}{n\bar{\sigma}^{-2}}\left(\frac{s_B}{1-b} - \frac{s_E}{b}\right)\right] f_B(b)\,db \Big/ \int_{\frac{s_E}{s_E+s_B}}^{1} f_B(b)\,db \quad (9.24)$$

Since, $Pr(R \le \bar{\sigma}^2 \mid R \le 0) = 1$, the unconditional probability of the event $\{ R \le \bar{\sigma}^2 \}$ exceeds γ, as desired. However, the resulting intervals can be somewhat conservative.

The Generalized Wald confidence interval

As before, the foregoing results easily extend to the unbalanced one-way random effect model. This is accomplished by using the weighted between group sum of squares $S_{WB}(\delta_\alpha)$ in place of S_B appearing in the above formulae. In comparing (9.17) and (9.23) the results can be deduced directly from (9.18). The 100γ% upper confidence bound for σ_α^2 found by this method is the value of $\bar{\sigma}^2$ satisfying the equation

$$\gamma = 1 - EG_{k-1}\left[\frac{\tilde{U}}{s_E} s_{WB}\left(\bar{\sigma}^{-2}\frac{\tilde{U}}{s_E}\right)\right],$$

where the expectation is taken with respect to the random variable $\tilde{U} \sim \chi^2_{N-k}$. In a similar manner (9.24) can also be extended to the unbalanced case. The software package XPro provides confidence intervals for both cases.

Example 9.4. A comparison

Williams (1962) constructed a confidence interval for a variance component using the data

$$k = 10 \quad n = 4, \text{ and } s_B = 81.27 \quad s_E = 30.60.$$

The confidence interval computed by applying the formula (9.20) with $\beta = .05$ is

$$[.76 \le \sigma_\alpha^2 \le 7.29]$$

and has length 6.55. The confidence coefficient of the interval is guaranteed to be at least 90%. According to Wang (1990), however, this is likely to be a 95% confidence interval. If the formula (9.23) or (9.24) (both formulas yield the same intervals up to two decimal points) is applied to find the equal-tail 90% and 95% generalized confidence intervals based on the same data, by employing XPro we get

$$[.92 \le \sigma_\alpha^2 \le 5.84] \quad \text{and} \quad [.78 \le \sigma_\alpha^2 \le 7.25],$$

respectively. Clearly both of these intervals are subsets of the former, a considerable improvement over the original Tukey-Williams interval and a small improvement with respect to Wang's confidence coefficient.

It should be emphasized that, as in Bayesian credible intervals (see, for instance Box and Tiao (1973)) , the latter intervals are based on exact probability statements with no special regard to repeated sampling properties. In the present context, there seems no point insisting on repeated sampling properties of little practical use when intervals with such properties can have undesirable features with the observed sample. Williams (1962) argued that in this situation the length of the equal-tail 95% confidence interval attainable with the exact theory with a knowledge of σ_ε^2 is 6.48. It should be noted, however, that this statement is valid only if one constructs interval estimates using a procedure based on the pivotal quantity $S_B/(\sigma_\varepsilon^2 + n\sigma_\alpha^2)$. If σ_ε^2 is known, shorter intervals can be obtained by the analogues of (9.23) or (9.24).

Next consider the data set considered by Healy (1961) in his illustration of the exact confidence interval for $\sigma_0^2 = 19\sigma_\alpha^2/300$:

$$k = 4 \quad n = 19, \text{ and } s_B = 139{,}997 \quad s_E = 33{,}048.$$

He also reported the statistics $z_1 = 0.628$ and $s_z = 62.72$ based on an auxiliary sample of size 72 from the standard normal distribution. The 90% confidence interval obtained by applying (9.21) is

$$[62 \le \sigma_0^2 \le 1514].$$

In contrast, the 90% equal-tail confidence interval obtained using (9.23) is

$$[58 \le \sigma_0^2 \le 1325],$$

which is much shorter than the former interval. Yet, according to some simulation studies the confidence intervals obtained using (9.23) guarantees the intended level. In this application, the 90% Tukey-Williams interval for σ_0^2 is [48,2161], which is much worse than the generalized and Healy intervals. However, it can be improved using the Wang's correction. It should also be mentioned that depending on the random numbers used by Healy's method it may yield shorter intervals than other method reported in this section.

9.5. Two-way Layout

Results in previous sections can be easily extended to models involving higher-way balanced designs. We consider only the two-way balanced layout with replication. Readers interested in results pertaining to higher-way balanced and unbalanced designs are referred to Khuri and Littell (1987), Khuri (1990), Lindman (1992), Burdick and Graybill (1992), Searle, Casella, and McCulloch (1992), and Zhou and Mathew (1994). Consider a random effect model or a mixed model of the form

$$X_{ijk} = \mu + \alpha_i + \beta_j + \gamma_{ij} + \varepsilon_{ijk}, \tag{9.25}$$
$$i=1,\ldots,I;\ j=1,\ldots,J;\ k=1,\ldots,K,$$

where α_i is either a random effect or a fixed effect, β_j is a random effect, γ_{ij} is the interaction random effect, and ε_{ijk} is the error term. Available data from this two-way design, x_{ijk}, $i=1,\ldots,I$, $j=1,\ldots,J$, $k=1,\ldots,K$ can be set out as in Table 8.9. Let us use the same notation as in Section 8.8 to denote the various sample means that we get from cells, columns and rows of the table. The sums of squares associated with α, β, γ and the error term of the model are

$$S_T = \sum_{i=1}^{I}\sum_{j=1}^{J}\sum_{k=1}^{K}(X_{ijk} - \bar{X})^2,$$

$$S_A = JK\sum_{i=1}^{I}(\bar{X}_{i.} - \bar{X})^2,$$

$$S_B = IK\sum_{j=1}^{J}(\bar{X}_{.j} - \bar{X})^2,$$

$$S_G = K\sum_{i=1}^{I}\sum_{j=1}^{J}(\bar{X}_{ij} - \bar{X}_{i.} - \bar{X}_{.j} + \bar{X})^2,$$

and

$$S_E = \sum_{i=1}^{I}\sum_{j=1}^{J}\sum_{k=1}^{K}(X_{ijk} - \bar{X}_{ij})^2 = (K-1)\sum_{i=1}^{I}\sum_{j=1}^{J}S_{ij}^2,$$

where S_{ij}^2 is the sample variance (unbiased) of the data in ijth cell of the two-way layout.

First consider the case where there are no fixed effects in the model. For example, suppose the experiment described in the introduction is carried out with one brand of tires. Then, α_i, β_j, γ_{ij} are the random effects associated with cars, drivers, and car-driver interactions. As in the one-way model we need to make further distributional assumptions in a full parametric setting. Assume that

$$\varepsilon_{ijk} \sim N(0,\sigma_\varepsilon^2),$$
$$\alpha_i \sim N(0,\sigma_\alpha^2),$$

$$\beta_j \sim N(0,\sigma_\beta^2),$$

$$\gamma_{ij} \sim N(0,\sigma_\gamma^2),$$

and that

α_i, β_j, and ε_{ijk} are jointly independent for all $i=1,...,I,\ j=1,...,J,\ k=1,...,K.$

These distributional assumptions yield results parallel to those in Section 9.2. The counterparts of equations (9.6) and (9.2) are

$$\frac{S_E}{\sigma_\varepsilon^2} \sim \chi^2_{IJ(K-1)} \tag{9.26}$$

$$\frac{S_A}{\sigma_\varepsilon^2 + K\sigma_\gamma^2 + JK\sigma_\alpha^2} \sim \chi^2_{I-1}, \tag{9.27}$$

$$\frac{S_B}{\sigma_\varepsilon^2 + K\sigma_\gamma^2 + IK\sigma_\beta^2} \sim \chi^2_{J-1}, \tag{9.28}$$

and

$$\frac{S_G}{\sigma_\varepsilon^2 + K\sigma_\gamma^2} \sim \chi^2_{(I-1)(J-1)}. \tag{9.29}$$

Inferences on the variance components can be based on these results. Their structure compared with those in the one-way design also indicates how the results can be extended to higher-way random effect models. As in previous designs, the elements needed in significance testing of hypotheses about variance components and their point estimates can be set out in an analysis of variance table. Table 9.3 shows the appropriate ANOVA table for the two-way random effect model.

Table 9.3. ANOVA of two-way random effect model

Source	D.F.	SS	MS	E(MS)
α	$I-1$	s_A	MSA	$\sigma_\varepsilon^2 + K\sigma_\gamma^2 + KJ\sigma_\alpha^2$
β	$J-1$	s_B	MSB	$\sigma_\varepsilon^2 + K\sigma_\gamma^2 + KI\sigma_\beta^2$
γ	$(I-1)(J-1)$	s_G	MSG	$\sigma_\varepsilon^2 + K\sigma_\gamma^2$
Error	$IJ(K-1)$	s_E	MSE	σ_ε^2
Total	$IJK-1$	s_T		

As usual, the mean sums of squares are obtained from the division of corresponding sums of squares by their degrees of freedom. For example, $MSA = s_A/(I-1)$. These mean sums of squares, which are displayed in the fourth column of the ANOVA table, provide unbiased estimates for the linear combinations of variance components displayed in the last column of the table. Unbiased estimates of individual variance components can be obtained by solving these simultaneous equations. Unbiased estimates of σ_ε^2, σ_γ^2, σ_β^2 and σ_α^2 obtained by this approach are

$$\hat{\sigma}_e^2 = MSE,\ \hat{\sigma}_\gamma^2 = \frac{(MSG - MSE)}{K},$$

$$\hat{\sigma}_\beta^2 = \frac{(MSB - MSG)}{IK},\ \text{and}\ \hat{\sigma}_\alpha^2 = \frac{(MSA - MSG)}{JK},$$

respectively.

Other inferences about the variance components can be easily deduced by comparing the last four equations with (9.6) and (9.7). Let σ^2 be any one of the three variance components and consider, for instance, the problem of testing right-sided null hypotheses of the form

$$H_0: \sigma^2 \geq \sigma_0^2. \tag{9.30}$$

The null hypothesis is to be tested against the natural alternative $H_1: \sigma^2 < \sigma_0^2$ based on the appropriate sums of squares. To present the general form of the testing procedure define

$$ss_x = \begin{cases} s_A & \text{if } \sigma^2 = \sigma_\alpha^2 \\ s_B & \text{if } \sigma^2 = \sigma_\beta^2, \\ s_G & \text{if } \sigma^2 = \sigma_\gamma^2 \end{cases}$$

$$ss_y = \begin{cases} s_G & \text{if } \sigma^2 = \sigma_\alpha^2 \\ s_G & \text{if } \sigma^2 = \sigma_\beta^2 \\ s_E & \text{if } \sigma^2 = \sigma_\gamma^2 \end{cases}$$

and

$$N = \begin{cases} JK & \text{if } \sigma^2 = \sigma_\alpha^2 \\ IK & \text{if } \sigma^2 = \sigma_\beta^2 \\ K & \text{if } \sigma^2 = \sigma_\gamma^2 \end{cases}$$

Let q and r be the degrees of freedom of the sums of squares ss_y and ss_x, respectively. Then, the null hypothesis in (9.30) can be tested on the basis of the p-value

$$p = E\left[G_r\left(\frac{ss_x}{N\sigma_0^2 + ss_y/U_q}\right)\right], \tag{9.31}$$

where G_r is the cdf of the chi-squared distribution with r degrees of freedom and the expectation is taken with respect to the chi-squared random variable

$$U_q \sim \chi_q^2.$$

Similarly, left-sided null hypotheses can be tested using the p-value $1 - p$.

Analogously, interval estimates for the three variance components can be deduced from equation (9.23). The $100\gamma\%$ lower (generalized) confidence bound is the solution of the equation

$$EG_r\left[\frac{1}{N\bar{\sigma}^2}\left(\frac{ss_x}{(1-B)} - \frac{ss_y}{B}\right)\right] = \gamma \tag{9.32}$$

for $\bar{\sigma}^2$, where the expectation is taken with respect the beta random variable

$$B \sim Beta\left(\frac{q}{2}, \frac{r}{2}\right)$$

and $\bar{\sigma}^2$ is taken to be zero if no positive value of the parameter satisfies equation (9.32). If σ^2 is known to be strictly positive, as in Section 9.4, the confidence bound is computed as

$$EG_r\left[\frac{1}{N\bar{\sigma}^2}\left(\frac{ss_x}{(1-B)} - \frac{ss_y}{B}\right) \mid B > \frac{ss_x}{ss_x + ss_y}\right] = \gamma$$

The $100\gamma\%$ upper confidence bound is found by replacing γ in equation (9.32) by $1 - \gamma$.

To illustrate the inference procedures consider the hypothetical data in Table 9.4 from a random effect model involving two factors A and B (e.g. operating temperature and operating pressure of an engine).

Table 9.4. Random factor levels and observed response

	A_1	A_2
B_1	23, 26, 28	29, 33, 35
B_2	24, 30, 32	26, 30, 37
B_3	22, 29, 33	28, 39, 32

The data can be summarized by the sample means and the sample variances (unbiased) of data in each cell. The table below presents these summary statistics as well as row means and columns means that are important in the calculation of necessary sums of squares.

	A_1		A_2		
	$\bar{x}_{i1}$	s_{i1}^2	$\bar{x}_{i2}$	s_{i2}^2	$\bar{x}_{i.}$
B_1	25.67	6.33	32.33	9.33	29.00
B_2	28.67	17.33	31.00	31.00	29.84
B_3	28.00	31.00	33.00	31.00	30.5
$\bar{x}_{.j}$	27.45		32.11		

Hence we get the following ANOVA table:

Source	D.F.	SS	MS	E(MS)
α	1	97.86	97.86	$\sigma_\varepsilon^2 + 3\sigma_\gamma^2 + 9\sigma_\alpha^2$
β	2	6.779	3.390	$\sigma_\varepsilon^2 + 3\sigma_\gamma^2 + 6\sigma_\beta^2$
γ	2	14.32	7.16	$\sigma_\varepsilon^2 + 3\sigma_\gamma^2$
Error	12	251.98	21.00	σ_ε^2
Total	17	s_T		

The unbiased estimator of σ_α^2 computed from MSA and MSG is $\hat{\sigma}_\alpha^2 = (97.86 - 7.16)/9 = 10.08$. In this application, the unbiased estimates of σ_β^2 and σ_γ^2 are both negative and therefore it is more sensible to report interval estimates rather than point estimates. The 90% equal-tail confidence limits for these parameters can be computed by setting $\gamma = .95$ and $\gamma = .05$ in formula (9.32). The 90% confidence intervals computed by employing XPro are

$$[\ 0.0 \le \sigma_\beta^2 \le 8.1\ ] \quad \text{and} \quad [\ 0.0 \le \sigma_\gamma^2 \le 37.9\ ].$$

Now suppose that the α_i's appearing in (9.25) are fixed effects due to I treatments being compared and β_j's and γ_{ij}'s are still random effects. The ANOVA table appropriate for this mixed model is shown in Table 9.5.

Table 9.5. ANOVA of two-way mixed model

Source	D.F.	SS	MS	E(MS)
α	$I-1$	s_A	MSA	$\sigma_\varepsilon^2 + K\sigma_\gamma^2 + \frac{KJ}{I-1}\sum(\alpha_i - \bar{\alpha})^2$
β	$J-1$	s_B	MSB	$\sigma_\varepsilon^2 + K\sigma_\gamma^2 + KI\sigma_\beta^2$
γ	$(I-1)(J-1)$	s_G	MSG	$\sigma_\varepsilon^2 + K\sigma_\gamma^2$
Error	$IJ(K-1)$	s_E	MSE	σ_ε^2
Total	$IJK-1$	s_T		

It is clear from the ANOVA table that inferences about variance components σ_β^2 and σ_γ^2 can still be made using the foregoing results of this section. A test of the equality of treatment effects, H_0: $\alpha_1 = \alpha_2 = \cdots = \alpha_I$, can be based on (9.29) and the result

$$\frac{S_A}{\sigma_\varepsilon^2 + K\sigma_\gamma^2} \sim \chi_{I-1}^2$$

under H_0; if the null hypothesis is false, then the distribution is noncentral chi-squared. Therefore, the desired hypothesis can be tested on the basis of the F-statistic

$$\frac{S_A/(I-1)}{S_G/(I-1)(J-1)} \sim F_{I-1,(I-1)(J-1)}. \tag{9.33}$$

By comparing the F-statistic in (8.41) with that in (9.33), the form of the p-value can now be deduced from (8.42). For balanced incomplete block designs, Cohen and Sackrowitz (1989) gave an exact test that utilizes both the interblock information and the intrablock information.

9.6. Comparing Variance Components

In some experiments, the problem of interest is the comparison of the variances of random effects of two forms of a certain factor. For example in an industrial experiment involving two competing production processes, the problem of comparing the two variance components (corresponding to the two processes) of each factor arises. Zhou and Mathew (1994) gave an interesting example of a situation where the problem is to compare a new tube against a control tube used for firing ammunition from tanks. They report some data from an experiment carried out at a U.S. army ballistic research laboratory. The response variable was the miss distance and the tube-to-tube variance due to the new tube is compared with the control tube.

To describe the canonical form of the problem, consider two independent balanced mixed models with variance components σ_x^2 and σ_y^2, among other variance components. It is of interest to compare these two variance components on the basis of the statistics S_x, S_y, S_G, and S_H with following distributions:

$$W_g = \frac{S_G}{\sigma_g^2} \sim \chi_g^2, \quad W_h = \frac{S_H}{\sigma_h^2} \sim \chi_h^2, \tag{9.34}$$

$$W_x = \frac{S_x}{\sigma_g^2 + M\sigma_x^2} \sim \chi_x^2, \quad \text{and} \quad W_y = \frac{S_y}{\sigma_h^2 + N\sigma_y^2} \sim \chi_y^2, \tag{9.35}$$

where σ_g^2 and σ_h^2 are nuisance parameters, M and N are some known positive constants. Notice that the variance components of the two-way model (9.25) as well as the variance component of the one-way model (9.1) each have this canonical form (see also Exercises 9.6 and 9.7). Consider the problem of testing hypotheses of the form

$$H_0: \frac{\sigma_x^2}{\sigma_y^2} \le \rho \text{ versus } H_1: \frac{\sigma_x^2}{\sigma_y^2} > \rho\,, \tag{9.36}$$

where ρ is a prespecified real number.

Zhou and Mathew (1994) described how unbiased tests can be obtained in this kind of situation. As suggested by their solution for the case $\rho = 1$, the test variable appropriate for constructing tests for (9.36) is

$$T = \frac{M(\sigma_h^2 + \frac{N}{\rho}\sigma_x^2)\frac{s_y}{S_y} + \frac{N}{\rho}\sigma_g^2\frac{s_G}{S_G}}{\frac{N}{\rho}(\sigma_g^2 + M\sigma_x^2)\frac{s_x}{S_x} + M\sigma_h^2\frac{s_H}{S_H}} \tag{9.37}$$

$$= \frac{\lambda_\sigma M(\sigma_h^2 + N\sigma_y^2)\frac{s_y}{S_y} + \frac{N}{\rho}\sigma_g^2\frac{s_G}{S_G}}{\frac{N}{\rho}(\sigma_y^2 + M\sigma_x^2)\frac{s_x}{S_x} + M\sigma_h^2\frac{s_H}{S_H}}$$

$$= \frac{\lambda_\sigma M\frac{s_y}{W_y} + \frac{N}{\rho}\frac{s_G}{W_g}}{\frac{N}{\rho}\frac{s_x}{W_x} + M\frac{s_H}{W_h}}, \tag{9.38}$$

where

$$\lambda_\sigma = \frac{\sigma_h^2 + N\sigma_x^2/\rho}{\sigma_h^2 + N\sigma_y^2} = 1 \text{ when } \sigma_x^2/\sigma_y^2 = \rho. \tag{9.39}$$

It is evident from the representations (9.37) and (9.38), and from (9.39) that, at the boundary of the null hypothesis, the observed value of $T = 1$ and that the distribution of T does not depend on unknown parameters. Zhou and Mathew (1994) gave a derivation of the test variable based on the principles of unbiasedness and invariance.

Since T tends to take larger values under the alternative hypothesis, the right tail of the distribution of T corresponds to the appropriate extreme region. Hence, the p-value for testing hypotheses in (9.36) is obtained as

$$p = Pr(T \ge 1 \mid \lambda_\sigma = 1)$$
$$= Pr(M\frac{s_y}{W_y} + \frac{N}{\rho}\frac{s_G}{W_g} \ge \frac{N}{\rho}\frac{s_x}{W_x} + M\frac{s_H}{W_h}). \tag{9.40}$$

In order to obtain a computationally more efficient formula, define the beta random variables

$$B_1 = \frac{W_x}{W_x + W_y} \sim Beta(\frac{x}{2}, \frac{y}{2}),$$

$$B_2 = \frac{W_x + W_y}{W_x + W_y + W_g} \sim Beta(\frac{x+y}{2}, \frac{g}{2}),$$

$$B_3 = \frac{W_x + W_y + W_g}{W_x + W_y + W_g + W_h} \sim Beta(\frac{x+y+g}{2}, \frac{h}{2}),$$

which are mutually independent. It is clear from the definitions of the beta random variables that equation can be expressed as

$$p = Pr\left[\frac{Ms_y}{(1-B_1)B_2B_3} + \frac{N}{\rho}\frac{s_G}{(1-B_2)B_3} \geq \frac{N}{\rho}\frac{s_x}{B_1B_2B_3} + \frac{Ms_H}{(1-B_3)}\right] \quad (9.41)$$

where the probability is evaluated with respect to the beta random variables. The computation of this p-value which involve a three dimensional numerical integration can be conveniently performed using the XPro software package.

Example 9.5. Comparing the effect of two musical programs

An experiment is carried out to study the effect of two musical programs on the efficiency of some factory workers. Program X is administered over ten days and the number of units produced per day by 8 randomly selected workers is recorded. Program Y is administered over 8 days and the productions of 8 different workers are observed. Table 9.6 exhibits the summary data (sample means and variances) from the experiment.

Table 9.6. Productivity under two musical programs

	Program X							
Worker	A	B	C	D	E	F	G	H
Mean	93.2	98.1	89.6	88.4	96.2	95.0	99.6	97.9
Variance	23.4	27.6	18.6	22.1	15.4	26.2	33.1	29.8
	Program Y							
Worker	I	J	K	L	M	N	O	P
Mean	90.3	85.1	99.4	98.4	86.2	82.5	103.9	96.7
Variance	32.4	26.3	16.8	23.7	18.4	25.6	34.1	28.3

Assuming a one-way random model for data from each program, let μ_x and μ_y be the two grand means, and σ_x^2 and σ_y^2 be the variances of random effects due to workers, under the programs X and Y, respectively. Let σ_g^2 and σ_h^2 be the error variances of the two models. In this application, comparison of these variances as well as means are important. Consider the null hypothesis H_0: $\sigma_x^2 \leq \sigma_y^2/2$ that the worker-to-worker variability of the productivity under Program X is at least half as small as that under Program Y. From the above data we can obtain the ANOVA table for each program using the steps illustrated by Example 9.1. The results are as follows:

ANOVA for Program X

Source	D.F.	SS	MS	E(MS)
Workers	7	1160.8	165.8	$\sigma_g^2 + 10\sigma_x^2$
Error	72	1765.8	24.5	σ_g^2
Total	79	2926.6		

ANOVA for Program Y

Source	D.F.	SS	MS	E(MS)
Between	7	3428.2	489.8	$\sigma_h^2 + 8\sigma_y^2$
Within	56	1439.2	25.7	σ_h^2
Total	63	4867.4		

Therefore, the p-value for testing the null hypothesis is

$$p = Pr\left[10\frac{3428.2}{(1-B_1)B_2B_3} + \frac{8}{.5}\frac{1765.8}{(1-B_2)B_3} \geq \frac{8}{.5}\frac{1160.8}{B_1B_2B_3} + 10\frac{1439.2}{(1-B_3)}\right],$$

$$= .79,$$

which was computed using XPro. Hence, there is no reason to doubt the null hypothesis.

Exercises

9.1. Consider the one-way random effect model defined by (9.1). Show directly (without invoking normality assumptions) that

$$E\left[\frac{\sum_{i=1}^{k}\sum_{j=1}^{n}(X_{ij}-\bar{X}_{i.})^2}{k(n-1)}\right]=\sigma_\varepsilon^2$$

and that

$$E\left[\frac{n\sum_{i=1}^{k}(\bar{X}_{i.}-\bar{X})^2}{k-1}\right]=\sigma_\varepsilon^2+n\sigma_\alpha^2.$$

9.2. Consider again the random effect model in (9.1), this time with normality assumptions. Find the maximum likelihood estimates of σ_ε^2 and σ_α^2 and show that the MLE of σ_α^2 is not necessarily positive.

9.3. Consider the one-way random effect model considered in Section 9.1. Do each of the following using the distributional results in Section 9.2.

(a) Show that the p-value appropriate for testing the null hypothesis $H_0;\ \sigma_\varepsilon^2 \le \delta$ against the alternative hypothesis $H_1;\ \sigma_\varepsilon^2 > \delta$ is

$$p = 1 - F_U(s_E/\delta),$$

where F_U is the cdf of chi-squared random variable U with $k(n-1)$ degrees of freedom.

(b) Establish procedures for significance testing of point null hypotheses about σ_ε^2 (Hint: apply Corollary 2.1).

(c) Find the 100γ% upper confidence bound for σ_ε^2.

(d) Establish procedures for significance testing of point null hypotheses about σ_α^2.

9.4. Consider the two-way random effect model discussed in Section 9.4. Find the distribution of the grand mean $\bar{X}$ if there are no fixed effects in the model. Establish procedures for testing point null hypotheses concerning the mean μ. Also construct a 100γ% lower confidence bound for μ.

9.5. Repeat Exercise 9.4 if α is a fixed effect term and β and γ are random effects.

9.6. Consider the following random effect model that arise in a certain hierarchical classification:

$$X_{ijk} = \mu + \varepsilon_i + \varepsilon_{ij} + \varepsilon_{ijk},$$
$$i=1,\ldots,I;\ j=1,\ldots,J;\ k=1,\ldots,K,$$

where the random error terms ε_i, ε_{ij}, and ε_{ijk} are all independently and normally distributed with zero means and certain variances. Construct the appropriate ANOVA and examine whether the variance components have the canonical form given by equations (9.34) and (9.35).

(a) Find the MLEs of all four unknown parameters of the model.

(b) Establish procedures for testing left-sided null hypotheses about each variance component.

(c) Establish procedures for constructing interval estimates for each variance component.

9.7. Consider the mixed model

$$X_{ijk} = \mu_i + \gamma_{ij} + \varepsilon_{ijk},$$
$$i=1,\ldots,I;\ j=1,\ldots,J;\ k=1,\ldots,K,$$

where μ_i, $i=1,\ldots,I$ are I fixed effects, $\gamma_{ij} \sim N(0,\sigma_\gamma^2)$, and ε_{ijk} are normally and independently distributed error terms. Construct the appropriate ANOVA and show that the variance components have the canonical form given by equations (9.34) and (9.35). Establish procedures for making inferences about the parameters μ_i's and the variance component σ_γ^2.

9.8. Consider the data in Table 9.2 concerning weight distribution at birth. Construct 95% confidence intervals for the variance component of the problem by each of the following methods and compare the results:

(a) Tukey-Williams method;

(b) Healy method;

(c) generalized confidence intervals method.

9.9. Consider the data in Table 9.4 and assume the two-factor random effect model in (9.25).

(a) Construct the equal-tail 95% confidence interval for σ_ε^2 based on the chi-squared distribution given by (9.26).

(b) Construct the left-sided 95% confidence interval for σ_α^2.

(c) Test the null hypothesis, H_0: $\sigma_\alpha^2 \leq 7$.

(d) Test the null hypothesis, H_0: $\sigma_\gamma^2 \leq 1$.

9.10. Consider the problem in Example 9.5 and carry out each of the following inferences.

(a) Construct 95% confidence intervals for each variance component.

(b) Construct 95% confidence intervals for the grand mean of each model.

(c) Compare the two grand means.

(d) Compare the error variances of the two models.

9.11. The data below represent the weights of sample bottles from five groups of vegetable oil (source: Swallow and Searle (1978)):

Group A:	15.75	15.82	15.75	15.71	15.84
Group B:	15.70	15.68	15.64	15.60	
Group C:	15.68	15.66	15.59		
Group D:	15.69	15.71			
Group E:	15.65	15.60			

Assuming a one-way random effects model, construct by Wald's method, a 95% confidence interval for ρ, the fraction of variation of bottle weights between groups out of the total variation. Also test the hypothesis that the variance of the random effect term is at least 1.

9.12. An experiment is conducted to study the effect of three heat treatments and the machine-to-machine variation in bolt production. A certain measurement is taken from a random sample of four bolts turned out by three randomly chosen machines in a factory. The table below exhibits the observed measurements:

Machine	Heat treatment H_1	H_2	H_3
M_1	13.2, 12.3, 9.80	14.4, 11.2, 13.4	8.5, 13.2, 11.6
M_2	12.7, 11.2, 11.8	13.7, 12.9, 14.2	12.6, 9.1, 12.5
M_3	13.8, 12.9, 13.1	13.9, 14.2, 12.9	11.8, 11.7, 12.8

Assume a mixed model of the form (9.25), where the fixed effects are due to the three heat treatments.

(a) Construct 99% confidence intervals for each variance component.

(b) Construct a 99% confidence interval for the among machine variation as a fraction of the total variation.

Chapter 10
Regression

10.1. Introduction

In the previous two chapters we dealt with models having linear structures in which the factors affecting the response variable take on a fixed number of levels. This is indeed the case in many designed experiments. In other designed experiments, and in many other situations, we encounter response variables that depend linearly on one or more continuous variables. Both of these are particular cases of what is known as *linear models*. The purpose of this chapter is to develop procedures for making inferences about general linear models when the response variable is normally distributed.

In many areas of scientific research, investigators often need to understand the association between two or more variables. Sometimes they need to quantify how one variable depends on the others and then make predictions concerning the dependent variable. For example, an agricultural research scientist may wish to establish how the yield of a crop, say Y, depends on the amount of fertilizer applied, say X. As another example, an economist may wish to establish the relationship between the sales of a product and the factors (price, promotions, level of the economy, etc.) that affect the sales. For this purpose they assume a model from a certain parametric family and then estimate the parameters based on some observed values of the variables. For instance, in the first example, in view of the fact that too little or too much fertilizer can be bad, and Y is a nonnegative random variable, the scientist may assume a model of the form $log(Y) = a + bX + cX^2$, where a, b, and c are the parameters of the model. In some applications making inferences about some of these parameters is itself the problem of interest and the purpose of modeling. In other applications, investigators need to compare parameters of models established under different conditions. In this chapter we shall address these kinds of problems and try to provide solutions when the underlying models are linear or when they can be transformed into linear models, as was the case in the previous example.

10.2. Simple Linear Regression Model

In Section 1.8, we studied the problem of fitting a straight line to a set of data from two variables. Further analysis of the regression line obtained by the method of least squares requires formulation of the problem as a statistical model. To do this, consider two variables, X and Y, which are either jointly distributed according to a bivariate normal distribution, or X is a *deterministic control variable* while Y at any given X is a normally distributed random variable. In both cases we can study the conditional distribution of Y given $X=x$. Then, the conditional expectation of Y given X, $E(Y|x)$, is called the *regression function of Y on X*. We consider only the situation where the regression function is a linear function of x; this is indeed the case if X and Y have a bivariate normal distribution. So, as in the previous two chapters, consider a linear model of the form

$$Y_i = \alpha + \beta x_i + \varepsilon_i, \tag{10.1}$$

$$\varepsilon_i \sim N(0,\sigma^2)\ ,\ i=1,\dots,n\ , \tag{10.2}$$

which is the stochastic counterpart of (1.20), where ε_i, $i=1,\dots,n$ are the independently distributed random error terms. Obviously, $E(Y_i|x_i) = \alpha + \beta x_i$ is the value of the regression line at $x = x_i$.

The likelihood function of the unknown parameters, given the data (x_i,y_i), $i=1,\dots,n$, is

$$L(\alpha,\beta,\sigma^2) = (2\pi\sigma^2)^{-n/2}\exp[-\frac{1}{2\sigma^2}\sum_{i=1}^{n}(y_i - \alpha - \beta x_i)^2]\ . \tag{10.3}$$

Clearly, regardless of the value of σ^2, the likelihood function can be maximized with respect to (α,β) by minimizing

$$q = \sum_{i=1}^{n}(y_i - \alpha - \beta x_i)^2 \tag{10.4}$$

with respect to the parameters of the regression line. Therefore, the MLEs of α and β are the same as their least squares estimates found in Chapter 1:

$$\hat{\beta} = \frac{\sum_{i=1}^{n} x_i y_i - n\overline{xy}}{\sum_{i=1}^{n} x_i^2 - n\overline{x}^2} \quad \text{and} \quad \hat{\alpha} = \overline{y} - \hat{\beta}\overline{x}. \tag{10.5}$$

Further, it can be shown by minimizing (10.4) with respect to σ^2 that the MLE of σ^2 is

$$\hat{\sigma}^2 = \frac{1}{n}\sum_{i=1}^{n}(y_i - \hat{\alpha} - \hat{\beta}x_i)^2. \tag{10.6}$$

It can be shown that this estimator is biased and that an estimator which corrects this bias is

$$\tilde{\sigma}^2 = \frac{1}{n-2}\sum_{i=1}^{n}(y_i - \hat{\alpha} - \hat{\beta}x_i)^2. \tag{10.7}$$

10.3. Inferences about Parameters of the Simple Regression Model

It is evident from their structure that $\hat{\alpha}$ and $\hat{\beta}$ are linear functions of Y_i, $i = 1,\ldots, n$. Therefore, the estimators are normally distributed. To specify the complete distribution we need to find the mean and the variance of each estimator; the covariance of $\hat{\alpha}$ and $\hat{\beta}$ is also required.

The least squares estimator of β can be rewritten as

$$\hat{\beta} = \frac{\sum_{i=1}^{n}(x_i - \bar{x})Y_i}{\sum_{i=1}^{n}(x_i - \bar{x})^2}. \tag{10.8}$$

Since $\sum_{i=1}^{n}(x_i - \bar{x}) = 0$, we get

$$E(\hat{\beta}) = \frac{\sum_{i=1}^{n}(x_i - \bar{x})E(Y_i)}{\sum_{i=1}^{n}(x_i - \bar{x})^2} = \frac{\sum_{i=1}^{n}(x_i - \bar{x})(\alpha + \beta x_i)}{\sum_{i=1}^{n}(x_i - \bar{x})^2}$$

$$= \beta\frac{\sum_{i=1}^{n}(x_i - \bar{x})x_i}{\sum_{i=1}^{n}(x_i - \bar{x})^2} = \beta.$$

That is, $\hat{\beta}$ is an unbiased estimator of β; as a matter of fact it is the best linear unbiased estimator of β. (See Exercise 10.1 for details.) Furthermore, $\hat{\alpha}$ is the best linear unbiased estimator of α. The variance of $\hat{\beta}$ is

$$Var(\hat{\beta}) = \frac{\sum_{i=1}^{n}(x_i - \bar{x})^2 Var(Y_i)}{[\sum_{i=1}^{n}(x_i - \bar{x})^2]^2} = \frac{\sigma^2}{\sum_{i=1}^{n}(x_i - \bar{x})^2}. \tag{10.9}$$

By taking a similar approach it can be shown (Exercise 10.2) that

$$Var(\hat{\alpha}) = \frac{\sigma^2 \sum_{i=1}^{n} x_i^2}{n\sum_{i=1}^{n}(x_i - \bar{x})^2} \tag{10.10}$$

is the variance of the MLE of α and that

$$Cov(\hat{\alpha},\hat{\beta}) = -\frac{\sigma^2 \bar{x}}{\sum_{i=1}^{n}(x_i - \bar{x})^2} \tag{10.11}$$

is the covariance of $\hat{\alpha}$ and $\hat{\beta}$.

The variance of a predicted value of Y corresponding to a particular value of x, say $\hat{Y} = \hat{\alpha} + \hat{\beta}x$, can be deduced from equations (10.9), (10.10), and (10.11). The result is

$$\begin{aligned} Var(\hat{Y}) &= Var(\hat{\alpha}) + x^2 Var(\hat{\beta}) + 2xCov(\hat{\alpha},\hat{\beta}) \\ &= \sigma^2\left[\frac{1}{n} + \frac{(x - \bar{x})^2}{\sum_{i=1}^{n}(x_i - \bar{x})^2}\right]. \end{aligned}$$

Since $E(\hat{Y}) = \alpha + \beta x = E(Y)$, $\hat{Y}$ is an unbiased predictor. In constructing confidence bands for predicted values of the dependent variable, mean squared error $E[(\hat{Y} - Y)^2] = Var(\hat{Y} - Y)$ is more important. Since $\hat{Y}$ is based on Y_i, $i = 1,\ldots, n$ and Y are independent, this can be obtained from the formula

$$\begin{aligned} E[(\hat{Y} - Y)^2] &= Var(Y) + Var(\hat{Y}) \\ &= \sigma^2\left[\frac{(n+1)}{n} + \frac{(x - \bar{x})^2}{\sum_{i=1}^{n}(x_i - \bar{x})^2}\right]. \end{aligned} \tag{10.12}$$

Moreover, $\hat{Y} - Y$ is normally distributed with zero mean.

Before we can construct confidence intervals and test hypotheses about these parameters we need one more result, because σ^2 appearing in the above equations is an unknown parameter. It can be shown that

$$\frac{(n-2)\tilde{\sigma}^2}{\sigma^2} \sim \chi^2_{n-2} \tag{10.13}$$

and that this random variable is independent of $\hat{\alpha}$ and $\hat{\beta}$. We shall prove this result in Section 10.4 in a more general setting. Hence, inferences about the parameters α and β can be based on the t-statistics

$$T_\alpha = \frac{(\hat{\alpha} - \alpha)}{\hat{\sigma}_{\hat{\alpha}}} \sim t_{n-2}$$

and

$$T_\beta = \frac{(\hat{\beta} - \beta)}{\hat{\sigma}_{\hat{\beta}}} \sim t_{n-2}$$

respectively, where $\hat{\sigma}^2_{\hat{\alpha}}$ is the estimated variance of $\hat{\alpha}$ obtained by replacing σ^2 in (10.10) by $\tilde{\sigma}^2$ and $\hat{\sigma}^2_{\hat{\beta}}$ is the estimated variance of $\hat{\beta}$ obtained by replacing σ^2 in (10.9) by $\tilde{\sigma}^2$. Obviously, each t-statistic is free of unknown parameters and both are stochastically increasing with respect to the underlying parameters of interest.

Inferences about the parameters of the regression can be based on these results. For example, the hypotheses

$$H_0: \beta = \beta_0 \quad \text{versus} \quad H_1: \beta \neq \beta_0,$$

can be tested based on the p-value

$$p = 2G_{n-2}\left(-\frac{|\hat{\beta} - \beta_0| \left[\sum_{i=1}^{n} (x_i - \bar{x})^2\right]^{1/2}}{\tilde{\sigma}} \right), \tag{10.14}$$

where G_{n-2} is the cdf of the t distribution with $n-2$ degrees of freedom. Furthermore, the shortest $100(1-\alpha)\,\%$ confidence interval for β is

$$\hat{\beta} \pm t_{\alpha/2} \frac{\tilde{\sigma}}{\left[\sum_{i=1}^{n} (x_i - \bar{x})^2\right]^{1/2}}, \tag{10.15}$$

where $t_{\alpha/2}$ is the $1 - \alpha/2$-th quantile of the t distribution with $n-2$ degrees of freedom. Confidence intervals of a predicted value of Y corresponding to a x value can be constructed in a similar manner. The shortest $100(1-\alpha)\,\%$ confidence interval for Y at $X = x$ is

$$(\hat{\alpha} + \hat{\beta}x) \pm t_{\alpha/2}\tilde{\sigma}\left[\frac{(n+1)}{n} + \frac{(x-\bar{x})^2}{\sum_{i=1}^{n}(x_i-\bar{x})^2}\right]^{1/2}.$$

Example 10.1. Association of test scores in Mathematics and Statistics

Consider again the regression of Y on X that we used in Chapter 1 to predict scores in statistics using scores in mathematics; that is the predictor $\hat{Y} = \hat{\alpha} + \hat{\beta}x$. The parameters are estimated based on the data in Table 10.1.

Table 10.1. Test scores in mathematics and in statistics

Mathematics	48 52 55 61 66 69 70 70 74 75 78 82 85 90 95
Statistics	54 55 50 67 65 70 74 68 65 73 85 80 85 89 90

Recall that the least squares estimates of α and β are $\hat{\alpha} = 8.80$ and $\hat{\beta} = .877$. The unbiased estimate of σ^2 computed by applying equation (10.7) is $\hat{\sigma}^2 = .339$. Estimated variances and the covariance of $\hat{\alpha}$ and $\hat{\beta}$ computed using equations (10.7), (10.9), (10.10), and (10.11) are

$$\hat{\sigma}^2_{\hat{\alpha}} = 39.24, \hat{\sigma}^2_{\hat{\beta}} = .00746, \text{ and } \hat{\sigma}_{\hat{\alpha}\hat{\beta}} = -.532,$$

respectively. Therefore, the observed t-statistics for testing $H_{01}: \alpha = 0$ and $H_{02}: \beta = 0$ are

$$t_\alpha = 8.80/6.26 = 1.41 \text{ and } t_\beta = .877/.086 = 10.2,$$

respectively. Therefore, while there is very strong evidence to reject H_{02}, H_{01} can not be rejected at the .05 level. Since $t_{.025}$ with 13 degrees of freedom is 2.16, the 95% confidence interval for β is

$$0.877 \pm 2.16 \times .086 = (0.691, 1.063).$$

Recall further that the predicted Statistics score for a student who scored 50 in Mathematics was $\hat{Y} = 52.6$. The estimated variance of this predictor is

$$\hat{Var}(\hat{Y}) = 39.24 + 2500 \times .00746 - 100 \times .532$$
$$= 4.69.$$

and therefore, $E[(\hat{Y} - Y)^2)] = 4.69 + .34 = 5.03$. Hence the 95% confidence interval for possible scores for students who obtained a score of 50 in Mathematics is

$$52.6 \pm 2.16 \times \sqrt{5.03} = (47.8, 57.4).$$

10.4. Multiple Linear Regression

Most practical problems of regressions involve more than one explanatory variable affecting the dependent variable. For example, the monthly expenditure of a household will depend on the household income, number of people living in the household, their average age, etc. Or, in estimating the gas consumption of a household during a month in which the gas meter was not read, the gas company needs to take into account a number of variables such as the average temperature during the month, the gas consumption of the household over previous months, and that during the same month in previous years. When a regression involves more than one explanatory variable it is said to be a multiple regression model. The ANOVA models that we studied in Chapter 8 are multiple linear regression models having a special structure. It will become clear that polynomial models with one or more independent variables can also be treated as multiple linear regression models.

Consider a variable Y which possibly depends on k independent variables $X_1, X_2, ..., X_k$. As in the previous section, given n observations from each variable, assume a linear model of the form

$$Y_i = \beta_1 x_{1i} + \beta_2 x_{2i} + \cdots + \beta_k x_{ki} + \varepsilon_i \tag{10.16}$$

$$\varepsilon_i \sim N(0,\sigma^2)\ ,\ \ i=1,..., n\ ,$$

where $\varepsilon_i,\ i=1,..., n$ are the random error terms. If β_1 stands for an intercept term, then

$$x_{1i} = 1 \quad \text{for all} \quad i=1,..., n\ .$$

Assume that there are more data from each variable than the number of unknown parameters; that is $k < n$. To rewrite (10.16) in a compact form, define the matrices

$$\mathbf{Y} = \begin{bmatrix} Y_1 \\ Y_2 \\ \cdot \\ \cdot \\ \cdot \\ Y_n \end{bmatrix} \qquad \mathbf{X} = \begin{bmatrix} X_{11} & X_{21} & \cdots & X_{k1} \\ X_{12} & X_{22} & \cdots & X_{k2} \\ \cdot & \cdot & & \cdot \\ \cdot & \cdot & \cdots & \cdot \\ \cdot & \cdot & & \cdot \\ X_{1n} & X_{2n} & \cdots & X_{kn} \end{bmatrix}$$

$$\boldsymbol{\beta} = \begin{bmatrix} \beta_1 \\ \beta_2 \\ \cdot \\ \cdot \\ \cdot \\ \beta_k \end{bmatrix} \qquad \boldsymbol{\varepsilon} = \begin{bmatrix} \varepsilon_1 \\ \varepsilon_2 \\ \cdot \\ \cdot \\ \cdot \\ \varepsilon_n \end{bmatrix}\ .$$

Assume that the columns of $\mathbf{X}$ are linearly independent so that $\mathbf{X}$ is of full rank k. With this notation all n equations in (10.16) can be set out compactly as

$$\mathbf{Y} = \mathbf{X}\boldsymbol{\beta} + \boldsymbol{\varepsilon},$$

where $\boldsymbol{\varepsilon}$ has a multivariate normal distribution with mean vector $\mathbf{0}$ and the variance covariance matrix $\sigma^2 I_n$. The distribution of $\boldsymbol{\varepsilon}$ is denoted as

$$\boldsymbol{\varepsilon} \sim N(\mathbf{0}, \sigma^2 \mathbf{I}_n),$$

where I_n is the n dimensional identity matrix with ones in the diagonal and zeros elsewhere.

Estimation of parameters

Since $\mathbf{Y} \mid \mathbf{X} \sim N(\mathbf{X}\boldsymbol{\beta}, \sigma^2 \mathbf{I}_n)$, the likelihood function of the unknown parameters given the data $(\mathbf{X}, \mathbf{y})$ is

$$L(\boldsymbol{\beta}, \sigma^2) = (2\pi\sigma^2)^{-n/2} \exp[-\frac{1}{2\sigma^2}(\mathbf{y} - \mathbf{X}\boldsymbol{\beta})'(\mathbf{y} - \mathbf{X}\boldsymbol{\beta})].$$

First of all, notice that, regardless of the value of σ^2, maximizing the likelihood function with respect to $\boldsymbol{\beta}$ is equivalent to choosing the parameter vector $\boldsymbol{\beta}$ so as to minimize the residual sum of squares

$$\begin{aligned} \mathbf{e}'\mathbf{e} &= (\mathbf{y} - \mathbf{X}\boldsymbol{\beta})'(\mathbf{y} - \mathbf{X}\boldsymbol{\beta}) \\ &= \mathbf{y}'\mathbf{y} - 2\boldsymbol{\beta}'\mathbf{X}'\mathbf{y} + \boldsymbol{\beta}'\mathbf{X}'\mathbf{X}\boldsymbol{\beta}, \end{aligned} \tag{10.17}$$

where $\mathbf{e} = (\mathbf{y} - \mathbf{X}\boldsymbol{\beta})$ is the observed value of the error term $\boldsymbol{\varepsilon}$. Therefore, the MLE of $\boldsymbol{\beta}$ is the same as its least squares estimator. To find this estimator, differentiating (10.17) with respect to $\boldsymbol{\beta}$ we get the vector of first partial derivatives

$$\frac{\partial}{\partial \boldsymbol{\beta}}(\mathbf{e}'\mathbf{e}) = -2\mathbf{X}'\mathbf{y} + 2\mathbf{X}'\mathbf{X}\boldsymbol{\beta} \tag{10.18}$$

and the Hessian matrix

$$\frac{\partial^2}{\partial \boldsymbol{\beta} \partial \boldsymbol{\beta}'}(\mathbf{e}'\mathbf{e}) = 2\mathbf{X}'\mathbf{X}.$$

Since the Hessian is positive definite, the least squares estimator can be obtained by equating the first derivatives in (10.18) to zero. Thus, the least squares estimator (MLE) $\hat{\boldsymbol{\beta}}$ satisfies the equation

$$\mathbf{X}'\mathbf{X} = \hat{\boldsymbol{\beta}}\mathbf{X}'\mathbf{y}.$$

Since, $\mathbf{X}$ is assumed to be of full rank, solving this equation we get

$$\hat{\boldsymbol{\beta}} = (\mathbf{X}'\mathbf{X})^{-1}\mathbf{X}'\mathbf{y}. \tag{10.19}$$

By maximizing the likelihood function with respect to σ^2 as well we obtain the MLE of σ^2 as

$$\hat{\sigma}^2 = \frac{(\mathbf{y} - \mathbf{X}\hat{\boldsymbol{\beta}})'(\mathbf{y} - \mathbf{X}\hat{\boldsymbol{\beta}})}{n} = \frac{\mathbf{e}'\mathbf{e}}{n},$$

where $\mathbf{e}'\mathbf{e}$ is the residual sum of squares due to the least squares regression. As in Section 10.3, the unbiased estimator which corrects the finite bias of this estimator is

$$\tilde{\sigma}^2 = \frac{\mathbf{e}'\mathbf{e}}{n-k}. \tag{10.20}$$

Although it is possible to establish this result directly from its definition and without any distributional assumptions (see Exercise 10.3), later we shall deduce this result from the distribution of $\mathbf{e}'\mathbf{e}$.

The ratio of the *explained sum of squares*, $\mathbf{y}'\mathbf{y} - \mathbf{e}'\mathbf{e}$, to the total sum of squares, $\mathbf{y}'\mathbf{y}$, is called the *coefficient of determination*; if the linear model (10.16) has an intercept, then each sum of squares must be corrected by subtracting $n\bar{y}^2$, where $\bar{y}$ is the sample mean of y data. This quantity is also called the *coefficient of multiple correlation* and is denoted by r^2. Since, the error sum of squares of the least squares regression can be expressed as $\mathbf{e}'\mathbf{e} = \mathbf{y}'\mathbf{y} - \hat{\boldsymbol{\beta}}'\mathbf{X}'\mathbf{y}$, r^2 of a regression line with an intercept term can be conveniently computed as

$$r^2 = \frac{\hat{\boldsymbol{\beta}}'\mathbf{X}'\mathbf{y} - n\bar{y}^2}{\mathbf{y}'\mathbf{y} - n\bar{y}^2}. \tag{10.21}$$

Example 10.2. Polynomial regressions

The foregoing formulae can be utilized to fit polynomial regressions as well. To illustrate this and the application of the above results, consider the data in Table 10.2 on the age (in weeks) of a certain breed of hens and the mean weight (in pounds) at that age.

Table 10.2. Age and mean weight of hens

Age	3	4	6	8	10	12	14
Weight	.42	.50	.76	.89	1.10	1.14	1.22

A scatter diagram of this data is shown in Figure 10.1. The scatter diagram suggests that a simple linear regression is not quite adequate to fit this data. Therefore, let us fit polynomials of degrees 2 and 3 using the formulae of multiple linear regression that we studied in this section. First assume the model

$$Y = \beta_1 + \beta_2 t + \beta_3 t^2 + \varepsilon,$$

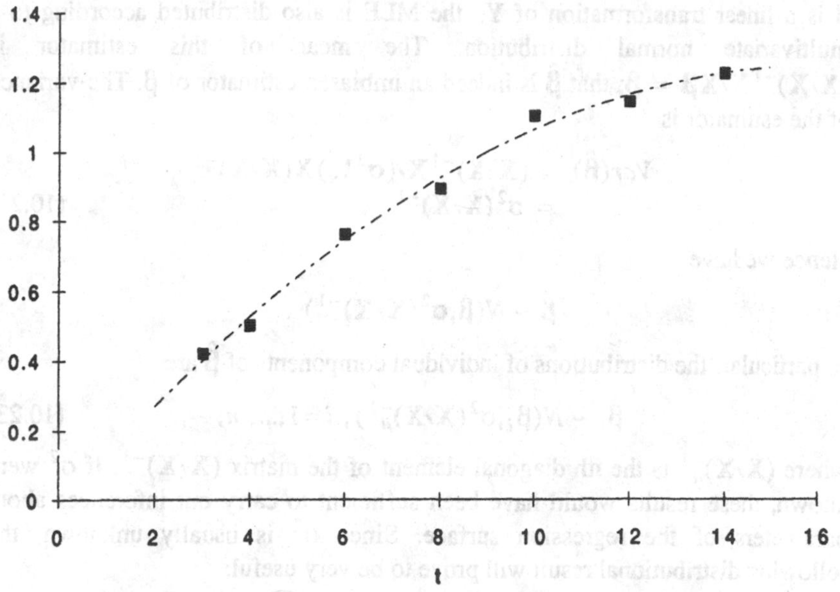

Figure 10.1. Scatter diagram of weights by age and the parabolic fit

where t is the age of hens and Y is the mean weight at age t. In the notation of equation (10.16), $x_1 = 1, x_2 = t, x_3 = t^2$. The least squares estimates of the parameters of the model found by applying the formula (10.19) are

$$\hat{\beta}_1 = -.0264, \quad \hat{\beta}_2 = .1578,$$

and

$$\hat{\beta}_3 = -.00049\,.$$

If a polynomial of degree 3 is fitted to the same data, we get the estimated model $\hat{Y} = .0373 + .1280t - .00098t^2 - .00015t^3$. The coefficients of determination for these two regressions estimated using formula (10.21) are .9929 and .9933, respectively. Perhaps, this indicates that the third order term t^3 has not significantly contributed to the regression. In the following section we shall develop procedures to answer questions of this nature.

10.5. Distributions of Estimators and Significance Tests

Recall that given **X**, $\mathbf{Y} \sim N(\mathbf{X}\boldsymbol{\beta}, \sigma^2\mathbf{I}_n)$. Since the MLE $\hat{\boldsymbol{\beta}} = (\mathbf{X}'\mathbf{X})^{-1}\mathbf{X}'\mathbf{Y}$ of $\boldsymbol{\beta}$ is a linear transformation of **Y**, the MLE is also distributed according to a multivariate normal distribution. The mean of this estimator is $(\mathbf{X}'\mathbf{X})^{-1}\mathbf{X}'\mathbf{X}\boldsymbol{\beta} = \boldsymbol{\beta}$; that $\hat{\boldsymbol{\beta}}$ is indeed an unbiased estimator of $\boldsymbol{\beta}$. The variance of the estimator is

$$\begin{aligned} Var(\hat{\boldsymbol{\beta}}) &= (\mathbf{X}'\mathbf{X})^{-1}\mathbf{X}'(\sigma^2\mathbf{I}_n)\mathbf{X}(\mathbf{X}'\mathbf{X})^{-1} \\ &= \sigma^2(\mathbf{X}'\mathbf{X})^{-1} \end{aligned} \tag{10.22}$$

Hence we have

$$\hat{\boldsymbol{\beta}} \sim N(\boldsymbol{\beta}, \sigma^2(\mathbf{X}'\mathbf{X})^{-1}) .$$

In particular, the distributions of individual components of $\hat{\boldsymbol{\beta}}$ are

$$\hat{\beta}_i \sim N(\beta_i, \sigma^2(\mathbf{X}'\mathbf{X})_{ii}^{-1}), \; i = 1, \ldots, n, \tag{10.23}$$

where $(\mathbf{X}'\mathbf{X})_{ii}^{-1}$ is the ith diagonal element of the matrix $(\mathbf{X}'\mathbf{X})^{-1}$. If σ^2 were known, these results would have been sufficient to carry out inferences about parameters of the regression surface. Since σ^2 is usually unknown, the following distributional result will prove to be very useful:

$$\frac{\mathbf{e}'\mathbf{e}}{\sigma^2} \sim \chi^2_{n-k}. \tag{10.24}$$

That $\tilde{\sigma}^2$ defined by (10.20) is an unbiased estimator of σ^2 immediately follows from this result, because the mean of a chi-squared random variable is equal to its degrees of freedom.

To derive equation (10.24), note that the residual $\mathbf{e} = (\mathbf{Y} - \mathbf{X}\boldsymbol{\beta})$ can be expressed as

$$\begin{aligned} \mathbf{e} &= (I_n - \mathbf{X}(\mathbf{X}'\mathbf{X})^{-1}\mathbf{X}')\mathbf{Y} \\ &= (I_n - \mathbf{X}(\mathbf{X}'\mathbf{X})^{-1}\mathbf{X}')(\mathbf{X}\boldsymbol{\beta} + \boldsymbol{\varepsilon}) \\ &= P\boldsymbol{\varepsilon}, \end{aligned} \tag{10.25}$$

where $P = (I_n - \mathbf{X}(\mathbf{X}'\mathbf{X})^{-1}\mathbf{X}')$ and

$$P\mathbf{X} = (I_n - \mathbf{X}(\mathbf{X}'\mathbf{X})^{-1}\mathbf{X}')\mathbf{X} = \mathbf{0} .$$

It is easily seen by direct multiplication that P is a symmetric idempotent matrix (a projection matrix); that is $P = P'$ and $PP = P$. Moreover, the trace of P is $n-k$ so that the Eigenvalues of P consist of $n-k$ ones and k zeros. Therefore, an orthogonal matrix Q can be formed using the Eigenvectors of P such that

$$Q'PQ = J_{n-k},$$

where J_{n-k} is a diagonal matrix with ones in the first $n-k$ diagonal elements, and zeros elsewhere. Consider the orthogonal transformation of the error vector

$$\delta = Q'\varepsilon \sim N(\mathbf{0},\sigma^2 I_n)$$

with its inverse transformation $\varepsilon = Q\delta$. Then, $e'e$ can be expressed as

$$\begin{aligned} e'e &= \delta' Q' P Q \delta \\ &= \delta' J_{n-k} \delta \\ &= \sum_{i=1}^{n-k} \delta_i^2 . \end{aligned}$$

Since δ_i, $i=1,\ldots,n-k$ are independently and identically distributed as $N(0,\sigma^2)$, (10.24) now follows.

Moreover, $\hat{\beta}$ and $e'e$ are independently distributed. To see this, first note that equation (10.25) implies that $E(\mathbf{e}) = \mathbf{0}$ and also recall that $P\mathbf{X} = \mathbf{0}$. Consequently, the covariance of the two random vectors in question becomes

$$\begin{aligned} Cov(\mathbf{e},\hat{\beta}) &= E(\mathbf{e}\hat{\beta}') \\ &= E(P\varepsilon[\beta' + \varepsilon'\mathbf{X}(\mathbf{X}'\mathbf{X})^{-1}]) \\ &= PE(\varepsilon\varepsilon')\mathbf{X}(\mathbf{X}'\mathbf{X})^{-1}]) \\ &= \sigma^2 P\mathbf{X}(\mathbf{X}'\mathbf{X})^{-1} \\ &= \mathbf{0}. \end{aligned}$$

Since underlying random vectors are normally distributed, the zero covariance of $\hat{\beta}$ and $e'e$ is sufficient to conclude that they are independently distributed.

In particular, equations (10.23), (10.24), and the preceding result imply that

$$t_i = \frac{\hat{\beta}_i - \beta_i}{\tilde{\sigma}\sqrt{(\mathbf{X}'\mathbf{X})_{ii}^{-1}}} \sim t_{n-k},\ i=1,\ldots,n . \qquad (10.26)$$

As in Section 10.3, inferences concerning a predicted value $\hat{Y} = \mathbf{x}'\hat{\beta}$ of the Y variable corresponding to a given value of $\mathbf{x}$ can also be made by taking a parallel approach. Noting that

$$Var(\hat{Y} - Y) = \sigma^2[1 + \mathbf{x}'(\mathbf{X}'\mathbf{X})^{-1}\mathbf{x}] ,$$

inferences about the predictor can be based on the t-statistic

$$t_y = \frac{\hat{Y} - Y}{\tilde{\sigma}[1 + \mathbf{x}'(\mathbf{X}'\mathbf{X})^{-1}\mathbf{x}]^{1/2}} .$$

These t-statistics can be employed to construct confidence intervals for predictions, individual regression coefficients, and to test their statistical significance. For instance, the shortest $100(1-\alpha)\%$ confidence interval for β_i is given by

$$\hat{\beta}_i \pm t_{\alpha/2}\tilde{\sigma}(\mathbf{X}'\mathbf{X})_{ii}^{-1/2} ,$$

where $t_{\alpha/2}$ is the $1 - \alpha/2$-th quantile of the t distribution with $n-k$ degrees of freedom.

In some applications, testing the significance of certain explanatory variables may itself be the main purpose of formulating a regression model. When there are many explanatory variables, repeated application of the t-test for testing individual regression coefficients can increase the Type-I error to prohibitive levels just as in the case of ANOVA that we discussed in Chapter 8. Therefore, joint tests for relevant coefficients or all coefficients (except perhaps the intercept term) need to be carried out before proceeding to t-tests. So, consider the problem of testing a sub-vector of coefficients of $\boldsymbol{\beta}$.

To describe an F-test appropriate for this purpose, consider the partitions

$$\boldsymbol{\beta} = \begin{bmatrix} \boldsymbol{\beta}_1 \\ \boldsymbol{\beta}_2 \end{bmatrix} \quad \text{and} \quad \mathbf{X} = [\mathbf{X}_1 \; \mathbf{X}_2]$$

so that the model $\mathbf{Y} = \mathbf{X}\boldsymbol{\beta}$ can be expressed as

$$\mathbf{Y} = \mathbf{X}_1\boldsymbol{\beta}_1 + \mathbf{X}_2\boldsymbol{\beta}_2, \tag{10.27}$$

where $\mathbf{X}_1$ is an $n \times k_1$ matrix, $\mathbf{X}_2$ is an $n \times k_2$ matrix, and $\boldsymbol{\beta}_1$ and $\boldsymbol{\beta}_2$ are vectors of length k_1 and k_2, respectively. Consider the problem of testing the hypotheses

$$H_0\colon \boldsymbol{\beta}_2 = \mathbf{0} \quad \text{versus} \quad H_1\colon \boldsymbol{\beta}_2 \neq \mathbf{0}\,. \tag{10.28}$$

Now consider the regression of $\mathbf{Y}$ on $\mathbf{X}_1$; that is

$$\begin{aligned} \mathbf{Y} &= \mathbf{X}_1\boldsymbol{\beta}_1 \\ &= \tilde{\mathbf{X}}\boldsymbol{\beta}, \text{ where } \tilde{\mathbf{X}} = [\mathbf{X}_1 \; \mathbf{0}]. \end{aligned} \tag{10.29}$$

Let $\mathbf{e}_1{}'\mathbf{e}_1$ be the residual sum of squares due to the least squares regression of this model and let $\mathbf{e}'\mathbf{e}$ be the error sum of squares due to the full regression in (10.27). In testing the effect of all explanatory variables of a linear model with an intercept term, we set $\boldsymbol{\beta}_1 = \beta_1$ and $\mathbf{e}_1{}'\mathbf{e}_1 = S_T = \sum_{i=1}^{n}(Y_i - \bar{Y})^2$. The reduction in the error sum of squares due to the addition of $\mathbf{X}_2$ is $\mathbf{e}_1{}'\mathbf{e}_1 - \mathbf{e}'\mathbf{e}$. This reduction will tend to be relatively small or large depending whether the null hypothesis in (10.28) is true or false. As a result, H_0 can be tested on the basis of the F-statistic

$$F = \frac{(\mathbf{e}_1{}'\mathbf{e}_1 - \mathbf{e}'\mathbf{e})/k_2}{\mathbf{e}'\mathbf{e}/(n-k)} \sim F_{k_2, n-k}\,. \tag{10.30}$$

We will show later in this section that, under H_0, the random variable F has an F-distribution with k_2 and $n-k$ degrees of freedom and that the tests based on F are unbiased. At level α, H_0 is rejected if $F > F_\alpha$, or equivalently if its p-value $p = Pr(F > F_{obs}) < \alpha$, where F_α is the $(1-\alpha)$th quantile of the underlying F-distribution.

Define the mean sums of squares

$$MSR = \frac{s_T - \mathbf{e}'\mathbf{e}}{k-1}, \; MSR2 = \frac{\mathbf{e}_1'\mathbf{e}_1 - \mathbf{e}'\mathbf{e}}{k_2}, \text{ and } MSE = \frac{\mathbf{e}'\mathbf{e}}{n-k}.$$

Then, as shown in Table 10.3, the computations involved in testing H_0 as well as the significance of the full regression can be set out in the form an analysis of variance table.

Table 10.3. ANOVA in regression

Source of Variation	D.F.	Sum of Squares	Mean Sum of Squares	F-Statistic
$\mathbf{X}_1$	$k_1 - 1$	$s_T - \mathbf{e}_1'\mathbf{e}_1$		
addition of $\mathbf{X}_2$	k_2	$\mathbf{e}_1'\mathbf{e}_1 - \mathbf{e}'\mathbf{e}$	MSR2	MSR2/MSE
$\mathbf{X}_1, \mathbf{X}_2$	$k - 1$	$s_T - \mathbf{e}'\mathbf{e}$	MSR	MSR/MSE
Error	$n - k$	$\mathbf{e}'\mathbf{e}$	MSE	
Total	$n - 1$	s_T		

Example 10.3. Polynomial regressions (continued)

Consider again the data in Table 10.2 and the least squares regression polynomials fitted to the data. The estimated standard errors of the parameters of the fitted parabola computed by applying the formulae (10.20) and (10.22) are

$$\hat{\sigma}_{\hat{\beta}_1} = .0655, \; \hat{\sigma}_{\hat{\beta}_2} = .0179, \; \hat{\sigma}_{\hat{\beta}_3} = .0011.$$

From (10.26), the t-statistics can be computed under $\boldsymbol{\beta} = \mathbf{0}$ (that is the ratios of estimated parameters and their estimated standard errors, $t_i = \beta_i / \hat{\sigma}_{\hat{\beta}_i}$) as

$$t_1 = -.403, \; t_2 = 8.82, \; t_3 = -4.67.$$

This suggests that the coefficients of both t and t^2 are highly significant. The F-statistic computed from (10.30) is 282.8, further confirming that the regression if highly significant. Let us now examine whether there is a need to add an additional term to the polynomial regression. Table 10.4 shows all necessary sums of squares including the total sum of squares (about the sample mean of y), the sum of squares explained by the full regression, and the sum of squares explained by the parabolic regression. The reduction of the residual sum of square due to the addition of the third power term is only .0002. The value of the F-statistic appropriate for testing the significance of this term is .154. Since the 95th quantile of the F-distribution with 1 and 3 degrees of freedom is 10.13, we can conclude that in the presence of the first and the second order terms, the third order term of the polynomial does not really make any contribution to the regression. It is of interest to note that inclusion of this term in the polynomial actually has an undesirable effect on the regression. For, when the full regression is fitted, the t-statistics of the estimated parameters of the polynomial become .20, 1.52, -.09, and -.36, thus suggesting that even the second order term is not statistically significant, a result that seems to contradict our previous conclusion. This example emphasizes the importance of careful significance testing in regression and the problem of over parameterization, especially when the explanatory variables are highly correlated. Moreover, in the presence of significant higher order terms (e.g. $\beta_4 \neq 0$) one should not test for the insignificance of lower order terms (e.g. $\beta_3 = 0$).

Source of Variation	D.F.	Sum of Squares	Mean Sum of Squares	F-Statistic
(t,t^2)	2	.5958		
addition of t^3	1	.0002	.0002	.154
(t,t^2,t^3)	3	.5960	.1987	152.8
Error	3	.0040	.0013	
Total	6	.6000		

Derivation of the F-test

The derivation of the F-statistic in (10.30) as well as other F-tests that we will encounter in the following sections are based on two very useful lemmas. Therefore, the lemmas are presented in a general notation rather than in the notation in the application of this section. These lemmas are based on some results in Fisher (1970) and an extension in Koschat and Weerahandi (1990). The proof of Lemma 10.1 is left as an exercise. (See also Fisher (1970)).

Lemma 10.1. Let $\mathbf{W}$ and $\tilde{\mathbf{W}}$ be $n \times m$ and $n \times \tilde{m}$ matrices of full rank with $m, \tilde{m} < n$ and such that every column vector of $\tilde{\mathbf{W}}$ is a linear combination of the column vectors of $\mathbf{W}$. Let $\mathbf{P} = \mathbf{I} - \mathbf{W}(\mathbf{W}'\mathbf{W})^{-1}\mathbf{W}'$ and $\tilde{\mathbf{P}} = \mathbf{I} - \tilde{\mathbf{W}}(\tilde{\mathbf{W}}'\tilde{\mathbf{W}})^{-1}\tilde{\mathbf{W}}'$. Then, $\mathbf{P}$ and $\tilde{\mathbf{P}}$ are symmetric idempotent matrices of rank $n-m$ and $n-\tilde{m}$ with $\tilde{\mathbf{P}}\mathbf{P} = \mathbf{P}$.

To present the second lemma, we use the notation $\chi^2_r(\lambda^2)$ to denote a noncentral chi-square distribution with non-centrality parameter λ^2 and r degrees of freedom.

Lemma 10.2. Let $\mathbf{P}_i$, $i = 1,\ldots,l$ be idempotent and symmetric matrices of rank r_i such that $\mathbf{P}_i\mathbf{P}_j = \mathbf{P}_j$ and $\mathbf{P}_i \neq \mathbf{P}_j$, for $i > j$. Let $\varepsilon \sim N(\mu,\mathbf{I})$ and define $\mathbf{e}_i = \mathbf{P}_i\varepsilon$. Then

$$\mathbf{e}'_{i+1}\mathbf{e}_{i+1} - \mathbf{e}'_i\mathbf{e}_i, \sim \chi^2_{r_{i+1}-r_i}\,(\mu'(\mathbf{P}_{i+1}-\mathbf{P}_i)\mu),\ \ i = 1,\ldots,l-1,$$

$$\mathbf{e}'_1\mathbf{e}_1 \sim \chi^2_{r_1}\,(\mu'\mathbf{P}_1\mu),$$

are jointly independent.

Proof. From $\mathbf{P}_i\mathbf{P}_j = \mathbf{P}_j$, for $i > j$, and the symmetry of the $\mathbf{P}_i$ it follows that $(\mathbf{P}_i\mathbf{P}_j)' = \mathbf{P}_j'$ and $\mathbf{P}_j\mathbf{P}_i = \mathbf{P}_j$. In particular, $\mathbf{P}_i\mathbf{P}_j = \mathbf{P}_j\mathbf{P}_i$. This means that the $\mathbf{P}_i, i = 1,\ldots,l-1$ are simultaneously diagonalizable, i.e., there is an orthogonal matrix $\mathbf{U}$ such that $\mathbf{P}_i = \mathbf{U}\Lambda_i\mathbf{U}'$, for $i = 1,\ldots,l$. The diagonal entries of Λ_i, λ_i^l, are either 0 or 1. Since $\Lambda_i\Lambda_j = \Lambda_j$, for $i > j$, $\lambda_j^l = 1$ implies that $\lambda_i^l = 1$ also. Therefore the λ_i^l may be assumed to be arranged in decreasing order, i.e., the first r_i diagonal entries of Λ_i are 1, the remaining are 0. Note that $r_1 < \ldots < r_l$. Denote by $\varepsilon^* = \mathbf{U}'\varepsilon$. Then $\varepsilon^* \sim N(\mu^*, \mathbf{I})$ and

$$\mathbf{e}'_{i+1}\mathbf{e}_{i+1} - \mathbf{e}'_i\mathbf{e}_i = \varepsilon^*(\Lambda_3 - \Lambda_2)\varepsilon^* = \sum_{l=r_i+1}^{r_{i+1}} \varepsilon_l^{*2},$$

$$\mathbf{e}'_1\mathbf{e}_1 = \varepsilon^*\Lambda_1\varepsilon^* = \sum_{l=1}^{r_1}\varepsilon_l^{*2}.$$

Since the ε_i^* are mutually independent and $\sum_{l=r_i+1}^{r_i} \mu_l^{*2} = \mu'(\mathbf{P}_{i+1} - \mathbf{P}_i)\,\mu$, the lemma follows.

Now we are in a position to establish the desired F-test. Setting $W = X$ and $\tilde{W} = \tilde{X}$ in Lemma 10.1 we see that

$$\mathbf{P} = \mathbf{I} - \mathbf{X}(\mathbf{X}'\mathbf{X})^{-1}\mathbf{X}' \quad \text{and} \quad \tilde{\mathbf{P}} = \mathbf{I} - \tilde{\mathbf{X}}(\tilde{\mathbf{X}}'\tilde{\mathbf{X}})^{-1}\tilde{\mathbf{X}}',$$

are symmetric idempotent matrices such that $\tilde{\mathbf{P}}\mathbf{P} = \mathbf{P}$. Consider the residual sums of squares

$$\mathbf{e} = \mathbf{P}\mathbf{y} \quad \text{and} \quad \mathbf{e}_1 = \tilde{\mathbf{P}}\mathbf{y},$$

where $\mathbf{y} \sim N(\mathbf{X}\boldsymbol{\beta}, \sigma^2 I_n)$. Since $P\mathbf{X} = \mathbf{0}$, it follows from Lemma 10.2 that

$$\frac{\mathbf{e}_1\prime\mathbf{e}_1 - \mathbf{e}\prime\mathbf{e}}{\sigma^2} \sim \chi_{k_2}(\boldsymbol{\beta}\prime\mathbf{X}\prime(\tilde{\mathbf{P}} - \mathbf{P})\mathbf{X}\boldsymbol{\beta}) \tag{10.31}$$

and

$$\frac{\mathbf{e}\prime\mathbf{e}}{\sigma^2} \sim \chi_{n-k} \tag{10.32}$$

are independently distributed. Moreover, if the null hypothesis specified by equation (10.28) is true, then the distribution in (10.32) reduces to a central chi-square distribution. In view of these results, consider the statistic

$$F = \frac{(\mathbf{e}_1\prime\mathbf{e}_1 - \mathbf{e}\prime\mathbf{e})/k_2}{\mathbf{e}\prime\mathbf{e}/(n-k)} \sim F_{k_2,n-k}(\boldsymbol{\beta}\prime\mathbf{X}\prime(\tilde{\mathbf{P}} - \mathbf{P})\mathbf{X}\boldsymbol{\beta}). \tag{10.33}$$

Under H_0, the distribution of this F-statistic reduces to a central F-distribution. It is now evident that F is a test statistic which can be employed to construct unbiased tests for H_0, as claimed.

10.6. Comparing Two Regressions with Equal Variances

A standard problem in statistical practice is the comparison of two or more regressions. Comparison of regressions often arises when an experimenter wishes to test whether the parameters of a regression of an important variable have changed due to a change in the structure on which the regression is based. The structural change could be a policy change, a change in strategy, a change of management, a war, and so on. As an example, consider the demand function for a certain product. Many factors affect the demand and this dependence is often adequately described by a multiple regression model. With changes in marketing strategy, such as a change in advertising copy, it is reasonable to expect changes in some of the parameters of the demand function. Comparisons of regressions also arise when comparing two brands of a product, two treatments, two kinds of fertilizers, etc., when factors affecting each response variable are controlled by a set of explanatory variables.

Here we consider only the problem of comparing two regressions. The readers interested in results concerning the comparison of more than two regressions are referred to Kullback and Rosenblatt (1957) and Koschat and Weerahandi (1992). We also assume here that there are enough data from each regression to estimate the parameters of the two models separately. Chow (1960) provides a solution to the case where there is not enough data to conduct two separate regressions, but there is sufficient data to carry out one regression and a combined regression. An extended version of the result is given by Koschat and Weerahandi (1990) and is presented here in the form of an exercise. (See Exercise 10.6).

Consider two linear regressions with possibly unequal numbers of parameters. In general, a practitioner might be interested in comparing only a subset of parameters. Some of the rest of the parameters might be common to the two regressions while others are different. So, consider two linear models with the structure

$$\mathbf{y}_i = \mathbf{U}_i\mathbf{c} + \mathbf{V}_i\mathbf{d}_i + \mathbf{W}_i\mathbf{t}_i + \varepsilon_i, \quad i = 1,2, \qquad (10.34)$$

$$\varepsilon_i \sim N(\mathbf{0},\sigma_i^2\mathbf{I}_i) ,$$

where the $\mathbf{I}_i$, $i = 1, 2$, are n_i dimensional identity matrices. In this section we assume that $\sigma_1^2 = \sigma_2^2$. The vectors $\mathbf{y}_1$ and $\mathbf{y}_2$ are of length n_1 and n_2, and comprise the sample data on the dependent variables. The sample data for the explanatory variables fill the matrices $\mathbf{U}_i$, $\mathbf{V}_i$, and $\mathbf{W}_i$. The variables corresponding to $\mathbf{U}_i$ and $\mathbf{W}_i$ are the same for both samples. The variables corresponding to $\mathbf{V}_1$ and $\mathbf{V}_2$ could potentially be different. For example, it may be known that certain variables affect the response for one sample but not the other, or it could happen that certain explanatory data had been collected for one sample but not for the other. The vector $\mathbf{c}$ is the vector of parameters that both samples have in common, and it is of length p_c. The vectors $\mathbf{d}_1$ and $\mathbf{d}_2$ are of length p_{d_1} and p_{d_2} respectively, They contain either parameters corresponding to sample specific variables or parameters that are expected to be different for the two samples. It is desired to test the equality of the vectors $\mathbf{t}_1$ and $\mathbf{t}_2$, both of length p_t; that is, the problem is to test the following hypotheses:

$$H_0: \mathbf{t}_1 = \mathbf{t}_2 \quad \text{versus} \quad H_1: \mathbf{t}_1 \neq \mathbf{t}_2. \tag{10.35}$$

To present a test similar to that in (10.30) consider the least squares regressions of the combined model with and without the equality of $\mathbf{t}_1$ and $\mathbf{t}_2$. Denote by $\mathbf{y}' = (\mathbf{y}_1', \mathbf{y}_2')$ the transpose of the concatenated vector of all responses, and consider the matrices

$$\mathbf{X}_{12} = \begin{bmatrix} \mathbf{U}_1 & \mathbf{V}_1 & \mathbf{0} & \mathbf{W}_1 \\ \mathbf{U}_2 & \mathbf{0} & \mathbf{V}_2 & \mathbf{W}_2 \end{bmatrix} \quad (n_1+n_2) \times (p_c+p_{d_1}+p_{d_2}+p_t), \tag{10.36}$$

and

$$\mathbf{X}_{1,2} = \begin{bmatrix} \mathbf{U}_1 & \mathbf{V}_1 & \mathbf{0} & \mathbf{W}_1 & \mathbf{0} \\ \mathbf{U}_2 & \mathbf{0} & \mathbf{V}_2 & \mathbf{O} & \mathbf{W}_2 \end{bmatrix} \quad (n_1+n_2) \times (p_c+p_{d_1}+p_{d_2}+2p_t) \tag{10.37}$$

We are using the notation with an index of 1,2 to refer to regressions with $\mathbf{t}_1 \neq \mathbf{t}_2$, an index of 12 to refer to regressions with $\mathbf{t}_1 = \mathbf{t}_2$, while an index of 1 or 2 will refer to regressions for samples 1 or 2 only. Assume that $\mathbf{X}_{12}$ and $\mathbf{X}_{1,2}$ are of full rank. Consider the residual sums of squares of the two regressions

$$S_{12}^2 = \mathbf{y}'\mathbf{P}_{12}\,\mathbf{y}, \text{ where } \mathbf{P}_{12} = \mathbf{I} - \mathbf{X}_{12}(\mathbf{X}_{12}'\mathbf{X}_{12})^{-1}\mathbf{X}_{12}' \tag{10.38}$$

and

$$S_{1,2}^2 = \mathbf{y}'\mathbf{P}_{1,2}\,\mathbf{y}, \text{ where } \mathbf{P}_{1,2} = \mathbf{I} - \mathbf{X}_{1,2}(\mathbf{X}_{1,2}'\mathbf{X}_{1,2})^{-1}\mathbf{X}_{1,2}', \tag{10.39}$$

respectively. If the two regressions have no common parameters, then $S_{1,2}^2 = S_1^2 + S_2^2$, the sum of residual sums of squares obtained by running separate regressions.

As in Section 10.5, an unbiased test of H_0 in (10.35) can now be obtained using the following result: Under H_0: $\mathbf{t}_1 = \mathbf{t}_2$,

$$F = \frac{(S_{12}^2 - S_{1,2}^2)/p_t}{S_{1,2}^2/l} \sim F_{p_t,l}, \tag{10.40}$$

where $l = n_1 + n_2 - (p_c + p_{d_1} + p_{d_2} + 2p_t)$. Under H_1: $\mathbf{t}_1 \neq \mathbf{t}_2$, F has a non-central F-distribution with the non-centrality parameter $\sigma^{-2}\mu'\mathbf{P}_{12}\mu$. The result is a consequence of Lemma 10.1 with $\mathbf{W} = \mathbf{X}_{1,2}$ and $\tilde{\mathbf{W}} = \mathbf{X}_{12}$, and Lemma 10.2. Hence large values of F discredit H_0. Furthermore, under the assumption of full rank of the design matrices, $\mu'\mathbf{P}_{12}\mu = 0$ if and only if $\mathbf{t}_1 = \mathbf{t}_2$. Hence the test is strictly unbiased.

The observed level of significance of this test is

$$p = 1 - H_{p_t,l}\left[\frac{(s_{12}^2 - s_{1,2}^2)/p_t}{s_{1,2}^2/l}\right], \tag{10.41}$$

where $H_{p_t,l}$ is the cdf of F-distribution with p_t and l degrees of freedom. At fixed level α, H_0 is rejected if $p < \alpha$, or equivalently if $F > F_\alpha$, where F_α is the $(1-\alpha)$th quantile of the F-distribution with p_t and l degrees of freedom.

Example 10.4. Comparing two drugs with controlled age effect

Two drugs, *A* and *B*, are administered to compare their effect on reducing the pulse rate of patients. Three dose amounts of each drug (measured in certain units) chosen at random are given to eight subjects and the decrease in pulse rates (*Y*) are recorded after one hour. Table 10.3 shows the results of the experiment.

Table 10.3. Age of subjects and decrease in pulse rates

	Drug A							
Age	28	32	35	41	45	52	56	58
Dose	1.5	2.0	1.5	1.0	2.0	1.5	1.0	2.0
Y	8	16	14	11	21	18	20	29
	Drug B							
Age	36	38	41	44	44	48	50	51
Dose	2.0	1.5	1.5	1.0	2.0	1.0	1.0	1.5
Y	16	17	14	8	22	14	7	23

For each drug consider a model of the form

$$Y = \alpha + \beta Dose + \gamma Age + \varepsilon$$

The least squares estimates of the parameters and their standard errors (shown in parentheses) of the two regressions estimated using the above data are

$$\hat{\alpha}_A = -17.45\ (3.77),\ \hat{\beta}_A = 8.42\ (1.60),\ \hat{\gamma}_A = 0.494\ (0.059),$$

and

$$\hat{\alpha}_B = -28.05\ (18.19),\ \hat{\beta}_B = 13.81\ (4.19),\ \hat{\gamma}_B = 0.530\ (0.319),$$

respectively. The residual sums of squares from these regressions are $s_1^2 = 15.45$ and $s_2^2 = 73.17$, respectively. Since the two regressions have no common parameters, $s_{1,2}^2 = s_1^2 + s_2^2 = 88.62$. Now consider the null hypothesis

$$H_0\colon \beta_A = \beta_B$$

that arises in comparing the two drugs. In order to test this hypothesis using the results of this section, we assume that the error variances of the two regressions are equal. Under the null hypothesis, the residual sum of squares s_{12}^2 is to be computed by the concatenated regression of Y on $\mathbf{X}_{12}$ as defined by (10.36). The estimated parameters of this regression are

$$\hat{\alpha}_A = -21.17,\ \hat{\alpha}_B = -17.44,\ \hat{\beta} = 10.62,\ \hat{\gamma}_A = 0.500,\ \hat{\gamma}_B = 0.393.$$

The residual sum of squares due to this regression is 102.93. The increase in error sum of squares due to the assumption of equal effects of the drugs is 14.31. Therefore, the observed value of the F-statistic appropriate for testing the null hypothesis can be computed as

$$F = \frac{14.31/1}{88.62/10}.$$

Its p-value is $p = 1 - H_{1,10}(1.615) = .233$, and therefore the data in Table 10.3 do not provide sufficient evidence to suspect the validity of the null hypothesis. The effects of the two drugs are not significantly different.

10.7. Comparing Regressions without Common Parameters

when the Variances are Unequal Let us now try to tackle the problem of comparing two regression lines without the assumption of equal variances. In many applications the assumption of equal variances is not reasonable. This assumption is usually made in the literature for convenience and mathematical tractability rather than anything else. Toyoda (1974) studied the consequences of this assumption, encouraging much work concerning the case of heteroscedasticity. Jayatissa (1977), Tsurumi (1984), and Thursby (1992) provided Scheffe (1943) type solutions to the problem and Weerahandi (1987) and Koschat and Weerahandi (1992) provided more powerful tests based on generalized p-values. Thursby (1992) and Griffiths and Judge (1992) studied the performance of a number of exact and approximate solutions to the problem.

As in most of the literature on this subject, let us first consider the simpler case where there are no common parameters to the two regressions being compared. In other words, we set $\mathbf{c} = \mathbf{0}$ in equation (10.34). In order to present an unbiased test based on generalized p-values, define $\tilde{\mathbf{y}}_i = \alpha_i^{-1}\mathbf{y}_i$, $\tilde{\mathbf{V}}_i = \alpha_i^{-1}\mathbf{V}_i$, and $\tilde{\mathbf{W}}_i = \alpha_i^{-1}\mathbf{W}_i$. Let

$$\tilde{\mathbf{X}}_{1,2} = \begin{bmatrix} \tilde{\mathbf{V}}_1 & \mathbf{O}_0 & \tilde{\mathbf{W}}_1 & \mathbf{O}_0 \\ \mathbf{O} & \tilde{\mathbf{V}}_2 & \mathbf{O} & \tilde{\mathbf{W}}_2 \end{bmatrix} \quad (n_1+n_2)\times(p_{d_1}+p_{d_2}+2p_t), \quad (10.42)$$

and

$$\tilde{\mathbf{X}}_{12} = \begin{bmatrix} \tilde{\mathbf{V}}_1 & \mathbf{O}_0 & \tilde{\mathbf{W}}_1 \\ \mathbf{O} & \tilde{\mathbf{V}}_2 & \tilde{\mathbf{W}}_2 \end{bmatrix} \quad (n_1+n_2)\times(p_{d_1}+p_{d_2}+p_t). \quad (10.43)$$

With this notation let $\tilde{S}_{1,2}^2(\alpha_1^2,\alpha_2^2)$ and $\tilde{S}_{12}^2(\alpha_1^2,\alpha_2^2)$ be the residual sums of squares obtained by regressing $\tilde{\mathbf{y}} = (\tilde{\mathbf{y}}_1{}',\tilde{\mathbf{y}}_2{}')'$ on $\tilde{\mathbf{X}}_{1,2}$ and $\tilde{\mathbf{X}}_{12}$, respectively. Denote by $\tilde{S}.^2$ the random variables representing sums of squares and by $\tilde{s}.^2$ the corresponding observed values. Note that for any positive c

$$\tilde{S}.^2(c\alpha_1^2,c\alpha_2^2) = c^{-1}\tilde{S}.^2(\alpha_1^2,\alpha_2^2)\ . \quad (10.44)$$

Let S_1^2 and S_2^2 be the residual sums of squares obtained by performing the two regressions in (10.34) separately.

With these definitions, tests of the hypotheses in (10.35) can be based on the p-value

$$p = 1 - E_R\left\{H_{k,l}\left[\frac{l}{k}\left[\tilde{s}_{12}^2\left(\frac{s_1^2}{R},\frac{s_2^2}{(1-R)}\right)-1\right]\right]\right\}, \quad (10.45)$$

where $\tilde{s}_{12}^2$ is the normalized residual sum of squares obtained by setting $\alpha_1^2 = s_1^2/R$ and $\alpha_2^2 = s_2^2/(1-R)$. Furthermore, $k = p_t$, $l = n_1+n_1-(p_{d_1}+p_{d_2}+2p_t)$, $H_{k,l}$ is the cumulative distribution function of the F distribution with k and l degrees of freedom, and the expectation is taken with respect to the beta random variable

$$R \sim Beta\left(\frac{n_1-p_t-p_{d_1}}{2},\frac{n_2-p_t-p_{d_2}}{2}\right).$$

The observed significance value, p given by (10.45) serves to measure how well the data supports the hypothesis H_0: $\mathbf{t}_1 = \mathbf{t}_2$. The expected value in (10.45) can be computed by numerical integration. If desired, at level α, the hypothesis can be rejected if $p < \alpha$. The software package XPro can be employed to compute the p-value. XPro computes the p-value by exact (up to 4 decimal places) numerical integration.

Example 10.5. Comparing two drugs with controlled age effect (continued)

Consider again the problem of comparing drugs A and B that we discussed in Example 10.4. In Example 10.4 we tested the null hypothesis of equal effects due to the two drugs under the assumption that the variances of the two regressions are equal. Now let us drop that assumption and retest the hypothesis H_0: $\beta_A = \beta_B$ by applying the results of this section. Recall that the residual sums of squares of separate regressions were $s_1^2 = 15.45$ and $s_2^2 = 73.17$. Therefore, the observed significance value computed from (10.45) is

$$p = 1 - E_R\left\{ H_{1,10}\left[10\, \tilde{s}_{12}^{\,2}\left(\frac{15.45}{R}, \frac{73.17}{1-R}\right) - 1\right]\right\},$$

$$= .295\ ,$$

as computed using XPro, where the expectation is taken with respect to the beta random variable

$$R \sim Beta(2.5, 2.5).$$

Therefore, without making the assumption of equal error variances we can come to the same conclusions as before, namely the difference between the effects of the two drugs is not statistically significant.

Derivation of the test

Noting that

$$S_1^2/\sigma_1^2 + S_2^2/\sigma_2^2 = \tilde{S}_{1,2}^{\,2}(\sigma_1^2, \sigma_2^2)$$

so that

$$R = \frac{S_1^2/\sigma_1^2}{\tilde{S}_{1,2}^{\,2}(\sigma_1^2, \sigma_2^2)} \sim Beta\left[\frac{n_1 - p_t - p_{d_1}}{2}, \frac{n_2 - p_t - p_{d_2}}{2}\right],$$

consider the potential test variable

$$T = \frac{\tilde{S}_{12}^{\,2}\,(\sigma_1^2, \sigma_2^2)}{\tilde{s}_{12}^{\,2}\,(s_1^2\sigma_1^2 / S_1^2,\ s_2^2\sigma_2^2 / S_2^2)}.$$

In terms of the beta random variable, T can be rewritten as

$$T = \left[\frac{\tilde{S}_{12}^{\,2}(\sigma_1^2, \sigma_2^2)}{\tilde{S}_{1,2}^{\,2}(\sigma_1^2, \sigma_2^2)}\right] \frac{1}{\left[\tilde{s}_{12}^{\,2}\left(s_1^2/R,\ s_2^2/(1-R)\right)\right]}$$

In this product the second factor does not depend on any of the unknown parameters and it follows from Lemma 10.2 that it is statistically independent from the first factor. Further, the first factor has the familiar form of the F-test statistic in (10.30). It attains its minimum under stochastic ordering at $\mathbf{t}_1 = \mathbf{t}_2$. Moreover, under H_0 its distribution does not depend on any of the unknown parameters. Thus T is indeed a test variable and it can be utilized to define unbiased tests. The observed value of the test variable is $t_{obs} = 1$. Hence, the p-value can be obtained as

$$\begin{aligned} p &= Pr(T \geq 1) \\ &= Pr\left\{ \frac{\tilde{S}_{12}^2}{\tilde{S}_{1,2}^2} (\sigma_1^2, \sigma_2^2) \geq \tilde{s}_{12}^2 \left(s_1^2/R,\, s_2^2/(1-R) \right) \right\} \\ &= Pr\left\{ \frac{\tilde{S}_{12}^2 - \tilde{S}_{1,2}^2}{\tilde{S}_{1,2}^2} (\sigma_1^2, \sigma_2^2) \geq \tilde{s}_{12}^2 \left(s_1^2/R,\, s_2^2/(1-R) \right) - 1 \right\} \\ &= 1 - E_R\left\{ H_{k,l}\left(\frac{l}{k} \left[\tilde{s}_{12}^2 \left(s_1^2/R,\, s_2^2/(1-R) \right) - 1 \right] \right) \right\}, \end{aligned}$$

as claimed.

10.8. Comparison of Two General Models

Finally, consider the two general models in (10.34) with $\sigma_1^2 \neq \sigma_2^2$. Assume that there are enough observations to carry out the two regressions separately, i.e., $n_i > p_c + p_{d_i} + p_{t_i}$, $i = 1, 2$. Using the tilde notation introduced before, define $\tilde{S}_{1,2}^2(\alpha_1^2, \alpha_2^2)$ and $\tilde{S}_{12}^2(\alpha_1^2, \alpha_2^2)$ as the residual sums of squares of regressions of $\tilde{\mathbf{y}}$ on

$$\tilde{\mathbf{X}}_{1,2} = \begin{bmatrix} \tilde{\mathbf{U}}_1 & \tilde{\mathbf{V}}_1 & \mathbf{O}_0 & \tilde{\mathbf{W}}_1 & \mathbf{O}_0 \\ \tilde{\mathbf{U}}_2 & \mathbf{O} & \tilde{\mathbf{V}}_2 & \mathbf{O} & \tilde{\mathbf{W}}_2 \end{bmatrix}, \quad (n_1+n_2)\times(p_c+p_{d_1}+p_{d_2}+2p_t), \tag{10.46}$$

and

$$\tilde{\mathbf{X}}_{12} = \begin{bmatrix} \tilde{\mathbf{U}}_1 & \tilde{\mathbf{V}}_1 & \mathbf{O}_0 & \tilde{\mathbf{W}}_1 \\ \tilde{\mathbf{U}}_2 & \mathbf{O} & \tilde{\mathbf{V}}_2 & \tilde{\mathbf{W}}_2 \end{bmatrix}, \quad (n_1+n_2)\times(p_c+p_{d_1}+p_{d_2}+p_t), \tag{10.47}$$

respectively. Then, tests for comparing $\mathbf{t}_1$ and $\mathbf{t}_2$ can be obtained on the basis of the test variable

$$T = \frac{\left[\tilde{S}_{12}^2 - \tilde{S}_{1,2}^2\right](\sigma_1^2,\sigma_2^2)}{\left[\tilde{s}_{12}^2 - \tilde{s}_{1,2}^2\right](s_1^2\sigma_1^2/S_1^2,\, s_2^2\sigma_2^2/S_2^2)}. \quad (10.48)$$

The p-value given by this test variable is

$$p = 1 - E_{R_1,R_2}\left\{H_{k,l}\left[\frac{l}{k}R_2\tilde{s}^2\left(\frac{s_1^2}{R_1},\frac{s_2^2}{(1-R_1)}\right)\right]\right\}, \quad (10.49)$$

where $\tilde{s}^2 = \tilde{s}_{12}^2 - \tilde{s}_{1,2}^2$, is the difference of the normalized residual sums of squares, each normalized using normalization factors s_1^2/R_1 and $s_2^2/(1-R_1)$. Furthermore, $k = p_t$, $l = n_1+n_1-(p_c+p_{d_1}+p_{d_2}+2p_t)$, $H_{k,l}$ is the cumulative distribution function of the F distribution with k and l degrees of freedom, and the expectation is taken with respect to the independent beta random variables

$$R_1 \sim Beta\left(\frac{n_1-p_c-p_t-p_{d_1}}{2},\frac{n_2-p_c-p_t-p_{d_2}}{2}\right),$$

$$R_2 \sim Beta\left(\frac{n_1+n_2-2p_c-2p_t-p_{d_1}-p_{d_2}}{2},\frac{p_c}{2}\right).$$

The observed significance value, p, given by (10.49) serves to measure how data supports or discredits the null hypothesis H_0: $\mathbf{t}_1 = \mathbf{t}_2$. At fixed level α, the null hypothesis is rejected if $p < \alpha$.

To establish this result by taking an approach similar to that in Section 10.7, define R_1 and R_2 as

$$R_1 = \frac{S_1^2/\sigma_1^2}{S_1^2/\sigma_1^2+S_2^2/\sigma_2^2}, \qquad R_2 = \frac{S_1^2/\sigma_1^2+S_2^2/\sigma_2^2}{\tilde{S}_{1,2}^2(\sigma_1^2,\sigma_2^2)}.$$

Let $\tilde{S}_i^2 = S_i^2/\sigma_i^2$, $i=1,2$ and consider the sums of squares $\tilde{S}_{12}^2$, $\tilde{S}_{1,2}^2$, $\tilde{S}_1^2+\tilde{S}_2^2$ and $\tilde{S}_1^2$. It follows from Lemma 10.1 and Lemma 10.2 that the random variables

$$\tilde{S}_{12}^2-\tilde{S}_{1,2}^2,\ \tilde{S}_{1,2}^2-(\tilde{S}_1^2+\tilde{S}_2^2),\ (\tilde{S}_1^2+\tilde{S}_2^2)-\tilde{S}_1^2 \text{ and } \tilde{S}_1^2$$

are all independently distributed with chi-squared distributions. It is easily verified that their non-centrality parameters are $\mu'\tilde{\mathbf{P}}_{12}\mu$, 0, 0, 0, respectively. From the result on the sum and ratio of independently distributed χ^2 distributions we see that

$$\tilde{S}_{12}^2-\tilde{S}_{1,2}^2,\ \tilde{S}_{1,2}^2-(\tilde{S}_1^2+\tilde{S}_2^2),\ \tilde{S}_1^2+\tilde{S}_1^2, \text{ and } R_1$$

are mutually independent. Applying the result to the resulting two center terms yields that $\tilde{S}_{12}^2-\tilde{S}_{1,2}^2$, $\tilde{S}_{1,2}^2$, R_1, R_2 are mutually independent. Consequently,

$$R_1 = \frac{\tilde{S}_1^2}{\tilde{S}_1^2+\tilde{S}_2^2} \sim Beta\left(\frac{n_1-p_c-p_t-p_{d_1}}{2},\frac{n_2-p_c-p_t-p_{d_2}}{2}\right),$$

$$R_2 = \frac{(\tilde{S}_1^2+\tilde{S}_2^2)}{\tilde{S}_{1,2}^2} \sim Beta\left(\frac{n_1+n_2-2p_c-2p_t-p_{d_1}-p_{d_2}}{2},\frac{p_c}{2}\right),$$

$$\text{and } F = \frac{(\tilde{S}_{12}^2-\tilde{S}_{1,2}^2)/k}{\tilde{S}_{1,2}^2/l} \sim F_{k,l}(\mu'\tilde{\mathbf{P}}_{12}\mu),$$

with $k = p_c$, $l = n_1+n_2-p_c-p_{d_1}-p_{d_2}-2p_t$ and $\mu' = (\mathbf{c}',\mathbf{d}'_1,\mathbf{d}'_2,\mathbf{t}')\tilde{\mathbf{X}}'_{12}$, are jointly independent.

To this end, notice that

$$S_1^2/\sigma_1^2 = R_1 R_2 \tilde{S}_{1,2}^2(\sigma_1^2,\sigma_2^2) \text{ and } S_2^2/\sigma_2^2 = (1-R_1)R_2\tilde{S}_{1,2}^2(\sigma_1^2,\sigma_2^2)\ .$$

Therefore, using formula (10.44), the test variable defined by (10.48), T, can be expressed as

$$T = \left(\frac{\tilde{S}_{12}^2-\tilde{S}_{1,2}^2}{\tilde{S}_{1,2}^2}(\sigma_1^2,\sigma_2^2)\right) R_2^{-1}\left[\tilde{s}^2\left(s_1^2/R_1, s_2^2/(1-R_1)\right)\right]^{-1}.$$

Only the first factor of this representation of T depends on the null hypothesis being tested and that term has a noncentral F-distribution. It is now evident that T yields unbiased tests. When the null hypothesis is true, the noncentral F-distribution reduces to a central F-distribution so that probabilities based on T are free of unknown parameters. Since the observed value of the test statistic is 1, the p-value in (10.49) now follows as in Section 5.7.

Example 10.6. Comparing two drugs with controlled age effect (continued)

Consider again the problem of comparing drugs A and B discussed in the previous two examples. Recall that the effect of age was controlled by taking the age as an explanatory variable in the regressions. In Example 10.5 we tested the null hypothesis of equal effects due to the two drugs without the assumption of equal error variances, but we allowed the age effect to be different for the two regressions. Now let us retest the hypothesis when the investigator believes that the effect of this variable is common to the two regressions. Then, the appropriate p-value for testing the null hypothesis can be computed from (10.49) as

$$p = 1 - E_{R_1,R_2}\left\{ H_{1,11}\left[11\, R_2 \tilde{s}^2 \left(\frac{15.45}{R_1}, \frac{73.17}{(1-R_1)} \right) \right] \right\}$$

$$= .253$$

correct up to three decimal places, where the expectation is taken with respect the beta random variables

$$R_1 \sim Beta(2.5,2.5), \text{ and } R_2 \sim Beta(5.0,0.5).$$

This p-value does not discredit the null hypothesis of equal effects of drugs *A* and *B*. Therefore, the assumption of common age effect has not affected the conclusion we made before.

Exercises

10.1. Consider the linear model in (10.1) and consider linear estimates of β of the form

$$\hat{\beta} = c_1 Y_1 + \cdots + c_n Y_n ,$$

where c_i, $i=1,\ldots, n$ are constants (weights). Find the weights such that (i) $\hat{\beta}$ is an unbiased estimator of β, and that (ii) the variance of $\hat{\beta}$ is as small as possible; that is the *best linear unbiased estimator* of β. Show that the resulting estimator is the same as the least squares estimator of β. Find the best linear unbiased estimator of α.

10.2. Considering the linear regression in (10.1), show that the variance of the best linear unbiased estimator $\hat{\alpha}$ is given by equation (10.10). Also establish equation (10.11).

10.3. In the notation of Section 10.4, show that the error sum of squares of the least squares regression line can be expressed as $\mathbf{e}'\mathbf{e} = \boldsymbol{\varepsilon}'\mathbf{P}\boldsymbol{\varepsilon}$, where $\mathbf{P} = I_n - \mathbf{X}(\mathbf{X}'\mathbf{X})^{-1}\mathbf{X}'$. Show that $\mathbf{P}$ is a symmetric idempotent matrix. Writing the expected value of $\mathbf{e}'\mathbf{e}$ as $E(\boldsymbol{\varepsilon}\boldsymbol{\varepsilon}')\, tr(\mathbf{P})$, or otherwise, find the expected value of $\mathbf{e}'\mathbf{e}$.

10.4. Consider the multiple linear regression model in (10.16). If $\mathbf{a}$ is a $k \times 1$ vector of constants, establish procedures for testing hypotheses of the form

$$H_0\colon \mathbf{a}'\boldsymbol{\beta} \le k \text{ versus } H_1\colon \mathbf{a}'\boldsymbol{\beta} > k .$$

10.5. Prove Lemma 10.1.

10.6. Consider the linear models defined by equation (10.34). Assume that $n_1 > p_c + p_{d_2} + p_t$ and $p_{d_2} < n_2 \le p_{d_2} + p_t$. Let S_1^2 be the residual sum of squares due to the least squares regression of the model corresponding to $i = 1$. Show that under $H_0\colon \mathbf{t}_1 = \mathbf{t}_2$,

$$F = \frac{(S_{12}^2 - S_1^2)/k}{S_1^2/l} \sim F_{k,l} ,$$

where S_{12}^2 is the residual sum of squares given by (10.38), $k = n_2 - p_{d_2}$ and $l = n_1 - (p_c + p_{d_1} + p_t)$.

What is the distribution of F if H_0 is not true? Describe how H_0 can be tested based on these results and discuss the properties of the test.

10.7. The following are heights (in inches) of eight athletes and the heights of their fathers:

Athlete's height:	68	69	70	70	71	72	73	74
Father's height:	69	69	66	70	71	68	71	75

(a) Calculate the coefficient of correlation between athletes' heights and their fathers' heights and discuss the nature of the association between the two sets of heights.

(b) Fit the two regression lines by the method of least squares and display them in a scatter diagram.

(c) Clearly specifying the assumptions you make, test the significance of parameters of each regression line.

(d) Estimate the height of the father of an athlete who is 75 inches tall; give a point estimate along with a 95% confidence interval.

10.8. Consider again the regression problem in Exercise 10.7. Fit a parabolic regression of the form $Y = a + bx + cx^2$, where Y and x represent athletes' heights and fathers' height, respectively. Test whether the addition of the second order term is statistically significant.

10.9. The following table presents the level of cholesterol of eight adults and their average daily intake (in grams) of saturated fat:

Fat intake:	24	33	39	42	44	51	55	60	62
Cho. Level:	150	145	160	150	175	215	170	195	210

(a) Calculate the correlation coefficient between the fat intake and the cholesterol level. Interpret the result.

(b) Fit the regression line for predicting the cholesterol level from the fat intake.

(c) Specifying the assumptions you make, construct 99% confidence intervals for the parameters of the regression line.

(d) Plot the regression line for the purpose of predicting the cholesterol level of an adult from his or her fat intake. Draw the 95% confidence bands on the same chart.

10.10. Consider again the regression problem in Exercise 10.7. Suppose the ages and the sex of eight individuals in that study later became available. The ages and the sex in the same order as in table of Exercise 10.7 are as follows:

Age:	37	32	55	48	35	53	46	51	50
Sex:	F	M	F	M	M	F	F	M	M

(a) Obtain the least squares regression of cholesterol level on fat intake, age and sex.

(b) Test the significance of the effect of each explanatory variable as well as their joint effect on the regression.

(c) Find the coefficient of determination.

(d) Predict (with a 95% confidence interval) the cholesterol level of a male who is 50 years old and has an average daily intake of 50 grams of saturated fat.

10.11. An investigator wishes to compare two hybrids of sugar beet using a set of readily available data. The data available (in certain units) to the investigator are shown in the following table.

	Hybrid A						
Yield	28	31	29	38	32	36	24
Mean rainfall	16	17	19	24	24	28	20
Mean temperature	64	66	63	70	65	74	62
	Hybrid B						
Yield	31	29	22	33	35	39	28
Mean rainfall	19	12	14	29	29	32	20
Mean temperature	60	68	68	76	60	78	60

(a) For each hybrid, regress the yield of sugar beet as a linear function of mean rainfall and mean temperature.

(b) Test for the significance of each regression.

(c) Assuming that the error variances of the two regressions are equal, compare the two hybrids.

(d) Assuming that the error variances of the two regressions are equal and the effects of rainfall and the temperature are common to the two regressions, compare the two hybrids.

10.12. Consider again the problem in Exercise 10.11. Repeat the comparisons in (c) and (d) without the assumption of equal error variances.

10.13. Consider the problem of comparing two drugs discussed in Example 10.4. Assume that the error variances of the two regressions are equal. Supposing that the age effect and the intercept term are common to the two regressions, test the hypothesis that the effect of the two drugs are equal.

10.14. Consider again the problem in Exercise 10.13. Repeat the test without the assumption of equal error variances.

Appendix A
Elements of Bayesian Inference

A.1. Introduction

In Bayesian inference, the parameter of interest, say θ, is treated as a random variable rather than as a constant. In some applications, θ could indeed be a random variable in the usual sense. This is the case, for instance, in the random effect models and mixed models addressed in Chapter 8. Whenever we consider θ as a random variable, we shall denote the cdf of X given θ by $F(x|\theta)$ instead of $F(x;\theta)$; similar conventions will be used for pdf or pmf of X as well.

To describe these ideas briefly, let θ be the mean weight of 10 year-old boys living in a certain city. Here θ is simply an unknown constant and there is nothing random about it as long as we talk about a well defined population at a fixed time point. If θ is the mean weight of 10 year-old boys at a future time point, then in some sense θ is random due to the uncertainty of the future. But even in this case we are talking about a one time event and there is a particular value that θ will take. Finally, if θ denote the mean weight of 10 year-old boys in a city chosen at random from a group of cities, then θ is a random variable in the usual sense; that is, the probability distribution of θ can be interpreted using the frequencies with which cities are chosen. The concept of ***subjective probability*** enables one to make probability statements about θ in any of these situations. A subjective probability of an event reflects the personal belief that an individual has about the relative likelihood that the event will occur. For in-depth investigations of this concept, formal definitions, and for methods of constructing subjective probabilities, the reader is referred to Jeffreys (1961), deFinetti (1972), and DeGroot (1970).

A.2. The Prior Distribution

Consider the parameter, θ, of a distribution $F(x|\theta)$ before any observations are taken from the distribution. Based on previous experience, knowledge about the underlying population, or any other information, suppose an experimenter has constructed a subjective probability distribution for θ. Such a distribution that reflects the uncertainty about the value of the parameter from the experimenter's point of view prior to observing the current sample is called the *prior distribution* of θ. The prior density function is often denoted by $\pi(\theta)$.

It should be mentioned in passing that the concept of prior distribution of a parameter is somewhat controversial in statistics. Some statisticians believe that it is not appropriate to talk about a probability distribution of θ when it is known that the parameter is not a random variable but rather an unknown fixed number. The treatment of a fixed number as a random variable may lead to different results and inconsistencies even when the experimenter's prior distribution reflects no prior knowledge about the parameter. Moreover, the use of prior distributions may increase the possibility for abuse of statistics. Often, the prior distributions that experimenters develop show bias towards favorable directions of the parameter. The reader can find in Berger (1985) a detailed discussion of these and other issues as well as various methods of finding prior distributions.

In many situations, the relative likelihoods that the parameter can take values in $\boldsymbol{\Theta}$ can be represented by a proper probability distribution. This is not possible if the relative likelihoods have an infinite mass and yet such prior distributions may not cause any difficulty in implementing Bayesian methods. A prior density that does not integrate into a finite number is known as an *improper prior*. One can employ Bayesian methods even when no prior information about the underlying parameters are available. Such prior distributions, known as *noninformative priors*, have been extensively studied by Jeffreys (1961) among others and are the basis of Bayesian methods developed by Box and Tiao (1973). There is no common agreement about how the noninformative prior should be specified. An undesirable fact about noninformative priors is that while different definitions seem appropriate under different arguments they often lead to different improper priors (see, for instance, Laplace (1812), Villegas (1977), and Zellner (1977)). We shall use the most popular method suggested by Jeffreys (1961) to find noninformative priors.

Let $f(x|\boldsymbol{\theta})$ be the pdf or the pmf of the underlying observable random variable X, where $\boldsymbol{\theta}$ is a vector of unknown parameters. The noninformative prior for $\boldsymbol{\theta}$ suggested by Jeffreys (1961) is

$$\pi(\boldsymbol{\theta}) = [\ |I(\boldsymbol{\theta})|\ ]^{1/2}, \tag{A.1}$$

where $I(\boldsymbol{\theta})$ is the Fisher information matrix (defined in terms of the second derivatives of the log likelihood function),

$$I(\boldsymbol{\theta}) = -E\left\{\frac{\partial^2}{\partial\boldsymbol{\theta}\partial\boldsymbol{\theta}'}\log(f(\mathbf{X}|\boldsymbol{\theta}))\right\}.$$

Example A.1.

Consider the exponential distribution with pdf

$$f_X(x|\alpha) = \alpha e^{-\alpha x} \quad \text{for } x > 0 ,$$

where $\alpha > 0$ is the parameter for which a prior is desired. Since α is a positive parameter,

$$\alpha \sim G(a,b) \tag{A.2}$$

is a proper prior distribution for the parameter of the exponential distribution, where a and b are some constants. If a and b are not specified, then they are called the *hyper parameters* of the distribution. We will study this prior later.

Let us now find noninformative priors for α and for the mean of the distribution $\mu = 1/\alpha$. When α is the parameter of interest, the Fisher information is

$$I(\alpha) = -E\left[\frac{\partial^2}{\partial\alpha^2}(log(\alpha) - \alpha X)\right] = \frac{1}{\alpha^2} .$$

Therefore, the (Jeffreys') noninformative prior for α is an improper prior with density function (proportional to)

$$\pi(\alpha) = \frac{1}{\alpha} . \tag{A.3}$$

On the other hand, if μ is the parameter of interest, the Fisher matrix becomes

$$I(\mu) = -E\left[\frac{\partial^2}{\partial\alpha^2}\left\{-log(\mu) - \frac{X}{\mu}\right\}\right] = -E\left[\frac{1}{\mu^2} - \frac{2X}{\mu^3}\right] .$$

$$= -\frac{1}{\mu^2} + \frac{2\mu}{\mu^3} = \frac{1}{\mu^2} .$$

Hence, the noninformative prior for μ, the mean of the exponential distribution, is $\pi(\mu) = 1/\mu$.

A.3. The Posterior Distribution

Let $\mathbf{X}$ be a random vector with joint pdf or pmf (given $\boldsymbol{\theta}$) $f(\mathbf{X}|\boldsymbol{\theta})$. Since $\boldsymbol{\theta}$ itself is considered as a random vector, this should formally be regarded as a conditional pdf or pmf of $\mathbf{X}$ given $\boldsymbol{\theta}$. Let $\pi(\boldsymbol{\theta})$ be the prior density of $\boldsymbol{\theta}$. Then, the joint pdf of $\mathbf{X}$ and $\boldsymbol{\theta}$ can be found as

$$h(\mathbf{x},\boldsymbol{\theta}) = f(\mathbf{x}|\boldsymbol{\theta})\pi(\boldsymbol{\theta}) . \tag{A.4}$$

Assuming that $\boldsymbol{\theta}$ is absolutely continuous, the marginal pdf or the marginal pmf of $\mathbf{X}$ can be found as

$$g(\mathbf{x}) = \int_{\Theta} h(\mathbf{x},\boldsymbol{\theta})\, d\boldsymbol{\theta} . \tag{A.5}$$

Finally, the conditional probability density function of $\boldsymbol{\theta}$, given that $\mathbf{X} = \mathbf{x}$, is

$$\pi(\boldsymbol{\theta}|\mathbf{x}) = \frac{h(\mathbf{x},\boldsymbol{\theta})}{g(\mathbf{x})} \quad \text{for } \boldsymbol{\theta} \in \Theta . \tag{A.6}$$

The probability distribution of $\boldsymbol{\theta}$ given by (A.6) is called the *posterior distribution* of $\boldsymbol{\theta}$. The conditional distribution of the parameter given the data, $\mathbf{x}$, is given this name because it reflects the experimenter's updated beliefs about the parameters after a sample taken from the underlying population has been observed. In other words, the posterior combines the information in the prior and the information in the data (the likelihood function) about $\boldsymbol{\theta}$; this will become clearer from the following equations. Notice that, ignoring a proportionality constant, Equation (A.6) can be rewritten as

$$\pi(\boldsymbol{\theta}|\mathbf{x}) \propto f(\mathbf{x}|\boldsymbol{\theta})\pi(\boldsymbol{\theta}) \quad \text{for } \boldsymbol{\theta} \in \Theta ,$$

where the proportionality constant can usually be found such that the posterior is a proper probability density function that integrates to 1; in some applications this is the case even with many improper priors. In other words,

$$\textit{posterior} \propto \textit{prior} \times \textit{likelihood} . \tag{A.7}$$

If the posterior distribution belongs to the same class of distributions as the prior distribution, then the class is said to be a *natural conjugate family of distributions* for the distribution of $\mathbf{X}$. This means that if a prior distribution is conjugate to a sampling distribution, then, in order to find the posterior distribution, we need only update the parameters of the prior distribution. As demonstrated below, this is, for instance, the case in sampling from an exponential distribution when the prior is Gamma.

Example A.2.

Let $X_1, \ldots, X_n$ be a random sample from the exponential distribution with parameter α. Let us find the posterior distribution of α under the priors given by (A.2) and (A.3). First suppose that the prior of α is a gamma distribution with the shape parameter a and the scale parameter b; the prior mean is ab. Ignoring proportionality constants, we have

$$\text{likelihood} = \prod_{i=1}^{n} \alpha e^{-\alpha x_i} = \alpha^n e^{-\alpha n\bar{x}}$$

and

$$\text{prior} \propto \alpha^{a-1} e^{-\alpha/b}$$

where $\bar{x}$ is the mean of the sample. Therefore, the posterior distribution of α given the sample information can be found using (A.7) as

$$\pi(\alpha|\,\mathbf{x}) \propto \alpha^{n+a-1} e^{-\alpha(n\bar{x}+1/b)} .$$

This can be easily recognized as a gamma distribution. Identifying the underlying shape and scale parameters, the posterior of α can be displayed as

$$\alpha|\,\mathbf{x} \sim G(n+a\,,\ \frac{b}{nb\bar{x}+1}\,). \tag{A.8}$$

Since the prior and the posterior both belong to the same family of distributions, it is now evident that the class of gamma distributions provides a natural conjugate family for the family of exponential distributions.

Next consider the noninformative prior defined by (A.3). Taking the product of the prior and the likelihood yields

$$\pi(\alpha|\,\mathbf{x}) \propto \alpha^{n-1} e^{-\alpha n\bar{x}} .$$

This implies the posterior distribution

$$\alpha|\,\mathbf{x} \sim G(\,n\,,\ 1/n\bar{x}\,)\ , \tag{A.9}$$

which is a proper probability distribution regardless of the fact that the prior is improper. Note that this is also the limiting distribution given by (A.8) as

$$a \to 0 \quad \text{and} \quad b \to \infty\,.$$

Example A.3.

Let $X_1, \ldots, X_n$ be a random sample from the geometric distribution with pmf

$$f(x) = p(1-p)^x \qquad \text{for } x = 0,\,1,\,2,\ \cdots$$

Suppose the parameter p has a beta prior of the form

$$p \sim B(\alpha,\beta)\ ,$$

where α and β are hyper parameters of the prior distribution. Since the likelihood function is given by

$$f_{\mathbf{X}}(\mathbf{x}|\,p) = p^n (1-p)^{n\bar{x}}\ ,$$

it follows from formula (A.7) that

$$\text{posterior} \propto p^{n+\alpha-1} (1-p)^{\beta+n\bar{x}-1} .$$

This can be recognized as being proportional to the probability density function of a beta distribution, thus implying that

$$p \mid \mathbf{x} \sim B(n+\alpha,\ n\bar{x}+\beta) \tag{A.10}$$

is the posterior distribution of the parameter p. Noting that both the prior and the posterior belong to the Beta family, we can conclude that the class of beta distributions forms a natural conjugate family for the family of geometric distributions.

A.4. Bayes Estimators

Let $\pi(\boldsymbol{\theta} \mid \mathbf{x})$ be the posterior distribution of a parameter, θ, based on observations, $\mathbf{x}$, of the random vector, $\mathbf{X}$. The distribution of $\mathbf{X}$ is $F(\mathbf{x} \mid \boldsymbol{\theta})$ and the objective here is to obtain point estimators and interval estimators for $\boldsymbol{\theta}$. In general $\boldsymbol{\theta}$ is a vector of parameters of interest. In this section we treat the problem from a Bayesian standpoint.

First consider the problem of point estimation. From a Bayesian point of view, the posterior distribution plays a role similar to that of the likelihood function as an expression that incorporates all information about the parameter $\boldsymbol{\theta}$; unlike likelihood functions, however, posterior distributions incorporate both the information in the prior and the information in the sample. A point estimator of the parameter can be obtained parallel to the definition of the maximum likelihood estimator. In the spirit of the classical MLE procedure, therefore, we define the *generalized maximum likelihood* estimator as the mode of the posterior distribution $\pi(\boldsymbol{\theta} \mid \mathbf{x})$. The estimate obtained by this method will also be referred to as the *posterior mode estimate* of $\boldsymbol{\theta}$.

It should be noted, however, that the posterior distribution is a probability distribution while the likelihood function, as a function of $\boldsymbol{\theta}$, does not necessarily yield a probability distribution. The mode of the posterior distribution is a particular measure of the location of the distribution. Therefore, we can just as well estimate $\boldsymbol{\theta}$ by other measures of the location such as the mean or the median of the posterior distribution. These estimates will be referred to as the *posterior mean estimate* and the *posterior median estimate*, respectively. In a decision-theoretic approach to the problem of point estimation, these alternative estimates can be shown to be optimal under certain *estimation loss functions*. For instance, the posterior mean and the posterior median are the Bayes estimates of θ under the *squared error loss function*

$$L(\theta,\hat{\theta}) = (\theta-\hat{\theta})^2 \tag{A.11}$$

and the *absolute error loss function*

$$L(\theta,\hat{\theta}) = |\theta-\hat{\theta}|, \tag{A.12}$$

respectively, where $\hat{\theta}$ is an estimate of θ.

To describe briefly the idea of Bayes estimators, consider a general loss function $L(\theta,\hat{\theta})$ whose expected value exists with respect to the posterior density $\pi(\theta \mid \mathbf{x})$. Then, an estimator $\hat{\theta}$ is said to be the *Bayes estimator* of θ under the loss function $L(\theta,\hat{\theta})$ if $\hat{\theta}$ minimizes the *posterior expected loss*

$$E[L(\theta,\hat{\theta}) \mid \mathbf{x}] = \int_{\Theta} L(\theta,\hat{\theta})\pi(\theta \mid \mathbf{x})d\theta . \tag{A.13}$$

For example, with the squared error loss function, the posterior expected loss can be expressed as

$$E((\theta-\hat{\theta})^2 \mid \mathbf{x}) = E\left[\left((\theta-E(\theta \mid \mathbf{x})) + (E(\theta \mid \mathbf{x})-\hat{\theta}\right)^2 \mid \mathbf{x}\right]$$

$$= Var(\theta \mid \mathbf{x}) + [E(\theta \mid \mathbf{x})-\hat{\theta}]^2 \geq Var(\theta \mid \mathbf{x}) .$$

Moreover, it follows from the last equality that if $\hat{\theta} = E(\theta \mid \mathbf{x})$, the posterior mean, then the posterior expected loss is equal to $Var(\theta \mid \mathbf{x})$. It is now clear that the posterior mean as an estimate of θ minimizes the posterior expected loss. In other words, the posterior mean is the Bayes estimator under the squared error loss function.

Example A.4.

Consider the posterior distribution given by (A.8) for the parameter α of an exponential distribution, namely the gamma distribution

$$\alpha \mid \mathbf{x} \sim G(n+a, \frac{b}{1+nb\bar{x}}) .$$

It immediately follows from the mean of the gamma distribution that

$$\hat{\alpha} = \frac{b(n+a)}{1+nb\bar{x}}$$

is the posterior mean estimator of α. Note in particular that as $a \to 0$ and $b \to \infty$, that is under the noninformative prior, the Bayes estimator (under the squared error loss) reduces to $\hat{\alpha} = 1/\bar{x}$, which is the same as the maximum likelihood estimate of α.

Let us now find the generalized maximum likelihood estimate of α based on the sample $X_1, \ldots, X_n$ and the gamma prior $G(a,b)$. Obviously, the mode of the posterior distribution occurs at the same point as the maximum of the function

$$g(\alpha) = (n+a-1)\log(\alpha) - \frac{\alpha(1+nb\bar{x})}{b} \tag{A.14}$$

because, except for a constant independent of α, $g(\alpha)$ is the same as the logarithm of the posterior density function. Differentiating (A.14) with respect to α we have

$$\frac{dg}{d\alpha} = \frac{(n+a-1)}{\alpha} - \frac{(1+nb\bar{x})}{b} \quad \text{and} \quad \frac{d^2 g}{d\alpha^2} = -\frac{(n+a-1)}{\alpha^2} .$$

Since the second derivative is negative for all $n+a>1$, the maximum of $g(\alpha)$ can be found by equating the first derivative to zero. Hence, the generalized maximum likelihood estimate of alpha is

$$\tilde{\alpha} = \frac{b(n+a-1)}{1+nb\bar{x}} .$$

Example A.5.

Recall that the posterior distribution of the geometric parameter, p, based on a beta prior and a random sample of size n is of the form

$$p|\mathbf{x} \sim B(n+\alpha, n\bar{x}+\beta) ,$$

where α and β are hyper parameters of the prior distribution. Using the formula for the mean of a beta distribution, the posterior mean estimator of p can be obtained as

$$\hat{p} = \frac{n+\alpha}{n(\bar{X}+1)+\alpha+\beta} .$$

The posterior mode estimator can also be easily obtained as before. However, an explicit expression for the posterior median estimator of p is not available. Nevertheless, for given values of hyper parameters and a specific sample, the posterior mode estimator can be computed by numerical methods.

A.5. Bayesian Interval Estimation

Suppose now that a confidence set for $\boldsymbol{\theta}$ is desired. Recall that, unlike in the classical setting, $\boldsymbol{\theta}$ is regarded as a random vector. Therefore, in contrast to the classical approach of making probability statements about a subset of the sample space in order to obtain a confidence region, here we can make probability statements concerning subsets of the parameter space directly. As a result, confidence sets we obtain in a Bayesian setting have a direct probability interpretation. The Bayesian analog of a confidence interval is referred to as a *Bayesian confidence interval* or as a credible interval.

Definition A.1. If $C(\mathbf{x})$ is a subset of the parameter space $\boldsymbol{\Theta}$ such that

$$Pr(\theta \in C(\mathbf{x})|\mathbf{x}) = \int_{C(\mathbf{x})} \pi(\boldsymbol{\theta}|\mathbf{x})\,d\boldsymbol{\theta} = \gamma , \tag{A.15}$$

then $C(\mathbf{x})$ is said to be a $100\gamma\%$ *credible set* for $\boldsymbol{\theta}$, where $\pi(\boldsymbol{\theta}|\mathbf{x})$ is the posterior density function of $\boldsymbol{\theta}$.

Bayesian confidence intervals are more general than classical confidence intervals in that the former exist even in situations where the latter do not exist. For instance, one can construct 95% Bayesian intervals even for parameters of discrete distributions such as Binomial distributions. As is the case in classical confidence regions concerning parameters of continuous random variables, there is usually a class of credible sets with the desirable confidence coefficient γ. Depending on the application, a particular set having additional properties may be preferable. In some applications, one-sided intervals may be desired for a parameter of interest. What is desired in many applications, however, is a HPD (highest posterior density) credible set of a specified confidence level. The HPD credible set with coefficient γ is found by minimizing the volume of the credible set subject to the probability coverage γ. If the posterior density function is symmetric about its mode, the HPD credible sets can be computed quite easily. In other situations it can be calculated with the aid of numerical methods.

Example A.6.

Recall that the posterior distribution of the geometric parameter, p, based on a beta prior and a random sample of size n is of the form

$$p|\mathbf{x} \sim B(\alpha(\mathbf{x}),\beta(\mathbf{x})) ,$$

where $\alpha(\mathbf{x}) = n+\alpha$ and $\beta(\mathbf{x}) = n\bar{x}+\beta$, α and β being the hyperparameters of the prior distribution. To illustrate the procedure for constructing Bayesian interval estimates, suppose a left-sided 95% credible interval for p is desired. In that case the appropriate probability statement is of the form

$$Pr(\, p \leq B_{.95}(\alpha(\mathbf{x}),\beta(\mathbf{x}))\ ) \mid \mathbf{x}\ ) = .95 ,$$

where $B_{.95}(\alpha(\mathbf{x}),\beta(\mathbf{x}))$ is the $.95th$ quantile of the beta distribution with parameters $\alpha(\mathbf{x})$ and $\beta(\mathbf{x})$. Then it immediately follows from Definition A.1 that $(\ 0\ ,\ B_{.95}(\alpha(\mathbf{x}),\beta(\mathbf{x})\ )$ is a 95% credible interval for p.

A.6. Bayesian Hypothesis Testing

Consider next the problem of hypothesis testing from a Bayesian point of view. We shall consider only the problem of testing one-sided hypotheses under prior distributions which are independent of the hypotheses being tested. Such priors may not quite be appropriate for testing point null hypotheses in particular (see, for instance, Berger and Sellke (1987) and Casella and Berger (1987)); an account of the Bayesian treatment of point null hypotheses can be found in Berger (1985). We are particularly interested in noninformative priors and natural conjugate priors because they provide direct Bayesian analogs of results related to p-values, especially in normal theory.

To fix ideas, consider the problem of testing a left-sided hypothesis

$$H_0: \theta \le \theta_0 \quad \text{against} \quad H_1: \theta > \theta_0 ,$$

where θ is the parameter of interest and θ_0 is a specified value of the parameter. The hypotheses are to be tested on the basis of a random sample $X_1, \ldots, X_n$ from the distribution $F(x|\theta)$. Let $\pi(\theta|\mathbf{x})$ be the posterior density function of θ based on a prior $\pi(\theta)$ and the data $\mathbf{x}$. Since θ is regarded as a random variable, the problem of deciding between the two hypotheses is quite easy in a Bayesian setting. One can simply calculate the posterior probability of each hypothesis and compare the two.

Definition A.2. The *posterior odds ratio* of H_0 to H_1 is defined as

$$\rho = \frac{\pi_0(\mathbf{x})}{\pi_1(\mathbf{x})} \tag{A.16}$$

where

$$\pi_0(\mathbf{x}) = Pr(\theta \le \theta_0) = \int_{\theta \le \theta_0} \pi(\theta|\mathbf{x})\, d\theta ,$$

and

$$\pi_0(\mathbf{x}) = Pr(\theta \le \theta_0) = 1 - Pr(\theta > \theta_0)$$

are the posterior probabilities of H_0 and H_1, respectively.

Of course, one can define the *prior odds ratio* of H_0 to H_1 in a similar manner. The posterior odds ratio divided by the prior odds ratio is called the *Bayes factor*, a quantity which, more or less, measures the odds of H_0 to H_1 suggested by the data. These notions are of little interest in this context, however.

Example A.7.

Let $X_1, \ldots, X_n$ be a random sample from an exponential population with mean θ. Let $t = \sum x_i$. Assuming the noninformative prior for θ, consider the problem of testing the following hypotheses:

$$H_0: \theta \le \theta_\theta, \quad H_1: \theta > \theta_0$$

Recall that the noninformative prior for θ is the improper prior

$$\pi(\theta) = \frac{1}{\theta} .$$

Therefore, using formula (A.7) we obtain the posterior as

$$\pi(\theta|\mathbf{x}) \propto \frac{1}{\theta} \times \frac{1}{\theta^n} e^{-t/\theta} .$$

In order to express the posterior distribution in terms of a standard distribution, define $\alpha = 1/\theta$. By this change of variables we get

$$\pi(\alpha | \mathbf{x}) \propto \alpha^{n-1} e^{-\alpha t}.$$

This can be recognized as the gamma distribution with shape parameter, n, and scale parameter, $1/t$. Consequently,

$$\frac{t}{\theta} | \mathbf{x} \sim G(n,1) \quad \text{and in turn} \quad \frac{2t}{\theta} | \mathbf{x} \sim \chi^2_{2n} .$$

Hence, the posterior probability of the null hypothesis is

$$\begin{aligned} Pr(\theta \le \theta_0 | \mathbf{x}) &= Pr(\frac{2t}{\theta} \ge \frac{2t}{\theta_0}) \\ &= 1 - C_{2n}(\frac{2t}{\theta_0}) , \end{aligned} \tag{A.17}$$

where C_{2n} is the cdf of the chi-squared distribution with 2n degrees of freedom. Notice that this is numerically the same as the p-value given by (2.6), which was established treating θ as a constant rather than as a random variable (therefore, the results are logically different). However, it should be pointed out that a similar result does not hold with many other applications including testing problems concerning parameters of normal distributions; Example 2.13 is such a situation. In other words, the posterior probability of the null hypothesis under Jeffreys' noninformative prior is not necessarily the same as the p-value. Of course, the result is not at all true in the cases of point null hypotheses.

Equations (A.16) and (A.17) also imply that

$$\rho = \frac{1}{C_{2n}(\frac{2t}{\theta_0})} - 1$$

is the posterior odd ratio of H_0 to H_1.

Example A.8.

Let X be an observation from the geometric distribution with probability mass function

$$f(x) = p(1-p)^x \quad \text{for } x = 0, 1, 2, \cdots$$

Consider the problem of testing the hypotheses

$$H_0: p \le p_0 \quad \text{versus} \quad H_1: p > p_0 , \tag{A.18}$$

based on X and the improper prior

$$\pi(p) = p^{-1} . \tag{A.19}$$

It can be deduced from the previous example concerning the geometric distribution, or shown directly, that the posterior distribution of p given $X = x$ is the beta distribution

$$p \mid x \sim B(1, x+1) \ .$$

Therefore, the posterior probability of the null hypothesis is

$$Pr(p \leq p_0) = \int_0^{p_0} (x+1)(1-p)^x dp$$

$$= 1-(1-p_0)^{x+1} \ . \qquad \text{(A.20)}$$

On the other hand, according to Example 2.6, this is also the p-value for testing (A.18). It is of interest to note at this juncture that the prior defined by (A.19) is not the same as the noninformative prior (see Exercise 2.13) given by the Fisher information. Although one may argue that the above prior is the natural noninformative prior, changing the definition from one application to another after revealing facts is no good scientific practice. The foregoing findings should be treated as results of an existence theorem rather than anything else. Since the Bayesian interpretation of the posterior probability is not the same as that of the p-value, it is not surprising that the two treatments can lead to different conclusions.

Appendix B

Technical Arguments

B.1. Proof of Corollary 1.1

To prove the result in the case of continuous random variables, the joint density function of (U,V) can be written as

$$f_{\mathbf{U},\mathbf{V}}(\mathbf{u},\mathbf{v};\boldsymbol{\theta}) = A(\boldsymbol{\theta})\, B(\mathbf{u},\mathbf{v})\, e^{\boldsymbol{\theta}_u'\mathbf{u} + \boldsymbol{\theta}_v'\mathbf{v}},$$

where $\boldsymbol{\theta}_w'\mathbf{w}$ is the vector (dot) product of $\boldsymbol{\theta}_w$ and $\mathbf{w}$, $\mathbf{w} = \mathbf{u}$ *or* $\mathbf{v}$. The marginal distribution of V can be obtained by integrating this joint density with respect to u. Therefore, the required conditional density function can be obtained as

$$f_{\mathbf{U}|\mathbf{V}}(\mathbf{u}|\,\mathbf{v};\boldsymbol{\theta}) = \frac{A(\boldsymbol{\theta})\, B(\mathbf{u},\mathbf{v})\, e^{\boldsymbol{\theta}_u'\mathbf{u} + \boldsymbol{\theta}_v'\mathbf{v}}}{\int_{\mathbf{V}} A(\boldsymbol{\theta})\, B(\mathbf{u},\mathbf{v})\, e^{\boldsymbol{\theta}_u'\mathbf{u} + \boldsymbol{\theta}_v'\mathbf{v}}\, d\mathbf{u}} \tag{B.1}$$

$$= \frac{B(\mathbf{u},\mathbf{v})\, e^{\boldsymbol{\theta}_u'\mathbf{u}}}{\int_{\mathbf{V}} B(\mathbf{u},\mathbf{v})\, e^{\boldsymbol{\theta}_u'\mathbf{u}}\, d\mathbf{u}}$$

which is an exponential family independent of θ_v as desired.

B.2. Proof of Theorem 1.2

If T is a function of the form $\mathbf{T}(\mathbf{x}) = h(\overline{\mathbf{T}}(\mathbf{x}))$, then, given any $g \in G$, we have

$$\begin{aligned} \mathbf{T}(g(\mathbf{X})) &= h(\overline{\mathbf{T}}(g(\mathbf{X}))) \\ &= h(\overline{\mathbf{T}}(\mathbf{X})) \quad \text{since } \overline{\mathbf{T}} \text{ is invariant} \\ &= \mathbf{T}(\mathbf{X})\,. \end{aligned}$$

Hence, T is also invariant. Now with T invariant, whenever $\overline{\mathbf{T}}(\mathbf{X}_1) = \overline{\mathbf{T}}(\mathbf{X}_2)$

we have for some $g \in G$, $\mathbf{X}_1 = g(\mathbf{X}_2)$, which in turn implies that $\mathrm{T}(\mathbf{X}_1) = \mathrm{T}(\mathbf{X}_2)$, thus completing the proof.

B.3. Proof of Theorem 2.1

Let $\delta = \mu_0 - \mu$. In terms of δ the probability of interest can be expressed as

$$Pr(-k(w) \leq T - \mu_0 \leq w) = Pr(\mu - k(w) + \delta \leq T \leq \mu + w + \delta).$$

Since $f(\mu - k(w)) = f(\mu + w) \leq f(\mu)$, the probability over an interval of fixed length $w + k(w)$ remains constant or decreases for any departure of δ from $\delta = 0$. Therefore, the probability is a nonincreasing function of $|\delta|$.

B.4. Proof of Corollary 2.1

Let M be the mode of the distribution of Y. Then the mode of X is $M + \theta$. In order to apply Theorem 2.1 to the distribution of X, set $\mu = \theta + M$, $w = x - (\theta + M)$, and $-k(w) = \tilde{x} - (\theta + M)$, in the notation of Theorem 2.1. Then, we have $f(\mu + w) = f_X(x)$ and $f(\mu - k(w)) = f_X(\tilde{x})$, where $f_X(\,)$ is the pdf of X. But the equality $f_Y(\tilde{x} - \theta) = f_Y(x - \theta)$ implies that $f_X(x) = f_X(\tilde{x})$. Hence, the requirement $f(\mu - k(w)) = f(\mu + w)$ is satisfied as required by Theorem 2.1. Consequently,

$$\begin{aligned} &Pr(\tilde{x} - (\theta + M) \leq X - (\theta + M) \leq x - (\theta + M) \mid \theta = \theta_0) \\ &\quad = Pr(\tilde{x} - \theta \leq Y \leq x - \theta \mid \theta = \theta_0) \end{aligned}$$

is a nonincreasing function of $|(M + \theta) - (M + \theta_0)| = |\theta - \theta_0|$, as claimed.

B.5. Proof of Theorem 2.3

Suppose the test is based on an extreme region, $C(\mathbf{x})$, and let $Pr(\mathbf{X} \in C(\mathbf{x}) | \boldsymbol{\theta}_0) = p$ be its p-value, where p does not depend on $\boldsymbol{\delta}$. From (2.27) $Pr(\mathbf{X} \in C(\mathbf{x}) | \boldsymbol{\theta}_0) = E(\,E(\,I_c \mid \mathbf{S}\,)\,)$ and therefore $E(\,E(\,I_c \mid \mathbf{S}\,)\,) = p$. Since $\mathbf{S}$ is complete sufficient for $\boldsymbol{\delta}$, the latter equality implies that $E(\,I_c \mid \mathbf{S}\,)\,) = p$. This means that $Pr(\mathbf{X} \in C(\mathbf{x}) | \mathbf{S}\colon \boldsymbol{\theta}_0) = p$ as claimed.

B.6. Proof of Theorem 3.1

Let $\tau = \sigma^{-2}$ and let $g_1(\mu | \tau)$ and $g_2(\tau)$, respectively, denote the conditional prior distribution of μ given τ and the marginal distribution of τ. From (3.45) we get,

$$g_1(\mu|\tau) = c_1\tau^{1/2}e^{-\frac{\eta_0\tau}{2}(\mu-\mu_0)^2}$$

and

$$g_2(\tau) = c_2\tau^{\alpha_0-1}e^{-\lambda_0\tau} ,$$

where c_1, c_2, etc. are constants independent of parameters. The joint posterior distribution satisfies the equation

$$g(\mu,\tau|\,\mathbf{x}) = c_x\, f_{\mathbf{X}}(\mathbf{x}|\,\mu,\tau)g_1(\mu|\tau)g_2(\tau) ,$$

where $f_{\mathbf{X}}$ is the joint pdf of **X** and c_x is a constant independent of parameters, but dependent of **x**. Using (3.26) $f_{\mathbf{X}}(\mathbf{x}|\,\mu,\tau)$ can be written as

$$f_{\mathbf{X}}(\mathbf{x}|\,\mu,\tau) = c_3\,\tau^{n/2}e^{-\frac{\tau}{2}[(n-1)s^2+n(\bar{x}-\mu)^2]} .$$

Hence, the posterior distribution can be expressed as

$$g(\mu,\tau|\,\mathbf{x}) = c_4\tau^{\frac{n}{2}+\alpha_0-\frac{1}{2}}\,e^{-\tau(\lambda_0+\frac{(n-1)}{2}s^2)}\,e^{-\frac{\tau}{2}[\eta_0(\mu-\mu_0)^2+n(\mu-\bar{x})^2]} \tag{B.2}$$

But

$$\eta_0(\mu-\mu_0)^2 + n(\mu-\bar{x})^2 = (n+\eta_0)(\mu-\frac{\eta_0\mu_0+n\bar{x}}{n+\eta_0})^2 + \frac{\eta_0 n(\bar{x}-\mu_0)^2}{n+\eta_0}$$

$$= \eta_1(\mu-\mu_1)^2 + n\eta_0(\bar{x}-\mu_0)^2/\eta_1 . \tag{B.3}$$

It now follows from (B.2) and (2.24) that

$$g(\mu,\tau|\,\mathbf{x}) = c_4\tau^{\frac{n}{2}+\alpha_0-1}\,e^{-\lambda_1\tau}\,\tau^{1/2}e^{-\frac{\eta_1\tau}{2}(\mu-\mu_1)^2} ,$$

thus implying the desired result.

References

Agresti, A. (1990). *Categorical Data Analysis*, New York: John Wiley.

Agresti, A. (1992). A Survey of Exact Inference for Contingency Tables. *Statistical Science*, **7**, 131-153.

Agresti, A., Mehta, C. R., and Patel, N. R. (1990). Exact Inference for Contingency Tables with Ordered Categories. *Journal of the American Statistical Association*, **85**, 453-458.

Ananda, M. M. A. (1995). Nested Designs with Unequal Cell Frequencies and Unequal Variances. *Annals of the Institute of Statistical Mathematics*. **47**, 731-742.

Ananda, M. M. A., and Weerahandi, S. (1996). Testing the Difference of Two Exponential Means Using Generalized p-Values. **25**, 521-532. *Communications in Statistics: Simulation and Computation.*

Ananda, M. M. A., and Weerahandi, S. (1997). Two-Way ANOVA with Unequal Cell Frequencies and Unequal Variances. To appear in *Statistica Sinica.*

Anderson, V. L., and McLean, R. A. (1974). *Design of Experiments: A Realistic Approach*, New York: Marcel Dekker.

Baker, R. J. (1977). Exact Distributions Derived from Two-way Tables. *J. Royal Statistical Society*, **26**, 199-206.

Balmer, D. W. (1988). Recursive Enumeration of $r \times c$ Tables for Exact Likelihood Evaluation. *J. Royal Statistical Society*, **37**, 290-301.

Barnard, G. A. (1976). Conditional Inference is Not Inefficient. *Scandinavian Journal of Statistics*, **3**, 132-134.

Barnard, G. A. (1984). Comparing the Means of Two Independent Samples.

Applied Statistics, **33**, 266-271.

Bartlett, M. S. (1937). Properties of Sufficiency and Statistical Test. *Proc. Royal Society, Series A*, **160**, 268.

Bartlett, M. S. (1947). The Use of Transformations. *Biometrics*, **3**, 39-52.

Bartlett, M. S., and Kendall, D. G. (1946). The Statistical Analysis of Variance-Heterogeneity and the Logarithmic Transformation. *J. Royal Statistical Society Suppl.*, **8**, 128-138.

Bartlett, R. C. (1953). Approximate Confidence Intervals II: More Than One Parameter. *Biometrika*, **40**, 206-217.

Basu, D. (1959). The Family of Ancillary Statistics. *Sankya*, **21**, 247-256.

Basu, D. (1977). On the Elimination of Nuisance Parameters. *Journal of the American Statistical Association*, **72**, 355-366.

Berger, J. O. (1985). *Statistical Decision Theory and Bayesian Analysis*, New York: Springer-Verlag.

Berger, J. O., and Sellke, T. (1987). Testing a Point Null Hypothesis: The Irreconcilability of P Values and Evidence. *Journal of the American Statistical Association*, **82**, 112-122.

Bickel, P. J., and Doksum, K. A. (1977). *Mathematical Statistics*, San Francisco: Holden-Day.

Bishop, Y. M. M., Fienberg, S. E., and Holland, P. W. (1975). *Discrete Multivariate Analysis: Theory and Practice*, Cambridge, Mass.: The MIT Press.

Boik, J. B. (1987). The Fisher-Pitman Permutation Test: A Non-robust Alternative to the Normal Theory F Test when Variances are Heterogeneous. *British Journal of Mathematical and Statistical Psychology*, **40**, 26-42.

Box, G. E. P., and Tiao, G. C. (1973). *Bayesian Inference in Statistical Analysis*, Reading, Mass.: Addison-Wesley.

Bradbury, I. S. (1988). Approximations to Permutation Distribution in the Completely Randomized Design. *Communications in Statistics - Theory and Methods*, **17**, 543-555.

Bulmer, M. G. (1957). Approximate Confidence Limits for Components of Variance. *Biometrika*, **44**, 159-167.

Burdick, R. K., and Graybill, F. A. (1984). Confidence Intervals on Linear Combinations of Variance Components in the Unbalanced One-Way Classification. *Technometrics*, **26**, 131-136.

Burdick, R. K., and Graybill, F. A. (1992). *Confidence Intervals on Variance Components*, New York: Marcel Dekker.

Casella, G., and Berger, R. (1987). Reconciling Bayesian and Frequentist Evidence in the One-Sided Testing Problem. *Journal of the American Statistical Association*, **82**, 106-111.

Chow, G. C. (1960). Tests of Equality between Subsets of Coefficients in Two Linear Regressions. *Econometrica*, **28**, 591-605.

Church, J. D., and Harris, B. (1970). The Estimation of Reliability from Stress-Strength Relationships. *Technometrics*, **12**, 49-54.

Cochran, W. G., and Cox, G. M. (1957). *Experimental Design*, New York: John Wiley.

Cohen, A., and Sackrowitz, H. B. (1989). Exact Tests That Recover Interblock Information in Balanced Incomplete Designs. *Journal of the American Statistical Association*, **84**, 556-559.

Cohen, A., and Strawderman, W. E. (1971). Unbiasedness of Tests for Homogeneity of Variances. *Annals of Mathematical Statistics*, **42**, 355-360.

Conover, W. J. (1980). *Practical Nonparametric Statistics*, New York: John Wiley.

Cox, D. R. (1977). The Role of Significance Tests. *Scandinavian Journal of Statistics*, **4**, 49-62.

Cox, D. R., and Hinkley, D. V. (1974). *Theoretical Statistics*, London: Chapman and Hall.

Crowder, M. J., Kimber, A. C., Smith, R. L., and Sweeting, T. J. (1991), *Statistical Analysis of Reliability Data*, London: Chapman and Hall.

de Finetti, B. (1972). *Probability, Induction, and Statistics*, New York: John Wiley.

DeGroot, M. (1970). *Optimal Statistical Decisions*, New York: McGraw-Hill.

DeGroot, M. (1985). *Probability and statistics*, Reading, Mass.: Addison-Wesley.

Dempster, A. P. (1967). Upper and Lower Probability Inference Based on a Sample from a Finite Univariate Population. *Biometrika*, **54**, 515-528.

Downton, F. (1973). The Estimation of the Pr (Y < X) in the Normal Case. *Technometrics*, **15**, 551-558.

Ferguson, T. S. (1967). *Mathematical Statistics*, New York: Academic Press.

Fisher, F. M. (1970). Tests of Equality between Sets of Coefficients in Two Linear Regressions: An Expository Note. *Econometrica*, **38**, 361-366.

Fisher, R. A. (1922). On the Mathematical Foundations of Theoretical Statistics.

Phil. Trans. Royal Society, Ser. A, **222**, 309-368.

Fisher, R. A. (1925). *Statistical Methods for Research Workers*, Edinburgh: Oliver and Boyd.

Fisher, R. A. (1935). *The Design of Experiments*, Edinburgh: Oliver and Boyd.

Fisher, R. A. (1956). *Statistical Methods and Scientific Inference*. Edinburgh: Oliver and Boyd.

Fraser, D. A. S. (1979). *Inference and Linear Models*. New York: McGraw-Hill.

Fujikoshi, Y. (1993). Two-Way ANOVA Models with Unbalanced Data. *Discrete Mathematics*, **116**, 315-334.

Gail, M., and Mantel, N. (1977). Computing the Number of Contingency Tables with Fixed Magins. *Journal of the American Statistical Association*, **72**, 859-862.

Gibbons, J. D., and Pratt, J. W. (1975). p-Values: Interpretation and Methodology. *The American Statistician*, **29**, 20-25.

Gibbons, J. D. (1985). *Nonparametric Statistical Inference*, New York: Marcel Dekker.

Green, J. R. (1954). A Confidence Interval for Variance Components. *Annals of Mathematical Statistics*, **25**, 671-686.

Griffiths, W., and Judge, G. (1992). Testing and Estimating Location Vectors When the Error Covariance Matrix is Unknown. *Journal of Econometrics*, **54**, 121-138.

Guttman, I., Johnson, R. A., Bhattacharyya, G. K., and Reiser, B. (1988). Confidence Limits for Stress-Strength Models with Explanatory Variables. *Technometrics*, **30** 161-168.

Healy, W. C. (1961). Limits for a Variance Component With an Exact Confidence Coefficient. *Annals of Mathematical Statistics*, **32**, 466-476.

Hollander, M., and Wolfe, D. A. (1973). *Nonparametric Statistical Methods*, New York: John Wiley.

Huzurbazar, V. S. (1976). *Sufficient Statistics*, New York: Marcel Dekker.

Iman, R. L., Quade, D., and Alexander, D. A. (1975). Selected Probability Levels for the Kruskal-Wallis Test. *Selected Tables in Mathematical Statistics*, **3**, 329-384.

Jayatissa, W. A. (1977). Tests of Equality Between Sets of Coefficients in Two Linear Regressions When Disturbance Variances are Unequal. *Econometrica*, **45**, 1291-1292.

Jeffreys, H. (1961). *Theory of Probability,* London: Cambridge University Press.

Johnson, R. A. (1988). Stress-Strength Models for Reliability. *Handbook of Statistics,* 7 eds. P. R. Krishnaiah and C. R. Rao, Amsterdam: North Holland.

Johnson, R. A., and Weerahandi, S. (1988). A Bayesian Solution to the Multivariate Behrens-Fisher Problem. *Journal of the American Statistical Association,* **83**, 145-149.

Kempthorne, O., and Folks, L. (1971). *Probability, Statistics, and Data Analysis,* Ames, Iowa: Iowa State University Press.

Khuri, A. I. (1990). Exact Tests for Random Models with Unequal Cell Frequencies in the Last Stage. *Journal of Statistical Planning and Inference,* **24**, 177-193.

Khuri, A. I., and Littell, R. C. (1987). Exact Tests for the Main Effects Variance Components in an Unbalanced Random Two-Way Model. *Biometrics,* **43**, 545-560.

Kiefer, J. (1977). Conditional Confidence Statements and Confidence Estimators. *Journal of the American Statistical Association,* **72**, 789-808.

Koschat, M. A., and Weerahandi, S. (1990). Chow-type Tests under Homoscedasticity and Heteroscedasticity. Technical Report, Bell Communications Research.

Koschat, M. A., and Weerahandi, S. (1992). Chow-type Tests under Heteroscedasticity. *Journal of Business Economics and Statistics,* **10**, 221-228.

Kruskal, W. H., and Wallis, W. A. (1952). Use of Ranks on One-criteria Variance Analysis. *Journal of the American Statistical Association,* **47**, 583-621 (corrections in **48**, pp. 907-911).

Krutchkoff, R. G. (1988). One-way Fixed Effects Analysis of Variance when the Error Variances May be Unequal. *Journal of Statist. Comput. Simul.,* **30**, 259-271.

Krutchkoff, R. G. (1989). Two-way Fixed Effects Analysis of Variance when the Error Variances May be Unequal. *Journal of Statist. Comput. Simul.,* **32**, 177-183.

Kullback, S., and Rosenblatt, H. M. (1957). On the Analysis of Multiple Regressions in *k* Categories. *Biometrika,* 44, 67-83.

Laplace, P. S. (1812). *Theorie Analytique des Probabilites,* Paris: Courcier.

Lawless, J. F. (1982). *Statistical Models and Methods for Lifetime Data,* New York: John Wiley.

Lee, A. F. S., and Gurland, J. (1975). Size and Power of Tests for Equality of Means of Two Normal Populations with Unequal Variances. *Journal of the American Statistical Association*, **70**, 933-941.

Lehmann, E. L. (1975). *Nonparametrics: Statistical Methods Based on Ranks*, San Francisco: Holden-Day.

Lehmann, E. L. (1983). *Theory of Point Estimation*, New York: John Wiley.

Lehmann, E. L. (1986). *Testing Statistical Hypotheses*, New York: John Wiley.

Lindman, H. R. (1992). *Analysis of Variance in Experimental Design*, New York: Springer-Verlag.

Linnik, Y. (1968). Statistical Problems with Nuisance Parameters. *Translation of Mathematical monograph*. No. 20, New York: American Mathematical society.

Mann, H. B., and Whitney, D. R. (1947). On a Test of Whether One of Two Random Variables is Stochastically Larger Than the Other. *Annals of Mathematical Statistics*, **18**, 50-60.

Mehta, C. R., and Patel, N. R. (1980). A Network Algorithm for the Exact Treatment of the $2\times k$ Contingency Table. *Communications in Statistics*, **B9**, 649-664.

Mehta, C. R., and Patel, N. R. (1983). A Network Algorithm for Performing Fisher's Exact Test in $r\times c$ Contingency Tables. *Journal of the American Statistical Association*, **78**, 427-434.

Mehta, C. R., and Patel, N. R. (1986). A Hybrid Algorithm for Fisher's Exact Test on unordered $r\times c$ Contingency Tables. *Communications in Statistics*, **15**, 387-403.

Mehta, C. R., Patel, N. R., and Senchaudhuri, P. (1992). Exact Stratified Linear Rank Tests for Ordered Categorical and Binary Data. To appear in *Journal of Computational and Graphical Statistics*.

Meng, X., (1994). Posterior Predictive p-Values. *Annals of Statistics*, **22**, 1142-1160.

Mood, M. M., Graybill, F. A., and Boes, D. C. (1974). *Introduction to the Theory of Statistics*, New York: McGraw-Hill.

Neyman, J. (1935). Su un teorema concernente le cosiddette statistiche sufficienti. *Giornale dell' Instituto degli Attuari*, **6**, 320-334.

Pagano, M., and Halvorsen, K. (1981). An Algorithm for Finding the Exact Significance Levels of $r\times c$ Contingency Tables. *Journal of the American Statistical Association*, **76**, 931-934.

Pagano, M., and Tritchler, D. (1983). On Obtaining Permutation Distributions in Polynomial Time. *Journal of the American Statistical Association*, **78**, 435-440.

Pearson, K. (1900). On the Criterion that a Given System of Deviations from the Probable in the Case of a Correlated System of Variables is Such that it Can Reasonably be Supposed to Have Arisen from Random Sampling. *Phil. Mag.*, Series 5, **50**, 157-175.

Phadke, M . S. (1989). *Quality Engineering Using Robust Design*. Englewood Cliffs, New Jersey: Prentice-Hall.

Pratt, J. W. (1959). Remarks on Zeros and Ties in the Wilcoxon Signed Rank Procedures. *Journal of the American Statistical Association*, **54**, 655-667.

Pratt, J. W. (1961). Review of Lehmann's Testing Statistical Hypotheses. *Journal of the American Statistical Association*, **56**, 163-166.

Prentice, R. L. (1973). Exponential Survival with Censoring and Explanatory Variables. *Biometrika*, **60**, 279-288.

Reiser, B., and Guttman, I. (1986). Statistical Inference for Pr(Y < X). *Technometrics*, **28**, 253-257.

Rice, W. R., and Gaines, S. D. (1989). One-way Analysis of Variance with Unequal Variances. *Proc. Nat. Acad. Sci.*, **86**, 8183-8184.

Samaranayake, V. A., and Bain, L. J. (1988). A Confidence Interval for Treatment Component of Variance with Applications to differences in Means of Two Exponential Distributions. *J. Stat. Comput. Simul.*, **29**, 317-332.

Santner, T. J., and Duffy, D. E. (1989). *The Statistical Analysis of Discrete Data*, New York: Springer-Verlag.

Satterthwaite, F. E. (1946). An Approximate Distribution of Estimates of Variance Components. *Biomet. Bulletin*, **2**, 110-114.

Scheffe, H. (1943). On Solutions of the Behrens-Fisher Problem Based on the t-Distribution. *Annals of Mathematical Statistics*, **14**, 35-44.

Scheffe, H. (1953). A Method for Judging All Contrasts in the Analysis of Variance. *Biometrika*, **40**, 87-104.

Scheffe, H. (1959). *Analysis of Variance*, New York: John Wiley.

Scheffe, H. (1970). Practical Solutions of the Behrens-Fisher Problem. *Journal of the American Statistical Association*, **65**, 1501-1508.

Searle, S. R. (1971). *Linear Models*, New York: John Wiley.

Searle, S. R. (1987). *Linear Models for Unbalanced Data*, New York: John Wiley.

Searle, S. R., Casella, G., and McCulloch, C. E. (1992). *Variance Componenents*, New York: John Wiley.

Swallow, W. H., and Searle, S. R. (1978). Minimum Variance Quadratic Unbiased Estimation (MIVQUE) of Variance Components. *Technometrics*, **20**, 265-272.

Taguchi, G. (1986). *Introduction to Quality Engineering*, Tokyo: Asian Productivity Center.

Taguchi, G., Elsayed, E. A., and Hsiang, T. C. (1989). *Quality Engineering in Production Systems*, New York: McGraw-Hill.

Thomas, J. D., and Hultquist, R. A. (1978). Interval Estimation for the Unbalanced Case of the One-Way Random Effects Model. *Annals of Statistics*, **6**, 582-587.

Thompson Jr., W. A. (1985). Optimal Significance for Simple Hypotheses. *Biometrika*, **72**, 230-232.

Thursby, J. G. (1992). A Comparison of Several Exact and Approximate Tests for Structural Shift Under Heteroscedasticity. *Journal of Econometrics*, **53**, 363-386.

Toyoda, T. (1974). Use of the Chow Test Under Heteroscedasticity. *Econometrica*, **42**, 601-608.

Tsui, K., and Weerahandi, S. (1989). Generalized p-Values in Significance Testing of Hypotheses in the Presence of Nuisance Parameters. *Journal of the American Statistical Association*, **84**, 602-607.

Tsurumi, H. (1984). On Jayatissa's Test of Constancy of Regressions under Heteroskedasticity and Some Alternative Test Procedures, *Economic Studies Quarterly*, **35**, 57-62.

Tukey, J. W. (1951). Components in Regression. *Biometrika*, **7**, 33-69.

Verbeek, A., and Kroonenberg, P. M. (1985). A Survey of Algorithms for Exact Distributions of Test Statistics in *r×c* Contingency Tables with Fixed Margins. *Comput. Statist. Data Anal.*, **3**, 159-185.

Villegas, C. (1977). On the Representation of Ignorance. *Journal of the American Statistical Association*, **72**, 651-654.

Vollset, S. E., Hirji, K. F., and Elashoff, R. M. (1991). Fast Computation of Exact Confidence Limits for the Common Odds Ratio in a Series of 2×2 Tables. *Journal of the American Statistical Association*, **86**, 404-409.

Wald, A. (1940). A Note on the Analysis of Variance with Unequal Class Frequencies. *Annals of Mathematical Statistics*, **11**, 96-100.

Wang, C. M. (1990). On the Lower Bound of Confidence Coefficients for a Confidence Interval on Variance Components. *Biometrika*, **46**, 187-192.

Weerahandi, S. (1987). Testing Regression Equality with Unequal Variances. *Econometrica*, **55**, 1211-1215.

Weerahandi, S. (1991). Testing Variance Components in Mixed Models with Generalized p-Values. *Journal of the American Statistical Association*, **86**, 151-153.

Weerahandi, S., and Johnson, R. A. (1992). Testing Reliability in a Stress-Strength Model when X and Y are Normally Distributed. *Technometrics*, **34**, 83-91.

Weerahandi, S. (1993). Generalized Confidence Intervals. *Journal of the American Statistical Association*, **88**, 899-905, correction in **89**, 726, 1994.

Weerahandi, S. (1994). ANOVA under Unequal Error Variances. *Biometrics*. **51**, 589-599.

Weerahandi, S., and Amaratunga, D. (1996). A Performance Comparison of Methods of Inference in Mixed Models. Technical Report, Bell Communications Research.

Welch, B. L. (1947). The Generalization of Students' Problem When Several Different Population Variances are Involved. *Biometrika*, **34**, 28-35.

Welch, B. L. (1951). On the Comparison of Several Means Values: An Alternative Approach. *Biometrika*, **38**, 330-336.

Welch, B. L. (1956). On Linear Combinations of Several Variances. *Journal of the American Statistical Association*, **51**, 132-148.

Wilcoxon, F. (1945). Individual Comparisons by Ranking Methods. *Biometrics*, **1**, 80-83.

Williams, J. S. (1962). A Confidence Interval for Variance Components. *Biometrika*, **49**, 278-281.

Woolson, R. F. (1987). *Statistical Methods for the Analysis of Biomedical Data*, New York: John Wiley.

Zellner, A. (1977). Maximal Data Information Prior Distributions. In *New Methods in the Applications of Bayesian Methods*, A. Aykac and C. Brumat (Eds.), Amsterdam: North Holland.

Zhou, L., and Mathew, T. (1994). Some Tests for Variance Components Using Generalized p-Values. *Technometrics*, **36**, 394-402.

Index

Springer Series in Statistics

(continued from p. ii)

Knottnerus: Sample Survey Theory: Some Pythagorean Perspectives.
Kolen/Brennan: Test Equating: Methods and Practices.
Kotz/Johnson (Eds.): Breakthroughs in Statistics Volume I.
Kotz/Johnson (Eds.): Breakthroughs in Statistics Volume II.
Kotz/Johnson (Eds.): Breakthroughs in Statistics Volume III.
Küchler/Sørensen: Exponential Families of Stochastic Processes.
Lahiri: Resampling Methods for Dependent Data.
Le Cam: Asymptotic Methods in Statistical Decision Theory.
Le Cam/Yang: Asymptotics in Statistics: Some Basic Concepts, 2nd edition.
Liu: Monte Carlo Strategies in Scientific Computing.
Longford: Models for Uncertainty in Educational Testing.
Manski: Partial Identification of Probability Distributions.
Mielke/Berry: Permutation Methods: A Distance Function Approach.
Pan/Fang: Growth Curve Models and Statistical Diagnostics.
Parzen/Tanabe/Kitagawa: Selected Papers of Hirotugu Akaike.
Politis/Romano/Wolf: Subsampling.
Ramsay/Silverman: Applied Functional Data Analysis: Methods and Case Studies.
Ramsay/Silverman: Functional Data Analysis.
Rao/Toutenburg: Linear Models: Least Squares and Alternatives.
Reinsel: Elements of Multivariate Time Series Analysis, 2nd edition.
Rosenbaum: Observational Studies, 2nd edition.
Rosenblatt: Gaussian and Non-Gaussian Linear Time Series and Random Fields.
Särndal/Swensson/Wretman: Model Assisted Survey Sampling.
Santner/Williams/Notz: The Design and Analysis of Computer Experiments.
Schervish: Theory of Statistics.
Shao/Tu: The Jackknife and Bootstrap.
Simonoff: Smoothing Methods in Statistics.
Singpurwalla and Wilson: Statistical Methods in Software Engineering: Reliability and Risk.
Small: The Statistical Theory of Shape.
Sprott: Statistical Inference in Science.
Stein: Interpolation of Spatial Data: Some Theory for Kriging.
Taniguchi/Kakizawa: Asymptotic Theory of Statistical Inference for Time Series.
Tanner: Tools for Statistical Inference: Methods for the Exploration of Posterior Distributions and Likelihood Functions, 3rd edition.
van der Laan: Unified Methods for Censored Longitudinal Data and Causality.
van der Vaart/Wellner: Weak Convergence and Empirical Processes: With Applications to Statistics.
Verbeke/Molenberghs: Linear Mixed Models for Longitudinal Data.
Weerahandi: Exact Statistical Methods for Data Analysis.
West/Harrison: Bayesian Forecasting and Dynamic Models, 2nd edition.